Understanding and Responding to Hazardous Substances at Mine Sites in the Western United States

edited by

Jerome V. DeGraff
U.S. Department of Agriculture Forest Service
Sierra National Forest
1600 Tollhouse Road
Clovis, California 93611
USA

THE GEOLOGICAL SOCIETY OF AMERICA®

Reviews in Engineering Geology XVII

3300 Penrose Place, P.O. Box 9140 ■ Boulder, Colorado 80301-9140

2007

Copyright © 2007, The Geological Society of America (GSA). All rights reserved. GSA grants permission to individual scientists to make unlimited photocopies of one or more items from this volume for noncommercial purposes advancing science or education, including classroom use. For permission to make photocopies of any item in this volume for other noncommercial, nonprofit purposes, contact the Geological Society of America. Written permission is required from GSA for all other forms of capture or reproduction of any item in the volume including, but not limited to, all types of electronic or digital scanning or other digital or manual transformation of articles or any portion thereof, such as abstracts, into computer-readable and/or transmittable form for personal or corporate use, either noncommercial or commercial, for-profit or otherwise. Send permission requests to GSA Copyright Permissions, 3300 Penrose Place, P.O. Box 9140, Boulder, Colorado 80301-9140, USA.

Copyright is not claimed on any material prepared wholly by government employees within the scope of their employment.

Published by The Geological Society of America.
3300 Penrose Place, P.O. Box 9140, Boulder, Colorado 80301-9140, USA
www.geosociety.org

Printed in U.S.A.

GSA Books Science Editor: Marion E. Bickford

Library of Congress Cataloging-in-Publication Data

Understanding and responding to hazardous substances at mine sites in the western
    United States / edited by Jerome V. DeGraff.
       p. cm. (Reviews in engineering geology ; 17)
    Includes bibliographical references.
    ISBN-13 9780813741178 (pbk.)
    1. Abandoned mines—Environmental aspects—West (U.S.). 2. Hazardous substances—Environmental aspects—West (U.S.). I. DeGraff, Jerome V.
    Reviews in engineering geology (Geological Society of America) ; 17.

TD428.M56 U53   2007
363.17—22

                                          2007061020

Cover: Top left: Acidic mine drainage flows from the adit past a gray pile of arsenic-contaminated tailings at the Williams Brothers mine on the Sierra National Forest, Mariposa County, California. Gold was discovered and developed at this historic mine in 1875. Top middle: An open pit at the Siskon mine, Siskiyou County, California, on the Klamath National Forest near the California-Oregon border. It is part of extensive above and underground workings developed to extract gold ore for processing at an on-site mill. Top right: Ninety acres of tailings on the Okanogan-Wenatchee National Forest, Chelan County, Washington, near the mill facilities for the Holden mine. The tailings at this historic copper mine impact both air and water quality along Railroad Creek and nearby Lake Chelan. Bottom left: White Caps mill site within the Inyo National Forest, Inyo County, California. A response action capped tailings ponds containing elevated levels of heavy metals creating a large flat next to McGee Creek. Bottom middle: Mining and processing of mercury ore took place at the Rinconada mine near San Luis Obispo County, California. It resulted in contamination of private land and public land administered by the Los Padres National Forest and the Bureau of Land Management. Bottom right: Excavation collapsed the earth berms forming tailings ponds at the White Caps mill site. Uncontaminated soil was imported to cap the reshaped area near Bishop, California.

10 9 8 7 6 5 4 3 2 1

Contents

Preface .... *v*

1. *Addressing the toxic legacy of abandoned mines on public land in the western United States* .. *1*
   Jerome V. DeGraff

2. *Characterizing infiltration through a mine-waste dump using electrical geophysical and tracer-injection methods, Clear Creek County, Colorado.* *9*
   Robert R. McDougal and Laurie Wirt

3. *Strategies to predict metal mobility in surficial mining environments* *25*
   Kathleen S. Smith

4. *The effects of acidic mine drainage from historical mines in the Animas River watershed, San Juan County, Colorado—What is being done and what can be done to improve water quality?* *47*
   Stanley E. Church, J. Robert Owen, Paul von Guerard, Philip L. Verplanck, Briant A. Kimball, and Douglas B. Yager

5. *Mining-impacted sources of metal loading to an alpine stream based on a tracer-injection study, Clear Creek County, Colorado* *85*
   David L. Fey and Laurie Wirt

6. *On-site repository construction and restoration of the abandoned Silver Crescent lead and zinc mill site, Shoshone County, Idaho* *105*
   Jeff K. Johnson

7. *Approaches to contamination at mercury mill sites: Examples from California and Idaho* .. *115*
   Jerome V. DeGraff, Michelle Rogow, and Pat Trainor

8. *Approaches to site characterization, reclamation of uranium mine overburden, and neutralization of a mine pond at the White King–Lucky Lass mines site near Lakeview, Oregon* *135*
   Kent Bostick, Norm Day, Bill Adams, and David B. Ward

9. *Passive treatment of acid rock drainage from a subsurface mine* *153*
   Martin Foote, Helen Joyce, Suzzann Nordwick, and Diana Bless

10. *Management of mine process effluents in arid environments* *163*
    Christopher Ross

11. *Sampling and monitoring for closure* *171*
    Virginia T. McLemore, Kathleen S. Smith, Carol C. Russell, and the Sampling and Monitoring Committee of the Acid Drainage Technology Initiative—Metal Mining Sector

Preface

It is somewhat ironic that a volume with papers on cleaning up hazardous substances from historic mining sites would have its origins in Nevada, a state noted for its mining interests. This *Reviews in Engineering Geology* volume was an outgrowth of a symposium I organized for the 2002 Annual Meeting of the Association of Environmental and Engineering Geologists held 23–29 September 2002 in Reno. The Association of Environmental and Engineering Geologists is an associated society of the Geological Society of America (GSA) and has many members who are also members of the Engineering Geology Division of GSA.

The symposium was titled "Addressing Hazardous Waste and Contamination Issues at Abandoned Mines in the Western United States." Presenters were drawn from a number of federal agencies and environmental consulting firms. While the symposium presentations focused on the engineering aspects of clean-up technology or construction for mining-related problems, I was struck by the number of geochemical, groundwater, and related geologic topics that were highlighted. As a result, I decided to propose a volume for the *Reviews in Engineering Geology* series that would more fully document some of the technologies and mine clean-up experiences presented at the symposium, along with papers exploring the issues faced in characterizing mine-site contamination.

The problems caused by release of hazardous substances from historic or abandoned mines on public lands in the western United States are not just of interest to public lands managers. The many members of the public driving ATVs, hiking, horseback riding, camping, fishing, hunting, or just enjoying the solitude of a National Forest or other public land are equally concerned about their personal welfare and the well being of the environment.

In this volume, the initial paper examines the dimensions of the abandoned mine problem in the western United States and some site conditions that influence how characterization and response action can or cannot be carried out. The next four papers explore specific methods for characterization, particular contaminant issues, and impacts from the release of hazardous substances from mine and mill sites. These papers are followed by four more papers describing successful response actions, technologies, or practical approaches for addressing contaminant releases to the environment. The last two papers serve as a reminder that mining continues and that the potential for mine-related problems continues today. One deals with processing fluids for a particular extraction process and the other details monitoring strategies for after mining has ended.

The volume documents interesting approaches, techniques, and practical scientific considerations associated with mine-site remediation. It is hoped that this volume will prove useful to those geologists just getting involved in these types of sites or issues. It also should highlight how many federal, state, and local agencies and organizations are bringing the best science possible to address this serious problem.

Jerome V. DeGraff

Addressing the toxic legacy of abandoned mines on public land in the western United States

Jerome V. DeGraff
U.S. Department of Agriculture Forest Service, Sierra National Forest, 1600 Tollhouse Road, Clovis, California 93611, USA

ABSTRACT

The development and exploitation of mineral resources in the western United States was important to both our economic development and our history. With continued population growth and economic development in this region, the impacts of our mining legacy are proving to be equally important to citizens in our modern society. By one estimate, 500,000 abandoned mine sites are scattered across the western landscape, largely on public land (state and federal), affecting 16,000 miles of streams. Federal land management agencies such as the U.S. Department of Agriculture Forest Service and U.S. Department of Interior Bureau of Land Management are able to use their authorities under the Comprehensive Environmental Response, Compensation, and Liability Act to respond to the release of hazardous substances from these abandoned mines. Although human health is a primary consideration in prioritizing site response, environmental issues such as the impact on terrestrial species, water quality, or aquatic species also may influence site response priorities. Challenges faced in reducing or preventing further release of hazardous substances at historic mines sites include limited available funds, difficult access, changing public land uses, and increasing populations in nearby areas.

Keywords: abandoned mines, CERCLA, hazardous waste, western United States

THE PROBLEM OF ABANDONED MINES

Mining was a major force shaping the early history of the western United States. The romantic view of the early American West usually focuses on the cowboys and cattle drives. Although ranching and farming were clearly central to the migration of people from the eastern United States to the western frontier, they were not the only economic incentives promoting this movement. Mining and the possibility of finding rich deposits of gold, silver, copper, and other metals was a driving force from the early 1800s to well into the 1900s.

One of the best known examples is the 1849 gold rush to California. Most people know of James Marshall's discovery of gold in the channel where Sutter's mill was being built near Coloma, California. The first printed notice appeared during March 1848 in San Francisco. Less well known are the gold discoveries several months later in the Feather River and Trinity River (Clark, 1970). At the end of 1848, the non-Native American population of California was said to be ~20,000 and included 5000 miners (Harden, 1998). By the end of 1849, that population had grown to 100,000 and then to 224,000 by the end of 1852.

Mining was a boom and bust industry through much of the West, especially in the mountainous areas. Familiar examples of historic mining areas such as Tombstone, Arizona, and Butte, Montana, may give the impression that mining was dominated by large mines at widely scattered locations in the western states. In

Figure 1. Tailings from early twentieth-century gold ore processing at the Bright Star mine in the Piute Mountains section of the Sierra Nevada east of Bakersfield, California. Arsenic concentrations are 2–10 times greater than the threshold for defining the material as hazardous under State of California criteria.

reality, there were few mountain ranges where hopeful prospectors did not explore or industrious miners did not develop identified deposits. It is this historic legacy of small mining operations that dots the landscape, primarily on land under federal jurisdiction, and now represents a significant problem either by exposing the public to hazardous substances or as a source for their release into the environment.

The release of hazardous substances at abandoned mine sites is a consequence of the extraction and separation of minerals or metals from the host rock. Only a small proportion of the total ore extracted from a mine consists of the mineral or metal that is needed for industrial or commercial use. The remaining rock material after this separation, or beneficiation, is the waste left at the mine site. The proportion of waste generated varies with the mineral or metal and can range from 10% (potash) to 99.99% (gold) (Environmental Protection Agency [EPA], 2004). While waste rock (low-grade ore) and unprocessed ore stockpiles can be significant sources of hazardous substances, it is often the fine-grained tailings and other waste generated from further separation and chemical treatment at the mill site where the most significant releases occur (Fig. 1). EPA (2004) reviewed 156 hardrock mine sites where clean-up needs were identified and found the most common contaminants were lead, arsenic, zinc, and cadmium. The underground works of the mine can also generate a release of hazardous substances. Exposure to radiation and production of radon gas can be found at some uranium mines. More common is acid mine drainage from mines where sulfide-bearing rocks are present. The impact of acid water being released into streams and other bodies of water is made worse by the associated dissolution and transport of heavy metals (Forest Service, 1993) (Fig. 2).

Figure 2. Blowing dust from 90-acre area of tailings at the Holden mine 12 miles from Lake Chelan in Washington. The tailings are adjacent to Railroad Creek, a tributary to Lake Chelan, a few miles from the community of Holden and adjacent to the Glacier Peak Wilderness Area.

RESPONDING TO THE PROBLEM

Unless specifically withdrawn, public lands have been and remain available for mineral exploration and extraction. These lands are administered by the Forest Service (U.S. Department of Agriculture) and the Bureau of Land Management (U.S. Department of the Interior). It is not always apparent how significant an area this represents. If you total the land area of Alaska, Arizona, California, Colorado, Idaho, Montana, Nevada, New Mexico, Oregon, Utah,

Washington, and Wyoming, ~38% is administered by these two federal agencies (National Research Council [NRC], 1999). It is difficult to accurately estimate the impact from abandoned mines over this vast area. A 1993 report identified more than 1,500 mining sites with significant acid and metal drainage on National Forest system lands in the western United States (Forest Service, 1993). By 2002, the estimate for National Forest system land had been revised upward to more than 2000 mine sites based on additional field evaluation and site characterization (Fliniau, 2002). The conservation group Trout Unlimited issued a report in 2006 titled "Settled, Mined, and Left Behind" that states that 500,000 abandoned mine sites are scattered across the western landscape, largely on public land (state and federal), affecting 16,000 miles of streams.

The primary legal basis for addressing the impacts from hazardous substances at abandoned mines is the Comprehensive Environmental Response, Compensation, and Liability Act (CERCLA). In addition to its well-publicized program for hazardous substance cleanup on private land, the Superfund program, CERCLA also authorizes the EPA to ensure that federal agencies conduct appropriate cleanup on public land under their administration. EPA is primarily responsible for enforcement and response actions under CERCLA. However, federal agencies, such as the Forest Service, are empowered under CERCLA and Executive Order No. 12580 to exercise many of the same authorities in responding to releases of hazardous substances at sites on land under their administration. Even on land administered by federal agencies, the EPA will be the lead agency for response actions that are of an emergency nature. They may also act as the lead agency for sites that are on the National Priorities List (NPL).

Many western states have strong programs for both addressing releases of hazardous substances and cleanup of abandoned mines. Often there is state legislation that grants appropriate clean-up authorities similar to provisions in CERCLA to designated state agencies. This sometimes facilitates efforts to coordinate both federal and state response actions within the same watersheds to achieve the maximum benefit from addressing mine-related hazardous releases on both public and private land. In Oregon, the Department of Environmental Quality (DEQ) is the state agency with responsibility for mine site cleanup. Their agency strategy includes prioritizing sites based on human and environmental risks and undertaking cleanup in partnership with federal agencies such as the U.S. Department of Agriculture (USDA) Forest Service, U.S. Department of Interior (USDI) Bureau of Land Management, and U.S. Environmental Protection Agency.

Like the Oregon DEQ, federal agencies typically use both human health and environmental impacts to prioritize responding to hazardous substance concerns at abandoned mine sites. Human health risk is clearly a significant factor. In the past, human health risk at some sites was not so compelling due to the generally sparse population in the western states. With urbanization and population growth increasing, even on the boundaries of National Forests and other public land, it is not unreasonable to expect human health risks to rise. Human health concerns for abandoned mine sites on public land are typically associated with areas of concentrated recreational use, proximity to rural communities, or impacts to drinking water supplies. The Amalgamated Mill site within the Willamette National Forest illustrates the human health threat to drinking water supplies. During the 1930s, the mill generated tailings while processing copper ore from several nearby mines. For many years, these tailings abutting Battle Ax Creek were held in place by an extensive wooden crib wall. In 1990, it became clear that the aging wall could soon fail, allowing tailings into Battle Ax Creek, which would threaten the drinking water supply for the city of Salem, Oregon. In the late 1990s, the sulfur-rich mill tailings, unprocessed mined ore, rock, and soil portions of the mill site debris and containerized ore concentrate were excavated and transported to an off-site repository.

Establishing priority based on environmental effects varies with both the hazardous substance and the potential environmental effect. Environmental effects are especially evident where contaminants enter water and impact fish populations. The impact on fish is well described by the Trout Unlimited 2006 report titled "Settled, Mined, and Left Behind." In some instances, the effect leads to fish having concentrations of arsenic or mercury above safe levels for human consumption. In others, the metals being released into the aquatic environment are causing fish kills in streams or reservoirs or habitat degradation that limits fish populations. Terrestrial species can also be adversely affected by the presence of heavy metals and other contaminants at mine sites. The number of species involved can make this effect difficult to assess (Fig. 3). One approach is to compare concentrations of heavy metals present at the mine site to the limited ecotoxicological literature for wildlife. The Bureau of Land Management provides a means to implement this approach through their identification of risk management criteria for metals at abandoned mine sites (Ford, 1996). Despite uncertainties associated with ecological values, it does provide a benchmark for comparison to site conditions that suggests the degree of risk to wildlife species living in the vicinity of a mine site.

LIMITATIONS AFFECTING RESPONSE ACTIONS ON PUBLIC LAND

Response actions at abandoned mines on public land involve a number of issues common to CERCLA sites regardless of their location or the source of the contamination. One common issue is funding. The expected cost for cleanup at identified sites compared to annual funding makes it likely that response actions will occur over decades, and prioritization of which sites to clean up first is critical to limiting the adverse impact on human health and the environment.

Another common issue is the viability of potentially responsible parties (PRPs) available to undertake the cleanup at abandoned mine sites. EPA (2004) in their examination of 156 large hardrock mine sites found that 52% of the sites either had no identifiable PRP or an identified PRP that was not financially viable. The situation is generally worse for smaller abandoned mines on public land where many sites are attributable to early mining

Figure 3. Acid drainage from the Williams Brothers mine within the South Fork of the Merced River. This site is 3 miles west of the El Portal entrance to Yosemite National Park.

booms. The individuals are no longer alive, or the companies involved are no longer in existence.

Abandoned mine sites on western public land are subject to some significant operational problems associated with their location. The occurrence of mineral deposits, especially metals, in mountainous areas is pervasive throughout the West. Consequently, a common physical condition at abandoned mine sites is steep terrain. Steep terrain means that heavy equipment is difficult to operate effectively due to slope limitations. The gentler sloping areas suitable for staging areas or encapsulation sites are limited in size or unavailable. Often, the steep slopes at a site are where the tailings piles are found. It is not unusual for the toe of the tailings to be located at or in a stream or channel at the bottom of the slope (Fig. 4).

Site access is often limited by topography, especially in areas with steep slopes. The location of abandoned mines, whether in the mountains or deserts, often involves a long drive over roads in poor condition or even air access. This travel-time factor can significantly increase costs. The low-volume roads associated with access to these sites also may limit the size of equipment. Trucks used for transporting heavy equipment or bulk hauling of contaminated materials may have difficulty going up grade on

Figure 4. A black bear crossing tailings adjacent to Copper Creek at the Siskon mine in northern California. Copper Creek is a steelhead fishery and flows into Dillon Creek, a tributary to the Klamath River. This abandoned gold mine and mill site is on the Klamath River roughly 30 miles from Orlean, California.

these roads or negotiating the sharp turn radius on switchback curves. These transport factors can necessitate using smaller equipment or trucks that, in turn, lower production rates and increase costs.

Where the roads cross streams or rivers, available bridges are not usually engineered to withstand the large loads of bin trucks or low-bed trucks hauling heavy equipment. In some instances, there may be no bridge. Reaching the site may require the additional cost of building a temporary bridge or using smaller equipment capable of fording the stream or river. Fording streams or rivers can mean an additional limitation—timing to permit seasonal flows to drop to safe levels before crossing. Fish, amphibian, and sediment issues may also complicate crossings. The removal action at Gibraltar mine (DeGraff et al., this volume) illustrates how stream crossings can complicate access. The only road suitable for larger bin trucks involved crossing the Santa Ynez River at four low-water fords. These fords were only useable after Gibraltar Dam reduced flows following spring releases. Also, the mine access road crossed a stream that was designated habitat for the California Red-Legged Frog (*rana aurora draytonii*), a federally listed threatened and endangered species. Modifications were made to the drive-through crossing to protect this habitat after consultation with the U.S. Fish and Wildlife Service.

Roads previously suitable for mining with horse-drawn wagons, early trucks, or railway may no longer exist for use during a response action. In the Tahoe National Forest near Grass Valley, California, this situation was encountered during the LaTrinidad mine response action (R. Weaver, pers. comm., 2005). Nearly 80 cubic yards of tailings with elevated concentrations of arsenic were eroding into Sailor Creek, a cold-water trout stream. The cost of building a suitable three-mile-long temporary road would be steep and adversely affect one mile of historic trail providing recreational access for many years after completion of the response action. Cost-effective alternative access was accomplished using a Hughes 500 helicopter to lift the tailings in one-cubic yard Super Sacks to a road suitable for truck transport to a disposal facility.

A similar problem was addressed during the initial response action to acid mine drainage from the Golinsky mine north of Redding, California, in the Shasta-Trinity National Forest. When active, the mine was served by a railway. This access is no longer useable because the subsequent creation of Lake Shasta Reservoir has submerged much of the railway and the connecting road access. Implementation of the pilot scale sulfate reducing bioreactor to treat the acid drainage from the mine required using boats and overcoming unloading difficulties at this site (Gusek et al., 2005). Using boats or barges may actually be more routine at some abandoned mine sites. Similarly, the ongoing response action for the Holden mine on the Okanogan-Wenatchee National Forests in Washington requires using crew boats and equipment barges from the town of Chelan to Lucerne (N. Day, pers. comm., 2006). It is about a 40-mile trip across Lake Chelan. From Lucerne, 12 miles of gravel road connect to Holden and the mine area (Fig. 2). Fortunately, these modes of transport were used at the time the mine was active and continue to operate for other commercial activities at present.

Reservoirs like Lake Shasta may be only one of several ways in which changing land use patterns can affect site access. As populations have increased, subdivisions and previously small communities have grown larger. The only road to a site may now pass through a more populous area. The public may be pleased to have the nearby abandoned mine cleaned up but not be enamored of the many trucks that may need to pass back and forth through their neighborhood. In other cases, the road may be privately owned and require negotiating a road agreement to ensure its condition is restored should heavy traffic degrade it. This was necessary for the response action at the Deertrail mine, where the last 3 miles of narrow, native-surfaced road crossed private ranch land (DeGraff et al., this volume). This agreement specified the daily watering to reduce the amount of dust from traffic on the road, repair of any damage due to operator error or truck weight to culverts or cattle guards, and post response blading of the road with a grader to restore it to the same surface condition as prior to the passage of many large vehicles.

SOME SUCCESSFUL RESPONSE ACTIONS

While abandoned mine cleanup is a high priority, it is a formidable undertaking in the West due to both the number of sites and their geographic distribution (EPA, 1995). Each site requires time and resources to adequately characterize the site or risk spending funds on response measures that prove inappropriate or insufficient to address the hazardous substances present. The five basic steps for characterization are: (1) understanding premining conditions at the site, (2) inventorying the materials deposited above ground including processing equipment, (3) mapping what occurred underground as it relates to the substances being released, (4) monitoring the movement of water on the site, and (5) establishing the impacts to human health and the environment from the mining activity (EPA, 1995). Both McDougal and Wirt (this volume) and Fey and Wirt (this volume) describe the particular difficulties associated with the critical factors of surface water infiltration through mine waste and determining the downstream movement of metals from a mine site. These are good examples of how the monitoring of water is critical to fully characterizing the site.

These individual site problems become particularly important in effectively addressing abandoned mine impacts within a given watershed. Church et al. (this volume) point out that the Animas River watershed of Colorado hosts 5400 mines, mills, and prospects, of which 80 sites accounted for 90% of the metal loading of the surface waters. This demonstrates that concentrated areas of historic mining may be better characterized over an entire watershed. An example of a watershed approach to characterization and response to abandoned mines is the American Fork Canyon project in Utah (M. Manderbach, written comm., 2005). Located east of Salt Lake City and near the communities of American Fork and Alpine, this area hosted historic mining of lead, silver, copper, and some gold at six separate mines and mill sites. These operations spanned from 1870 through the 1950s and left tailings piles, waste rock dumps, and smelter wastes. The impact from these sites resulted in a 2002 state-issued advisory for high arsenic levels detected in trout. Its proximity to the Salt Lake Valley had made it popular with ATV users and motorcyclists, creating hazards through possible inhalation of toxic dust. By 2005, the Forest Service sites with the highest contaminant levels were characterized and encapsulated in engineered repositories. This isolation of hazardous wastes and associated restoration effectively addressed 75% of the identified problem. Monitoring will continue to determine what additional measures may be needed.

The Bear and South Yuba Rivers in California's northwestern Sierra Nevada are among a number of watersheds where the gold rush mining legacy affects people and the environment in the present day. Once the stream gravels and floodplain deposits had yielded the readily found gold, miners recognized that gravels preserved under Tertiary-aged volcanic rock on the adjacent ridges might be a source of gold. This discovery gave rise to extensive hydraulic mining in these watersheds. From 1853 to 1884, gold was mined by eroding steep faces of buried Tertiary gravels containing gold with high-pressure water cannon. This created large pits as the gravels were washed into a network of sluices. Frequently, tunnels were drilled through bedrock underlying the gravels to permit drainage when washing out the lower levels of gravel. Along the sluices, mercury was added to the finer materials to recover the gold. It is estimated that several hundred pounds of mercury would be lost to the local streams per year (Alpers and Hunerlach, 2000). The modern-day problem is associated with hydraulic pit drain tunnels and other locations where this mercury exists together with conditions suitable for bacteria that convert it from elemental form to methylmercury. Methylmercury is both toxic and soluble.

A watershed approach for the Bear and South Yuba is being jointly carried out by the Forest Service, Bureau of Land Management, EPA, U.S. Geological Survey, the California Department of Conservation, and the California Water Resources Control Board. The U.S. Geological Survey studies have demonstrated how methylmercury is bioaccumulating within these watersheds (May et al., 2000). Water quality studies are helping to identify particular hydraulic mine sites where methylmercury was being generated. Human exposure to the methylmercury comes from eating fish caught in local streams and reservoirs and from individuals involved in small-scale panning and dredging operations.

The EPA in 2000 undertook the first response action under this project in the Bear River watershed. Methylmercury in water discharged from a drain tunnel was measured by the U.S. Geological Survey (Hunerlach et al., 1999) where the tunnel discharge was readily accessed by the public. EPA focused on removal of mercury-contaminated gravels from a drain tunnel at the Polar Star mine near Dutch Flat, California (EPA, 2000). Once removed, the gravels were separated from the mercury. The recovered mercury, fine material from which mercury could not be separated, and mercury-contaminated wood from sluice boxes in the tunnel were transported to a treatment and disposal facility. The drain tunnel bottom was covered with concrete to ensure any remaining mercury was inaccessible to bacteria in this environment.

A second response action in the Bear River watershed was carried out by the USDA Forest Service at the Sailor Flat hydraulic mine within the Tahoe National Forest. This historic mine drains into Greenhorn Creek, a major tributary of the Bear River. Like the Polar Star mine, the U.S. Geological Survey studies had determined that methylmercury was in water discharging from the single drain tunnel present (Fig. 5). Their data also demonstrated that the local food chain was being affected through bioaccumulation of methylmercury through on-site exposure. In 2003, work began at the site with excavation down to the tunnel. This removed the tunnel roof exposing the entire 40-foot length of the tunnel floor and allowing access to the gravels on the tunnel floor. A concrete mixture was injected into the gravels to immobilize them and the mercury within them. Once completed, nearly 20 feet of soil was compacted along the length of the tunnel. A graded stream channel was engineered to drain the regraded floor of the tunnel. This action isolated the mercury from the bacteria capable of converting it to methylmercury form and also removed environmental components necessary for their presence.

The Bureau of Land Management and other cooperating agencies within the Bear-Yuba Project will be continuing response actions at other former hydraulic mine sites to reduce the effect of past mercury contamination. It represents one of many cooperative programs across the western United States that will need to continue for decades to successfully address the deleterious effect of hazardous substances from historic mining.

CONCLUSIONS

The legacy from historic mining in the western United States includes a considerable and widespread impact from heavy metals and other hazardous substances. A high proportion of historic mining sites are found on the significant amount of public land in western States. On public land, problems from release of hazardous substances commonly occur at numerous small historic mine sites rather than large ones. Impacts to the environment and human health are associated with many of these locations. The number of locations where human health is a consideration is likely to increase as population increases and urbanization concentrate that population closer to public land. Consequently, prioritization for response actions must be ongoing to ensure that the limited funds remain focused on the mine sites adversely impacting the environment and human health. The cost of response actions will continue to be influenced by both the on-site conditions and the limitations of access. Focusing response actions in watersheds with docu-

Figure 5. The discharging end of the drain tunnel in the Sailor Flat hydraulic mine pit. During the removal action, a leader line was shot by an arrow to the far end. The line helped place a plastic cover over the mercury-contaminated gravels during excavation of the tunnel roof.

mented problems and coordinating actions among federal, state, and local agencies are likely to produce the greatest benefit for the public. The imposing task of addressing hazardous substances at historic mine sites reinforces the importance of regulatory agencies diligently monitoring current operations.

REFERENCES CITED

Alpers, C.N., and Hunerlach, M.P., 2000, Mercury contamination from historic gold mining in California: U.S. Geological Fact Sheet FS-061-00.

Church, S.E., Owen, J.R., von Guerard, P., Verplanck, P.L., Kimball, B.A., and Yager, D.B., 2007, this volume, The effects of acidic mine drainage from historical mines in the Animas River watershed, San Juan County, Colorado: What is being done and what can be done to improve water quality?, in DeGraff, J.V., ed., Understanding and Responding to Hazardous Substances at Mine Sites in the Western United States: Geological Society of America Reviews in Engineering Geology, v. XVII, doi: 10.1130/2007.4017(04).

Clark, W.B., 1970, Bulletin 193, Gold District of California, California Division of Mines and Geology, Sacramento, California, 186 p.

DeGraff, J.V., Rogow, M., and Trainor, P., 2007, this volume, Approaches to contamination at mercury mill sites: Examples from California and Idaho, in DeGraff, J.V., ed., Understanding and Responding to Hazardous Substances at Mine Sites in the Western United States: Geological Society of America Reviews in Engineering Geology, v. XVII, doi: 10.1130/2007.4017(07).

EPA, 1995, Historic hardrock mining: The West's toxic legacy: EPA Publication 908-F-95-002, Region 8, Environmental Protection Agency, 8 p.

EPA, 2000, Polar Star Mine Tunnel: U.S. EPA removal action field work to start in May: U.S. Environmental Protection Agency, Region 9, San Francisco, California.

EPA, 2004, Nationwide identification of hardrock mining sites, Evaluation Report No. 2004-P-00005: Office of the Inspector General, Environmental Protection Agency, 52 p.

Fey, D., and Wirt, L., 2007, this volume, Mine-related sources of metal loading to an alpine stream, based on a tracer-injection study, Clear Creek County, Colorado, in DeGraff, J.V., ed., Understanding and Responding to

Hazardous Substances at Mine Sites in the Western United States: Geological Society of America Reviews in Engineering Geology, v. XVII, doi: 10.1130/2007.4017(05).

Fliniau, H.L., 2002, A whirlwind tour of USDA AML cleanup projects: Association of Engineering Geologists, Program with Abstracts, v. 45, p. 64.

Ford, K.L., 1996, Risk management criteria for metals at BLM mining sites, USDI Bureau of Land Management Technical Note 390 rev., Denver, Colorado, 26 p.

Forest Service, 1993, Acid drainage from mines on the national forests: A management challenge, Program Aid 1505: U.S. Department of Agriculture, Washington, D.C., 12 p.

Gusek, J., Shipley, B., and Lindsay, D., 2005, Overcoming access issues at a remote passive treatment site near Lake Shasta, CA: Proceedings ASMR 2005 National Meeting, American Society for Mining and Reclamation, Lexington, Kentucky, p. 443–453.

Harden, D.R., 1998, California geology: Prentice-Hall, Inc., Upper Saddle River, New Jersey, 479 p.

Hunerlach, M.P., Rytuba, J.J., and Alpers, C.N., 1999, Mercury contamination from hydraulic placer-gold mining in the Dutch Flat mining district, California, *in* Morganwalp, D.W., and Buxton, H.T., eds., U.S. Geological Survey Toxic Substances Hydrology Program, Proceedings of the Technical Meeting, Charleston, South Carolina, March 8–12, 1999: U.S. Geological Survey Water-Resources Investigations Report 99-4018B, p. 179–189.

May, J.T., Hothem, R.L., Alpers, C.N., and Law, M.A., 2000, Mercury bioaccumulation in fish in a region affected by historic gold mining: The South Yuba River, Deer Creek, and Bear River Watersheds, California, 1999: U.S. Geological Survey Open-file Report 00-367, 30 p.

McDougal, R.R., and Wirt, L., 2007, this volume, Characterizing infiltration through a mine waste dump using electrical geophysical and tracer injection methods, Clear Creek County, Colorado, *in* DeGraff, J.V., eds., Understanding and Responding to Hazardous Substances at Mine Sites in the Western United States: Geological Society of America Reviews in Engineering Geology, v. XVII, doi: 10.1130/2007.4017(02).

NRC, 1999, Hardrock mining on federal lands: National Academy Press, Washington, D.C., 247 p.

Manuscript Accepted by the Society 28 November 2006.

Characterizing infiltration through a mine-waste dump using electrical geophysical and tracer-injection methods, Clear Creek County, Colorado

Robert R. McDougal
Laurie Wirt*
U.S. Geological Survey, Denver, Colorado 80225, USA

ABSTRACT

Infiltration of surface water through mine waste can be an important or even dominant source of contaminants in a watershed. The Waldorf mine site in Clear Creek County, Colorado, is typical of tens of thousands of small mines and prospects on public lands throughout the United States. In this study, electromagnetic (EM) conductivity and direct current (dc) resistivity surveys were conducted in tandem with a NaCl tracer study to delineate ground-water flow paths through a mine-waste dump and adjacent wetland area. The tracer was used to tag adit water infiltrating from braided channels flowing over the top of the dump to seeps at the base of the dump. Infiltration from the braided channels had a maximum flow rate of 92 m/day and a hydraulic conductivity of 1.6×10^4 cm^3/s. After rerouting of adit flow around the waste dump, discharge at some of the largest seeps was reduced, although not all seepage was eliminated entirely.

Integrating results of the tracer study with those of the EM and dc geophysical surveys revealed two main flow paths of ground water, one beneath the dump and one through the dump. The main source of water to the first flow path is deeper ground water emerging from the fault zone beneath the collapsed adit. This flow path travels beneath the waste dump and appears to have been unaffected by rerouting of the adit discharge around the waste dump. The source of the second flow path is infiltration of adit water from braided channels flowing over the top of the dump, which is intermediate in depth and flows through the center of the waste dump. Following rerouting of adit flow, discharge to seeps at the toe of the dump along this flow path was reduced by as much as two-thirds, although not eliminated entirely. Improved understanding of ground-water flow paths through this abandoned mine site is important in developing effective remediation strategies to target sources of metals emanating from the adit, waste dump, and contaminated wetland area.

Any use of trade, firm, or product names is for descriptive purposes only and does not imply endorsement by the U.S. government.

Keywords: Waldorf mine, geophysics, tracer study, ground water

---

*Deceased

McDougal, R.R., and Wirt, L., 2007, Characterizing infiltration through a mine-waste dump using electrical geophysical and tracer-injection methods, Clear Creek County, Colorado, in DeGraff, J.V., ed., Understanding and Responding to Hazardous Substances at Mine Sites in the Western United States: Geological Society of America Reviews in Engineering Geology, v. XVII, p. 9–24, doi: 10.1130/2007.4017(02). For permission to copy, contact editing@geosociety.org. ©2007 The Geological Society of America. All rights reserved.

INTRODUCTION

Infiltration of rain, snowmelt, and adit drainage through solid mine waste into receiving waters and the resulting transport of dissolved metals in shallow ground and surface waters are dependent on a wide range of factors unique to each site. These factors include the geology, type of mineral deposit, climate, topography, and mine-waste characteristics. Most investigations of mining contamination focus on surface runoff, largely because of the expense of installing monitoring wells or the difficulty of measuring subsurface flow. Yet results from previous abandoned mine studies show that recharge that enters the subsurface through mine waste can be an important or even dominant factor in contaminant transport in a watershed (Cannon et al., 2005; Church et al., this volume). Remedial efforts that ignore subsurface loading and transport are likely to be ineffective. Consequently, there is a need for multidisciplinary studies that can assimilate the many factors involved in the subsurface transport of metals.

The Waldorf mine site in Clear Creek County, Colorado (Fig. 1) was selected for a suite of integrated studies because the adit, or horizontal shaft that serves as a drain beneath the main mine workings, delivered a steady stream of mine water discharge across the waste dump. The flow from the adit was measured above and below the waste dump, allowing the rate of infiltration passing through the dump to be accurately determined. In addition, water levels within the dump and characteristics of the mine-waste material have been determined using a preexisting network of shallow boreholes (Malem et al., 2000). The Waldorf mine site is typical of tens of thousands of small mines and prospects on public lands throughout the United States that have been abandoned (Moore and Luoma, 1990; Fields, 2003; and Church et al., this volume).

This report is the first of a two-part investigation of the near-surface transport of trace metals from a mine site to a nearby stream and was conducted by members of the U.S. Geological Survey Mineral Resources Team as part of the Process Studies of Contaminants Associated with Mineral Deposits project. In this first paper, electromagnetic (EM) conductivity and direct current (dc) resistivity surveys were conducted in tandem with a NaCl tracer study to delineate ground-water flow paths through a mine-waste dump and adjacent wetland area. In the second paper (Fey and Wirt, this volume), tracer studies and synoptic sampling were used to evaluate metal loading from different parts of the mine site to a nearby stream over a 2-year period (2002–2003). Immediately following the tracer studies in 2002, the adit flow was routed around the mine-waste dump in an attempt to reduce metal loading by limiting the amount of infiltration through the dump. The rerouting of the adit flow was seen as an opportunity to study the effect of changes in the amount of infiltration on metal transport and loading through the waste dump to the wetland below.

In this study, electrical geophysical methods were used to delineate saturated flow paths through the waste dump and wetland area. Electrical conductivity, the inverse of electrical resistivity, is a fundamental property of all earth materials. The degree of conductance varies with rock or sediment type, porosity, clay mineral type and content, and the quantity and quality of moisture. Poorer quality ground water (that is, water with higher concentrations of dissolved solids) or sediments with higher clay content are usually more conductive (Zohdy et al., 1974). Conductivity can be expressed in units of millisiemens/m and is a measure of the magnetic field resulting from an induced current. Resistivity is expressed in ohm-meters, and is an estimate of the earth's resistivity calculated using the relationship between resistivity, an electric field, and current density (Ohm's law), and the geometry and spacing of the current and potential electrodes. When the earth is not homogeneous and isotropic, which is usually the case, this estimate is called the apparent resistivity, which is an average of the true resistivity in the measured section of the earth (Dobrin and Savit, 1988). Interpretations of geophysical surveys can sometimes result in more than one nonunique interpretation; therefore, the addition of the NaCl tracer helped to distinguish both ground-water flow paths originating from surficial infiltration and those originating from other ground-water sources.

Purpose and Scope

The objectives of this study are to: (1) identify subsurface pathways of water movement through the mine-waste dump and adjacent wetland area using geophysical methods, and (2) determine how much of the adit water infiltrates through the mine-waste dump using discharge measurements and a NaCl tracer study. Prior to diversion of the adit flow in 2002, an EM survey was conducted concurrently with a tracer study to delineate the subsurface pathways of adit flow through the dump. The NaCl tracer was used to tag adit water infiltrating downward from braided channels flowing over the top of the dump to seeps at the base of the dump. In addition, discharge measurements were used to measure infiltration through the mine-waste dump and to calculate the residence time of subsurface water and the hydraulic conductivity of the dump material.

Direct current resistivity surveys were conducted across the waste dump and the wetland area below in order to provide cross-sectional profile information about the location and depth of conductive anomalies. The use of both EM contour maps and dc profiles allows the electrical properties of the dump and underlying material to be analyzed in three dimensions. Four dc surveys were completed, three transecting the dump and one down gradient along an old railroad grade.

Environmental Setting and Mining History

The environmental setting, described here for both papers, is a high-altitude mountain watershed draining the Continental Divide in the Colorado Mineral Belt of Clear Creek County, Colorado. The Waldorf study site is located ~10 km southwest of Georgetown, Colorado, near the headwaters of Leavenworth

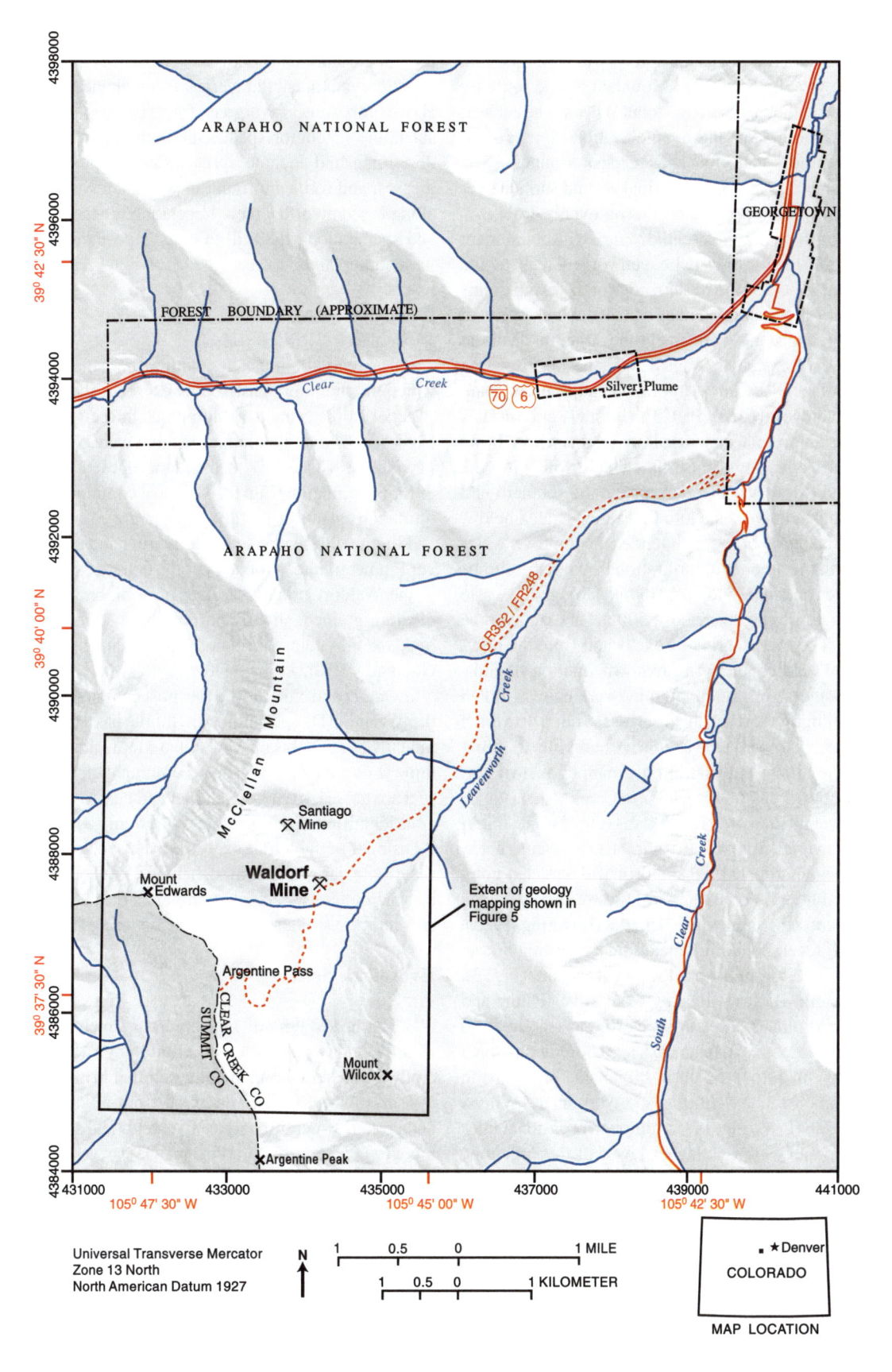

Figure 1. Location map of the Waldorf mine site and surrounding area.

Creek, a tributary of Clear Creek (Fig. 1). The headwaters of Leavenworth Creek drain a moderately steep-walled cirque that is bounded on the west by McClellan Mountain, to the south by Argentine Peak, and to the southeast by Mount Wilcox. The adit and mine dump are just above tree line at an elevation of ~3540 m, on a natural break in slope at the foot of McClellan Mountain. The mine dump stands ~12 m above the original ground surface over a thin layer of darkly colored, organic-rich soil over bedrock.

The surface of the dump is essentially flat, with three main lobes that drop off steeply toward Leavenworth Creek to the southeast. A collapsed adit on the western edge of the dump drains the Wilcox tunnel (aka Waldorf tunnel or adit), which drains a large complex of tunnels and workings in the Argentine Mining District. During the tracer study in August 2002, the adit discharge flowed over the mine-waste dump, through an adjacent wetland area, and into Leavenworth Creek. Part of the adit water infiltrates through the waste dump to become shallow ground water flowing to the wetlands down gradient. Immediately following the tracer study, the adit flow was diverted around the southern end of the dump through a culvert and into the wetlands area below. This low-cost remedial action was intended to improve water quality by reducing the amount of infiltration through the dump.

The waste pile contains ~51,000 m^3 of mill tailings and waste rock, based on an approximate area of 8500 m^2 and an average thickness of 6 m. The rock debris was excavated from the now-collapsed adit and also contains an unknown fraction of mill-tailings material. Some ore from the upper mine workings was milled on-site, and the tailings were discharged across the dump to a ditch running down gradient toward Leavenworth Creek (Fig. 2). Fine- to medium-grained tailings extend from the main pile down gradient along the western bank of Leavenworth Creek for a distance of ~1 km and cover an area of ~125,000 m^2 (Fig. 2). The thickness of the tailings varies from a few centimeters to over a meter.

Production began around 1900 and primarily targeted gold, silver, and lead, although copper and zinc were also extracted from the Waldorf and other mines in the district (Lovering, 1935). By the late 1920s, several buildings including a boarding house, a post office, and a 50-ton mill were located at the site (Fig. 3). The Argentine Central Railway, running from Silver Plume and Georgetown, carried miners and tourists as well as supplies and ore to and from the Argentine Mining District. The Wilcox tunnel was driven ~1200 m into McClellan Mountain, primarily to intersect the Commonwealth vein (Fig. 4). Work in the Wilcox tunnel ceased in 1926 (Lovering, 1935), although occasional mining and mill processing of ore in higher levels of the Argentine Mining District continued into the 1970s.

Climate and Vegetation

Average annual precipitation in the area is ~60–80 cm/year, with temperatures ranging from an average annual minimum of −15 °C to an average annual maximum of 17 °C (Western Regional Climate Center, 2005). Stream flow, resulting from snow melt, usually peaks in June or July. Afternoon thunderstorms, which can provide significant amounts of precipitation, are common in the summer months.

Ecosystem habitat varies from alpine meadow or tundra above timberline to evergreen forest below timberline. Alpine and subalpine vegetation species are found in the vicinity of the Waldorf mine and include spruce, fir, and pine. Willows, alders, grasses, and forbs are found in wetland areas near the mine and along Leavenworth Creek. Vegetation is absent on the waste pile and sparse along the mill-tailings deposits located between the waste dump and the creek.

Geologic Setting

The Waldorf mine lies within the Colorado Mineral Belt, which extends northeast-southwest from the Front Range near Boulder to the San Juan Mountains in the southwestern part of Colorado. The predominant rock in the area of the headwaters of Leavenworth Creek is the metamorphosed Precambrian Idaho Springs Formation (Fig. 5). The most common lithologies for this formation are quartz-biotite schist, quartz-biotite-sillimanite schist, and biotite-sillimanite schist (Lovering and Goddard, 1950). Silver Plume granite is found near the entrance of the Wilcox tunnel at the Waldorf mine. This late pre-Cambrian rock is generally medium-grained, slightly porphyritic, biotite granite, containing pink and gray feldspars, smoky quartz, and biotite (Lovering and Goddard, 1950).

Quartz and ankerite are the most common gangue minerals in the Argentine District, although fluorite has been reported to occur in many of the veins on McClellan Mountain west of the Waldorf mine (Lovering, 1935). The predominant ore minerals in the district are galena, pyrite, sphalerite, and chalcopyrite. Primary minerals identified in the Waldorf waste dump by hand specimen and by using a portable field spectrometer include quartz, chalcopyrite, pyrite, jarosite, goethite, hematite, and other amorphous iron bearing minerals. Secondary minerals include kaolinite, gypsum, and smectite (Malem et al., 2000; Lovering and Goddard, 1950).

Hydrologic Setting

The headwaters of Leavenworth Creek begin at the Continental Divide of the Colorado Front Range. Patches of snow generally are present throughout the summer in most years. Base flow to Leavenworth Creek is derived from infiltration of snowmelt and ground-water discharge to wetlands and alpine streams. Shallow ground water is stored in surficial veneers of unconsolidated colluvial or fluvial sediments containing a mixture of boulders, gravel, cobbles, and soil. Generally less than a few meters thick, these surficial deposits are moderately to highly permeable. At the valley floor and adjacent to streams, the water table is generally near or just below land surface. Streams may be gaining or losing within any given reach, but overall they tend to gain in proportion to drainage area.

The Waldorf mine dump apparently was deposited on top of a fen, or ground-water fed wetland area, which is exposed south

Figure 2. Digital orthophoto quad (DOQ) showing the Waldorf mine dump and extent of mill tailings along Leavenworth Creek.

and east of the dump. The dump is underlain by bedrock covered by a layer of native soil that ranges in thickness from ~0.5 m to 1 m (Malem et al., 2000). On the basis of water-level measurements from two transects of ten nested wells (Fig. 6), most of the dump is unsaturated (Malem et al., 2000). Each nested well consists of an upper and lower piezometer. The screened interval of the upper well is in the unsaturated zone, and that of the lower well is near the base of the dump beneath the water table. The water table was usually within a meter of the base of the mine waste (Malem et al., 2000).

The fen water is characterized by low specific conductance (30–150 ΩS/cm), an indication that its source is the shallow infiltration of snowmelt runoff. In contrast, discharge from the Waldorf adit is much higher in specific conductance (600–650 ΩS/cm) and is presumably deeper-circulating ground water in contact with crystalline rocks. Igneous and metamorphic rocks have little to no primary permeability; thus, water in bedrock aquifers is stored in fractures within the rocks (Topper et al., 2003). A role of the tunnel was to intercept water-bearing fractures in order to lower the water table surrounding the upper mine workings. Bedrock aquifers often produce higher concentrations of total dissolved solids than ground water from shallow unconsolidated deposits, owing to greater water-rock interaction along longer flow paths having a longer residence time.

METHODS OF INVESTIGATION

Field activities and methods described in this section include: (1) tracer-injection methods and discharge measurements, (2) an EM survey, and (3) dc resistivity surveys. Because part of this study focused on water movement through the waste dump, a NaCl tracer was used to tag water infiltrating into the waste dump from

Figure 3. Historical photos of the Waldorf mine site showing the mill and out buildings. The Vidler mine and Argentine Peak are seen in the background (top). Rail lines and aerial tram are shown in bottom photo. (Photos by Louis Charles McClure, ca. 1906).

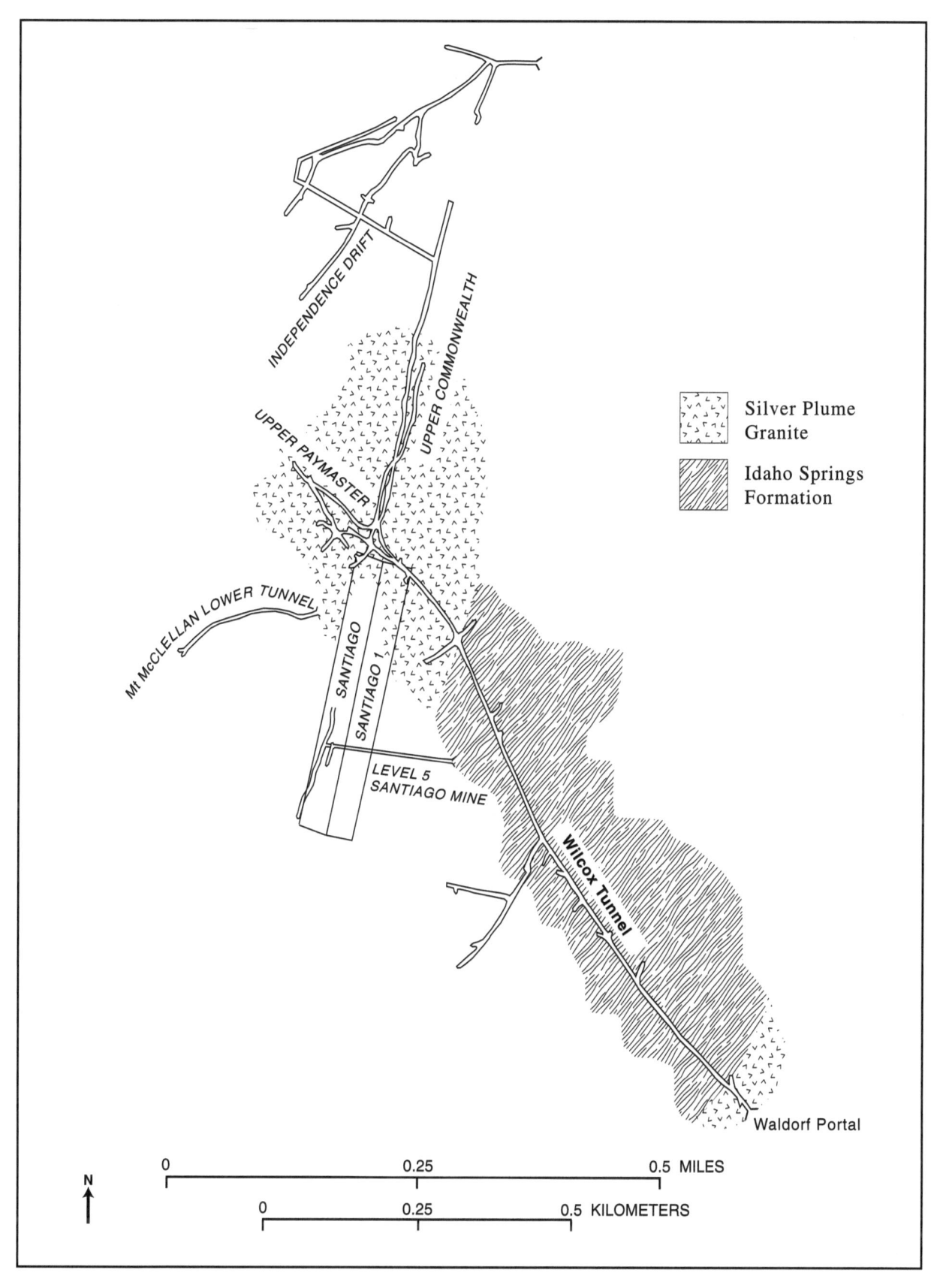

Figure 4. Generalized map of workings associated with the Waldorf mine and mill (after Lovering, 1935).

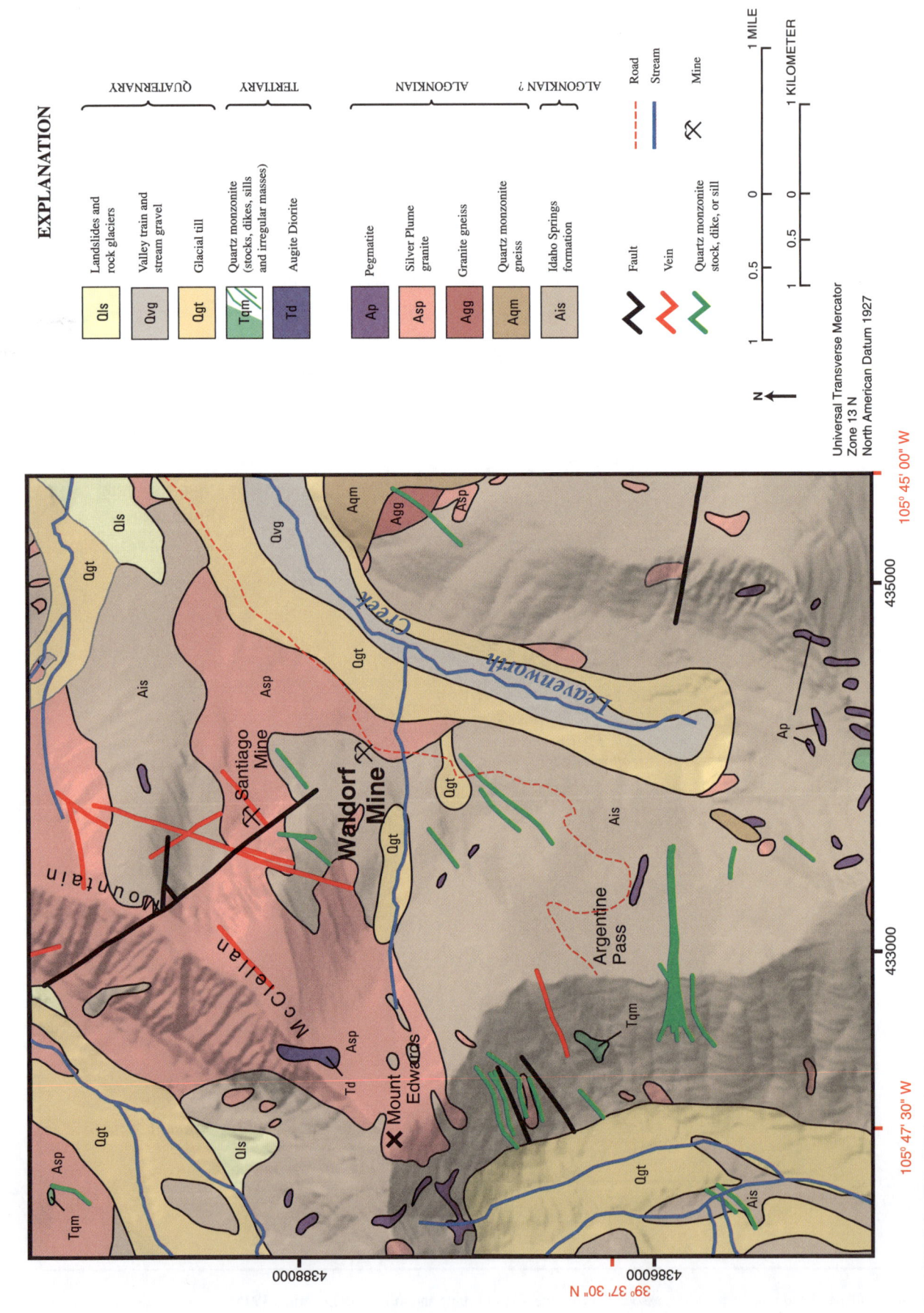

Figure 5. Generalized geology in the vicinity of the Waldorf mine (after Lovering, 1935).

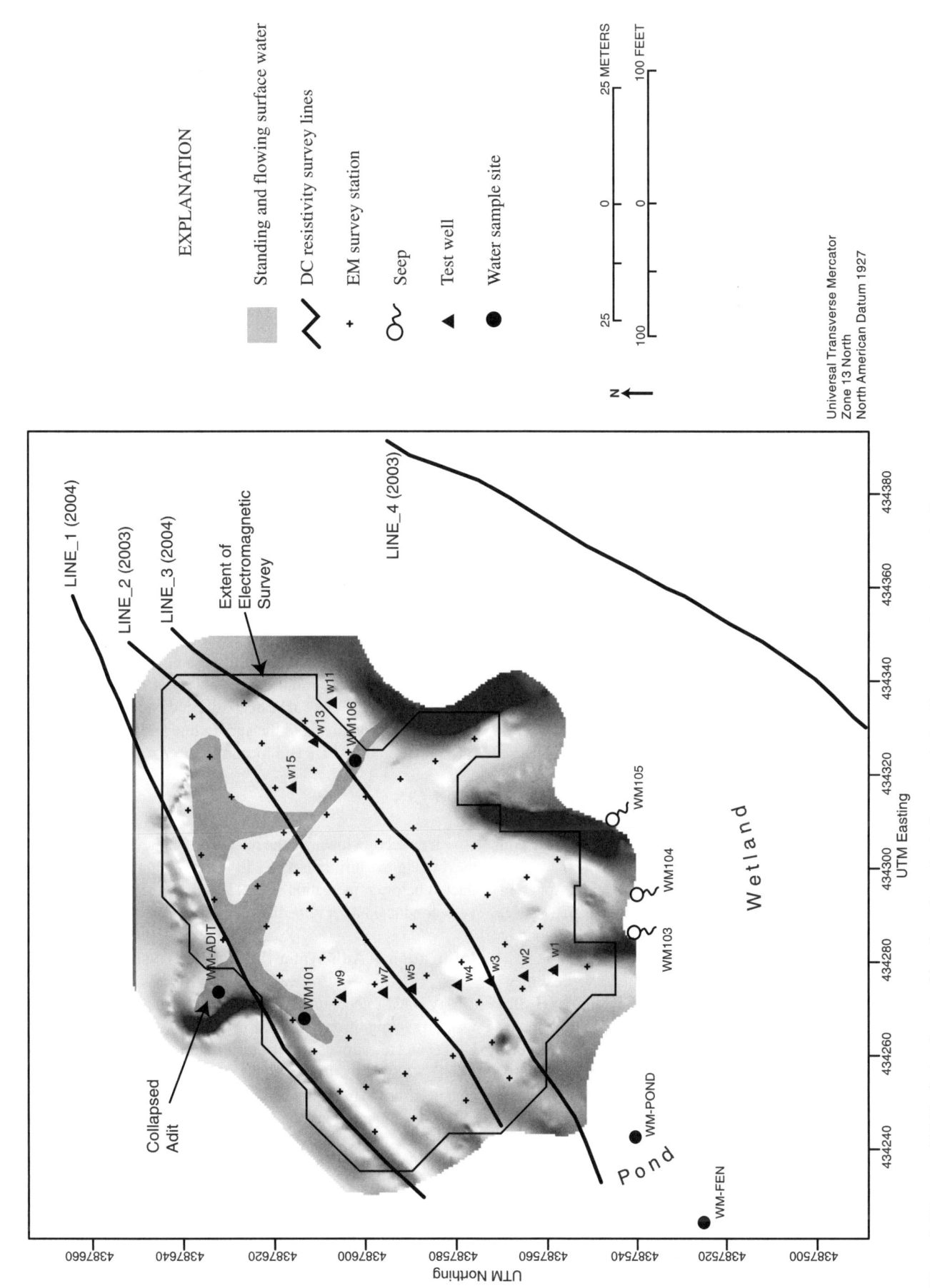

Figure 6. Site map of Waldorf mine dump and location of geophysical surveys, test wells, seeps, and water sample sites.

the collapsed adit. In addition, a LiBr tracer study was conducted to determine metal loading from different parts of the Waldorf mine site to nearby Leavenworth Creek (Fey and Wirt, this volume). The EM survey was conducted concurrently with two tracer studies in late July 2002 in order to delineate subsurface pathways prior to rerouting of the adit flow. Following diversion of water flowing from the adit, dc transects were made in 2003 and 2004 on the waste pile and in the wetland area down gradient from the waste dump. Discharge measurements were used to develop a water budget for the infiltration of adit flow through the waste dump.

Tracer-Injection and Discharge Methods

In July 2002 a saturated NaCl solution was injected into the adit stream up-gradient from the waste dump over a 3-day timeframe. The NaCl tracer experiment was timed to begin before and end after the LiBr tracer study in Leavenworth Creek (Fey and Wirt, this volume). Water from the adit was mixed with stock salt in a 55-gallon reservoir until the solution had reached a chloride concentration of ~150,000 mg/L. (For comparison, background chloride concentrations were ~0.2 mg/L.) The tracer solution was pumped from the reservoir through plastic tubing into the adit stream.

Discharge measurements using either a Baski cutthroat flume or a graduated cylinder and stopwatch approach were made at the adit on braided channels flowing over the surface of the waste pile and at seeps at the base of the dump. Measurements were made in 2002 before the adit water was diverted around the dump and again in 2003, one year after remediation efforts were completed.

Electromagnetic Survey Methods

Before adit flow was rerouted in August 2002, an EM survey was conducted to map the subsurface electrical conductivity of the mine dump. Measurements were made with a Geonics EM34–3 ground conductivity meter using a transmitter and receiver coil spacing of 10 m. This instrument measures the apparent conductivity of the ground through electromagnetic induction. Apparent conductivity is a volume average of a heterogeneous half-space. Only when the earth is a homogeneous half-space is the apparent conductivity the same as the true conductivity.

Measurements were made with the coils in the vertical magnetic dipole (VMD) and the horizontal magnetic dipole (HMD) orientations. With this instrument and given coil spacing, a maximum depth of investigation of ~15 m is possible in the VMD orientation and of ~7.5 m in the HMD orientation (McNeill, 1980). The survey was conducted on a grid with survey stations spaced at 10-m intervals. The survey grid was constructed by first selecting an origin point located on an existing benchmark.

Direct Current Resistivity Methods

In July 2003, one year after the adit flow was rerouted, two dc resistivity profiles were conducted to expand coverage and improve understanding of ground-water flow paths beneath the waste dump and through the wetland area. Two additional lines were surveyed in 2004. An Advanced Geosciences, Inc. (AGI) SuperSting R8/IP eight-channel multielectrode instrument was used to record apparent resistivity data to determine lateral and vertical resistivity variations within the mine-waste pile. Multielectrode systems have the capability of recording many channels of data simultaneously and allow for the collection of very dense dc resistivity data in a relatively short amount of time. For this study, the surveys were conducted using an inverse Schlumberger array (Telford et al., 1990).

The survey lines collected in 2003 consisted of 32 electrodes with a 4-m spacing between each electrode. Also included in the 2003 work was a high-resolution Global Positioning System (GPS) survey of the waste-pile surface. These data were used to produce a detailed digital elevation model (DEM) of the dump and to geometrically correct the apparent resistivity data. Survey lines completed in 2004 used 76 electrodes with 2-m spacing to improve data resolution (Fig. 6).

The apparent resistivity data are converted to true resistivity at depth using a numerical inversion routine. The inversion routine used for this study was EarthImager 2D, a dc resistivity inversion program from Advanced Geosciences, Inc. The result of the inversion process is a highly detailed cross section of resistivity corrected for topography along the profile.

It is important to note that two-dimensional lines of resistivity data are collected in three-dimensional space and must be interpreted bearing in mind the possibility of off-line anomalies. The magnitude of this "3D effect" depends upon the lateral distance of the anomaly from the survey line and can result in misinterpretation of its size and location.

RESULTS

This section includes the results of (1) the NaCl tracer study used to tag the infiltration of adit water, (2) the EM survey conducted before remediation in 2002, and (3) the dc resistivity surveys conducted after remediation in 2003 and 2004.

Infiltration of Adit Water through the Waste Dump

A NaCl tracer study was used to tag the infiltration of adit water traveling through the waste dump to create a simple water budget and to estimate rates of infiltration. Before remediation on July 29, 2002, flow from the adit was split between three braided channels shown in Figure 6. Flow was measured below the opening of the collapsed adit and at seeps and channels near the base of the mine-waste dump. Surficial discharge at the opening of the collapsed adit was 4.8 L per second (L/s). One braided channel with a discharge of ~1 L/s flowed to the southwest for ~10 m before completely infiltrating into the dump. The other two braids, merging above sampling site WM106, had a combined discharge of 2.7 L/s measured at the toe of the waste dump. Thus, by subtraction, 43% of the surficial adit water infiltrated into the waste dump. In addition, some of the adit water leaving the mine tunnel may directly infiltrate the rock rubble zone beneath the opening of the collapsed adit without discharging to the land surface. As will be evident from the geophysical surveys, subsurface adit water may provide a significant source of ground water moving beneath the western half of the waste dump.

Ground-water samples were collected from the base of the unsaturated mine-waste dump 23–24 h after the NaCl tracer-injection started (Fey and Wirt, this volume, Table 2 therein). Samples at sites WM104 and WM105 contained elevated chloride, indicating breakthrough of the tracer through the dump in less than 24 h. This confirms a hydraulic connection between surficial adit flow to seeps WM104 and WM105 (Fig. 6). Concentrations of Cl in the seeps (9.8 and 29 mg/L) were less than those in the adit flow (39–43 mg/L), indicating mixing of the tracer solution with either subsurface adit or background water containing little chloride along this flow path or indicating that the tracer had not reached steady-state conditions at the base of the dump at the time of sampling. The lower concentration of chloride at site WM104 suggests a slower rate of flow and greater dispersion along this flow path. In contrast, a chloride concentration of 0.2 mg/L at site WM102 indicates that none of the tracer had reached this flow path at the time of sampling, suggesting either a primary source other than infiltration of surficial adit discharge or a much slower rate of ground-water movement.

Hydraulic conductivity for the waste dump along the largest ground-water flow path was determined using Darcy's law. Based on the calculated infiltration rate of 2.06 L/s, a vertical distance of 13 m, and a horizontal distance of 100 m, the hydraulic conductivity along flow paths emerging at sites WM104 and WM105 is 1.6×10^4 cm^3/s. The rate of water movement through the waste dump is ~92 m/day or 3.8 m/hr, which is thought to represent a maximum rate.

Approximately 1.1 L/s of adit water emerged from seeps near sites WM104 and WM105. The remaining 1.6 L/s of infiltrated adit water that is unaccounted for presumably discharges to the wetland area between the waste dump and Leavenworth Creek (Fig. 2). This wetland area probably receives other sources of water, including recharge from rain runoff and snowmelt and background water in the shallow alluvium that is relatively uncontaminated by mining. At the time of the tracer study, other sources of water infiltrating through the waste dump were negligible but could be substantial during other times of the year, such as the spring snowmelt period.

Ground water may also enter the waste dump from bedrock beneath the collapsed adit. The Waldorf tunnel is oriented along a fault in the Silver Plume granite, which is evident by comparing trends of the tunnel in Figure 4 with the fault in Figure 5. Ground-water movement through bedrock is likely to be enhanced by a greater occurrence of connected fractures in the vicinity of the fault. The fen and the waste dump conceal any surface expression of the fault, although ground-water discharge in this area is consistent with this interpretation. Water chemistry of deeper ground water discharging along the fault zone is expected to be similar to that of the adit, although it was not possible to sample such water directly.

Delineation of Flow Paths Based on EM Surveys before Remediation

Data collected from the EM surveys prior to remediation were used to map apparent conductivity for the vertical and horizontal magnetic dipoles (Fig. 7). From top to bottom, the three horizontal layers are the DEM of the dump surface, the HMD contour map, and the VMD contour map. The HMD contour map represents relatively shallow apparent conductivity data to a depth midway through the pile (~7.5 m). In the HMD orientation, the instrument response is more sensitive to near surface material, whereas in the VMD orientation, the response is less sensitive to changes in near surface conductivity (McNeill, 1980). The VMD contour map shows relatively deep apparent conductivity data through the pile near the bedrock surface (~15 m).

Conductivity increases with depth along the contact between the dump material and the break in slope at the base of McClellan Mountain (site A in the HMD and the VMD layers in Fig. 7). This conductive zone extends toward site A' (Fig. 7), which coincides with seeps WM104 and WM105 (Fig. 6). Conductive anomalies A and A' appear to be connected at depth, as seen in the VMD data, but not in the middle of the waste dump measured by the HMD data. Therefore, we conclude that ground water and/or adit water infiltrates through the dump material and travels vertically downward along the waste pile–bedrock interface along flow path A, exiting at seeps and also in the wetland area farther down gradient.

A second conductive zone is evident in the HMD contour map (Fig. 7) as flow path B. The conductive zones at B, B', and B" are probably wet or saturated ground water overlying clay layers within the waste pile. The source of the water creating these conductive zones could be adit water flowing through fractures beneath the Wilcox tunnel near its outlet or could be derived from nonmining related ground water draining the colluvial slope of McClellan Mountain. In either case, underflow at the toe of the mountain slope creates a conductive saturated zone at the base of the waste pile near site B. As would be expected, the high conductivity areas in the HMD data generally correlate well with seeps at the base of the dump (B to B', at WM105) and beneath the braided surface channels shown in Figure 6.

Two other deep conductive anomalies are shown in the VMD contour map at C and D (Fig. 7). Zone C does not appear to have an upslope ground-water source, as in the case of flow paths A and B, but does lie beneath a wetter zone. Therefore, this anomaly may be caused by wet clay or organic-rich soil that is perched on top of relatively impermeable bedrock or by downward flow from flow path B above. The anomaly shown at D begins at the break in slope of McClellan Mountain but does not extend completely

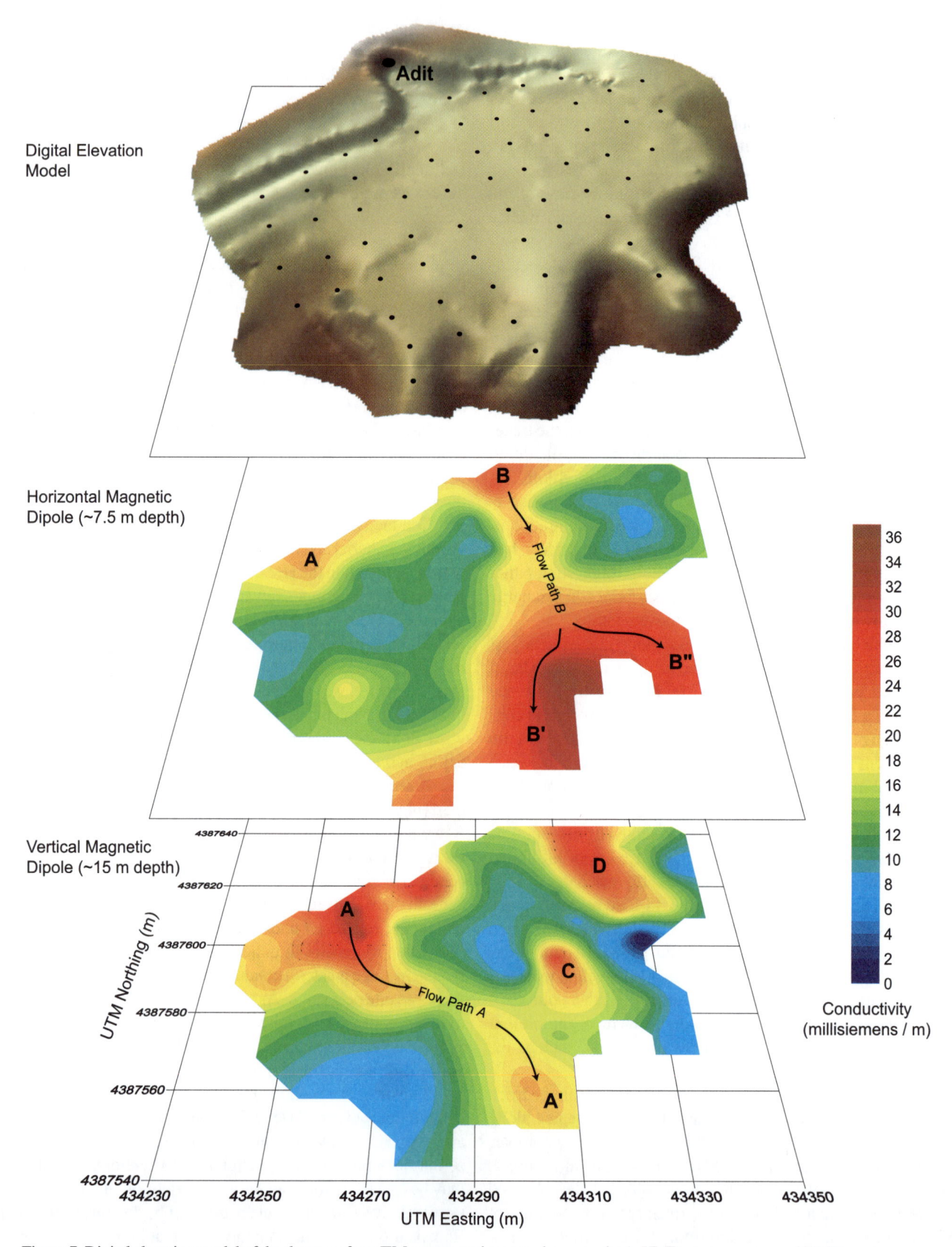

Figure 7. Digital elevation model of the dump surface (EM survey stations are shown as dots), HMD contour map, and VMD contour map.

across the VMD contour map toward the down-gradient end of the dump. In the shallower HMD data this same zone is relatively resistive, indicating that the source of this deeper anomaly is not from above. This presents two possibilities for source water. First, the conductive anomaly at D could be caused by shallow alluvial ground water draining the slope of McClellan Mountain up-gradient from the dump. Second, the source water could be related to flow path *B*, infiltrating vertically downward at a slight angle to the east. In either case, the native soil between the bedrock and the dump material appears to be saturated, but heterogeneities in the waste-dump material should be considered.

The nature of the waste-dump materials and the degree of saturation is partly known from two transects of nested wells (Fig. 6) that were drilled by Malem et al. (2000). The deepest well logs penetrating the full vertical thickness of the waste dump indicate that a 0.5 to 1-m-thick layer of native soil is present between the dump material and the underlying bedrock. As mentioned earlier, water-level measurements by Malem et al. (2000) indicate that the water table is at or near the base of the waste dump, near the original ground surface, which is believed to have been a fen. Thus, the dump is largely unsaturated, except where infiltration of adit water is occurring beneath the braided channels on the surface. The conductive zones suggest that infiltration occurs predominantly as vertical flow driven by gravity, exploiting preferential flow paths through the nonhomogeneous waste rock. Where wet or saturated, clays and organic material are thought to account for increased conductivity, particularly near the base of the waste-dump material.

Delineation of Flow Paths Based on DC Surveys after Remediation

Direct current resistivity surveys in 2003 and 2004 following remediation provide vertical profiles of electrical resistivity beneath the dump and wetland area (Fig. 8). Resistivity values in ohm-meters obtained from the inversion program were converted to conductivity values in millisiemens per meter so that the profiles could be easily compared to the apparent conductivity maps from the previous EM surveys. Values of apparent conductivity are not directly comparable with true conductivity, and therefore should be compared only in a relative sense.

Resistivity Line 1 (Fig. 8) is oriented along the break in slope between the waste pile and McClellan Mountain (Fig. 6). Several conductive zones are seen near the surface in this profile. Zone A is in an area of ground-water discharge extending from the slope of McClellan Mountain into the willow-dominated wetland down-gradient. Ground water sampled in this area is of good quality and has not been significantly affected by the adit or waste dump (Fey and Wirt, this volume). The high conductivity here is interpreted as saturated organic rich soils near the surface. Zone B is down-slope from a stand of willows, indicating the presence of shallow ground water. The resistive areas beneath Zones A and B are probably bedrock of Silver Plume granite.

Zones C and D in Line 2 and Zone E in Line 3 are associated with the occurrence of weathered pyritic ore and jarosite, observed during data collection and confirmed in the laboratory using a portable field spectrometer. Weathering of pyrite to jarosite and associated acid production from this process are likely to be the source of these conductive zones. Typically, as acid is produced in the pyrite weathering process, ion concentration increases resulting in an increase in conductivity (Baird and Cann, 2004). Conductive Zone E in Line 3 occurred near a pool of standing rain water on the surface of the dump. This anomaly coincides with a small depression on the waste pile surface where rain water would pond and then infiltrate. It was observed that the rate of infiltration beneath these pools was highly variable. In instances where the surface material was relatively permeable, rain water would infiltrate into the waste pile material within minutes or hours. In other instances where rinds resistive to surface infiltration had formed on the dump, the standing puddles would last for days until the water had evaporated.

Conductive anomalies associated with flow paths identified in the EM contour maps (Fig. 7) appear in Lines 1, 2, and 3 of the resistivity profiles and are shown in Figure 8 as flow paths *A* and *B*. Flow path *A* begins in the waste dump (Line 1) as a well-defined conductive zone but becomes less conductive and more diffuse down-gradient in Line 2 and Line 3. The conductive anomaly at F in Line 3 is probably associated with this flow path and correlates in lateral position with anomalies A' and B' in the EM VMD and HMD conductivity contour maps (Fig. 7). Due to possible 3D effects, as previously discussed, the exact vertical position of the anomaly may not be accurate in the profile.

The anomalies associated with flow path *B* in Lines 1, 2, and 3 are more conductive and well defined than those seen in flow path *A*. The lateral position of the flow path correlates well with flow path *B* shown in Figure 7. However, conductive anomaly G may be shallower than shown, again considering 3D effects in the profile.

Because adit flow had been rerouted around the dump at the time these resistivity data were collected, the most likely source of water creating the conductive zones in both flow paths *A* and *B* is ground water originating from McClellan Mountain. In Line 1, the anomalies appear to occur in bedrock. Therefore, it is possible that both flow paths discharge to the dump through faults, fractures, or veins in the Silver Plume granite.

The lower dc profile in Figure 8, Line 4, follows an old railroad grade through the wetland area down gradient from the waste pile (the location relative to the waste pile is shown in Fig. 6). Broad, shallow conductive anomalies at Zones H and I suggest that conductive waters containing high dissolved-solids content extend between the waste dump and the wetlands area. Zone F is associated with the main flow path of diverted adit water. The source of this higher conductivity is adit water and ground-water discharge from the seeps at the base of the waste pile. Some adit water probably flows underneath the dump above the bedrock surface. The lack of conductive zones at depth in this profile indicates either that resistive bedrock occurs at a depth of ~5 m

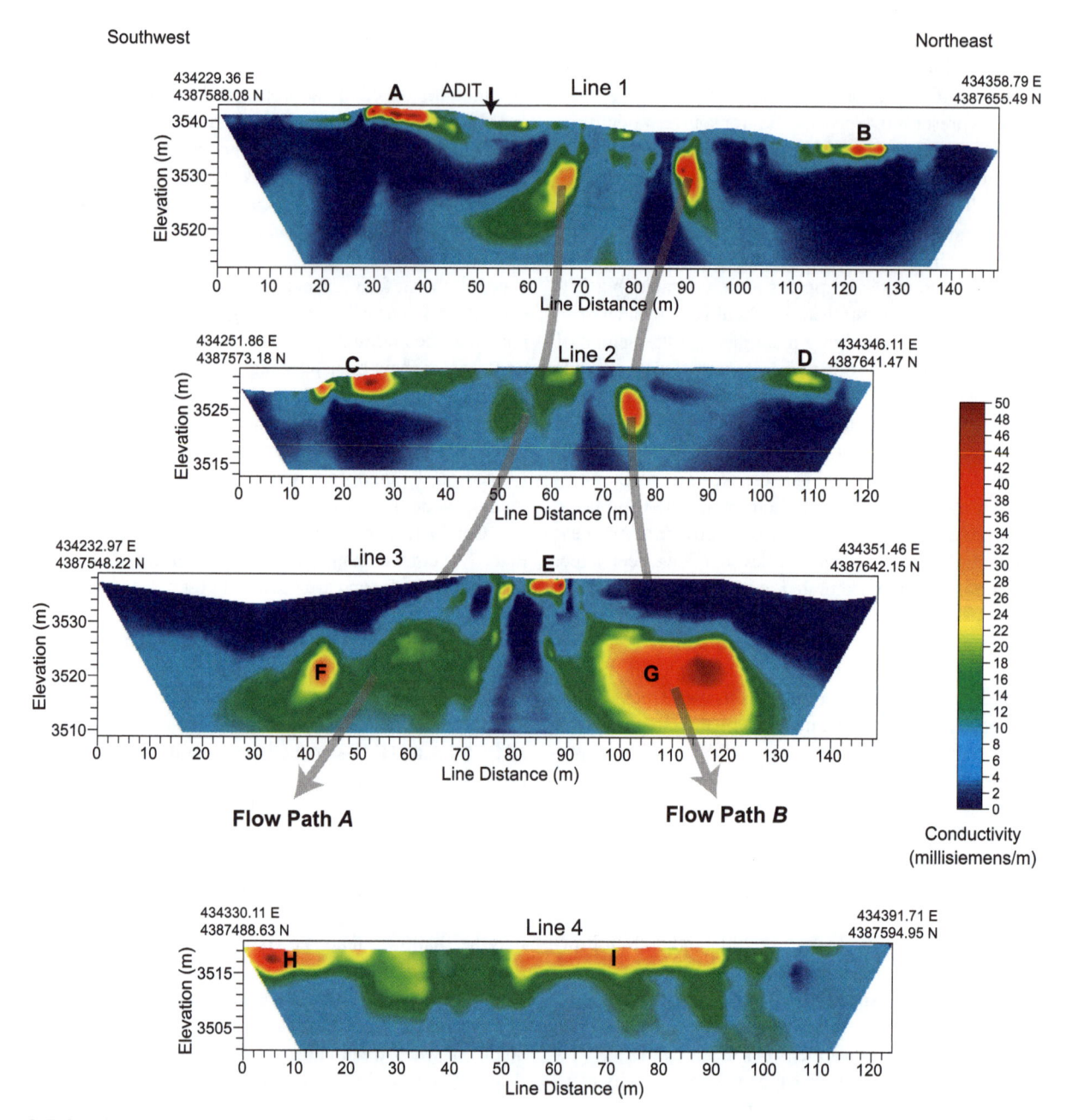

Figure 8. DC resistivity profiles collected in 2003 and 2004. Line locations and year of collection are shown in Figure 6. Resistivity values of Ohm-m have been converted to conductivity values of millisiemens/m. The end points of each line are given in universal transverse Mercator (UTM) coordinates (North American datum 1927).

or that ground water below this depth is sharply contrasting in quality, or both. Additional sampling of the wetland area using piezometers is planned to confirm this hypothesis.

CONCLUSIONS

By integrating the results of the tracer study with those of the EM and dc geophysical surveys, two main flow paths of groundwater movement are evident through the dump and beneath the dump. The flow paths are represented by arrows between A and A' (path A) and between B and B' (path B) in Figure 7. The EM and dc surveys provide a three-dimensional image of groundwater movement through the waste dump.

The source of water to path A appears to be deeper ground water that emerges from the fault zone beneath the collapsed adit and interpreted as predominantly adit water that could have mixed with a smaller fraction of non-mining-affected ground water. The unaffected ground water is probably similar in water chemistry to that discharging to the fen (Fey and Wirt, this volume). Ground water appears to flow downward along the northwest interface between the waste dump and underlying bedrock. The main flow path for path A continues horizontally beneath the waste dump and discharges to the wetland area rather than to seeps at the toe of the dump. Seeps at the base of the waste dump along path A, such as WM103, are small but persisted one year after the adit water had been rerouted, indicating that the primary source of water to this flow path is unchanged.

The source of water to path B is infiltration from discharge from the collapsed adit portal that traveled through the waste dump to seeps near WM104 and WM105. Ground water along this larger flow path was tagged by the tracer solution with a maximum flow rate of 92 m/day and a hydraulic conductivity of 1.6×10^4 cm^3/s. After remediation, discharge to WM105 ceased and discharge at WM104 decreased from 0.32 to 0.14 L/s. Although as much as two-thirds of infiltration along flow path B was eliminated by rerouting of the adit discharge, this did not appear to have a measurable impact on metal loading of mine-impacted inflows to nearby Leavenworth Creek in the following year (Fey and Wirt, this volume).

The dc profiles from 2003 and 2004 show additional conductive zones that are likely to be caused by wet or saturated clay, or native soil beneath the dump. The source of the water is either adit water that does not exit at the portal or ground water draining from McClellan Mountain. In either case, the unconsolidated material is saturated above the bedrock interface. The dc profile crossing the wetland area indicates that the contaminated apron of the waste dump is shallow and probably does not extend far into the underlying bedrock beneath the contaminated wetland.

This study has important implications for remediation of abandoned mine sites because it demonstrates that integrated studies using geophysical and geochemical methods can identify subsurface flow paths and thereby improve our understanding of sources of metal contaminants. It is hoped that the information provided here will be useful in developing future strategies for reducing loads of dissolved metals from the adit, waste dump, and contaminated wetland area that reach Leavenworth Creek. The proportions of metal loads from mining-impacted sources to Leavenworth Creek will be addressed in the following chapter.

ACKNOWLEDGMENTS

This project was funded by the U.S. Geological Survey Mineral Resources Team's Process Studies of Contaminants Associated with Mineral Deposits Project. The author would like to thank Douglas B. Yager, Rhonda Driscoll, Murray Beasley, and Ted Asch for fieldwork assistance and Robert Horton for assistance in the field and data analysis.

REFERENCES CITED

Baird, C., and Cann, M., 2004, Environmental chemistry: W.H. Freeman & Company, New York, 652 p.

Cannon, M.R., Church, S.E., Fey, D.L., McDougal, R.R., Smith, B.R., and Nimick, D.A., 2005, Understanding trace-element sources and transport to upper Basin Creek in the vicinity of the Buckeye and Enterprise Mines, chap. E1 of Nimick, D.A., Church, S.E., and Finger, S.E., eds., Integrated investigation of environmental effects of historical mining in the Basin and Boulder Mining Districts, Boulder River watershed, Jefferson County, Montana: U.S. Geological Survey Professional Paper 1652, p. 401–456.

Church, S.E., Owen, J.R., von Guerard, P., Verplanck, P.L., Kimball, B.A., and Yager, D.B., 2007, this volume, The effects of acidic mine drainage from historical mines in the Animas River watershed, San Juan County, Colorado: What is being done and what can be done to improve water quality?, in DeGraff, J.V., ed., Understanding and Responding to Hazardous Substances at Mine Sites in the Western United States: Geological Society of America Reviews in Engineering Geology, v. XVII, doi: 10.1130/2007.4017(04).

Dobrin, M.B., and Savit, C.H., 1988, Introduction to geophysical prospecting (4th edition): McGraw-Hill, New York, 630 p.

Fey, D.L., and Wirt, L., 2007, this volume, Mining-impacted sources of metal loading to an alpine stream, based on a tracer-injection study, Clear Creek County, Colorado, in DeGraff, J.V., ed., Understanding and Responding to Hazardous Substances at Mine Sites in the Western United States: Geological Society of America Reviews in Engineering Geology, v. XVII, doi: 10.1130/2007.4017(05).

Fields, S., 2003, The earth's open wounds: Abandoned and orphaned mines, environmental health perspectives: Journal of the National Institute of Environmental Health Sciences, v. 111, no. 3, p. A155–A161.

Lovering, T.S., 1935, Geology and ore deposits of the Montezuma quadrangle, Colorado: U.S. Geological Survey Professional Paper, 119 p.

Lovering, T.S., and Goddard, E.N., 1950, Geology and ore deposits of the Front Range, Colorado: U.S. Geological Survey Professional Paper, 319 p., 28 plates.

Malem, F., Wanty, R., Viellenave, J.H., and Fontana, J.V., 2000, Probe sampling and geophysics applied to ground water evaluation of mine dumps, in Tailings and mine waste '00: Balkema, Rotterdam, p. 223–230.

McNeill, J.D., 1980, Electromagnetic terrain conductivity measurement at low induction numbers, Technical Note TN-6: Geonics Limited, Ontario, Canada.

Moore, J.N., and Luoma, S.N., 1990, Hazardous wastes from large-scale metal extraction: Environmental Science & Technology, v. 24, no. 9, p. 1278–1284, doi: 10.1021/es00079a001.

Telford, W.M., Geldart, L.P., and Sheriff, R.E., 1990, Applied geophysics (2nd edition): Cambridge University Press, Cambridge, 770 p.

Topper, R., Spray, K., Bellis, W.H., Hamilton, J.L., and Barkmann, P.E., 2003, Ground Water Atlas of Colorado: Colorado Geological Survey, Division of Minerals and Geology, Special Publication 53, p. 191–202.

Western Regional Climate Center, 2005, Division of Atmospheric Sciences Desert Research Institute, Reno, Nevada (http://www.wrcc.dri.edu).

Zohdy, A.A.R., Eaton, G.P., and Mabey, D.R., 1974, Application of surface geophysics to ground-water investigations, *in* Techniques of water-resources investigations of the United States Geological Survey, Book 2, Chapter D1: United States Geological Survey, Denver, CO, 116 p.

MANUSCRIPT ACCEPTED BY THE SOCIETY 28 NOVEMBER 2006

Strategies to predict metal mobility in surficial mining environments

Kathleen S. Smith
U.S. Geological Survey, M.S. 964D, Denver Federal Center, Denver, Colorado 80225-0046, USA

ABSTRACT

This report presents some strategies to predict metal mobility at mining sites. These strategies are based on chemical, physical, and geochemical information about metals and their interactions with the environment. An overview of conceptual models, metal sources, and relative mobility of metals under different geochemical conditions is presented, followed by a discussion of some important physical and chemical properties of metals that affect their mobility, bioavailability, and toxicity. The physical and chemical properties lead into a discussion of the importance of the chemical speciation of metals. Finally, environmental and geochemical processes and geochemical barriers that affect metal speciation are discussed. Some additional concepts and applications are briefly presented at the end of this report.

Keywords: metals, transport, speciation, prediction, bioavailability

INTRODUCTION

The purpose of this report is to present some strategies to predict metal mobility at mining sites. Some metals found in mining-influenced waters include Ag, Al, As, Ba, Be, Cd, Co, Cr, Cu, Fe, Hg, Mn, Mo, Ni, Pb, Sb, Se, Sr, Tl, V, and Zn. Metals are different from regulated organic substances because they cannot be destroyed by biological or chemical processes. Instead, metals can only be reduced by physical removal (e.g., leaching, biological uptake). Hence, once released, metals persist in the environment. However, factors such as metal speciation can influence metal distribution and bioavailability within the environment. Consequently, the forms, transformations, and geochemical environment of metals need to be considered when evaluating potential effects of metals on the environment. In addition to the forms and concentrations of metals themselves, many synergisms or antagonisms involve interaction of other chemical elements and environmental factors with the metals. Once these other elements and factors are recognized and addressed, a more accurate assessment of metal mobility can be made. This report discusses some of the chemical and physical factors of metals and the geochemical processes that can influence metal mobility, distribution, and bioavailability in surficial mining environments. Understanding the ways that these factors and processes can influence metals can aid in forecasting the potential ecological effects of metals. Figure 1 illustrates some geochemical processes and conditions that can redistribute dissolved metals in the environment.

Spatial and Temporal Scales

In this report, chemical and physical properties and processes are discussed for a variety of scales, both spatial and temporal. Success in forecasting metal behavior in surficial environments depends on using an appropriate spatial scale (Fig. 2A), which can range from atomic scale to regional scale or larger. At a regional scale, generalizations often can be used to understand broad trends in metal mobility (e.g., Wanty et al., 2001). As the scale becomes increasingly finer, however, estimating metal behavior at an appropriate scale becomes increasingly difficult (e.g., Smith et al., 2000).

Figure 2B illustrates the rates of several types of reactions. The rates of geochemical and biological reactions can affect

---

*E-mail: ksmith@usgs.gov

Smith, K.S., 2007, Strategies to predict metal mobility in surficial mining environments, *in* DeGraff, J.V., ed., Understanding and Responding to Hazardous Substances at Mine Sites in the Western United States: Geological Society of America Reviews in Engineering Geology, v. XVII, p. 25–45, doi: 10.1130/2007.4017(03). For permission to copy, contact editing@geosociety.org. All rights reserved.

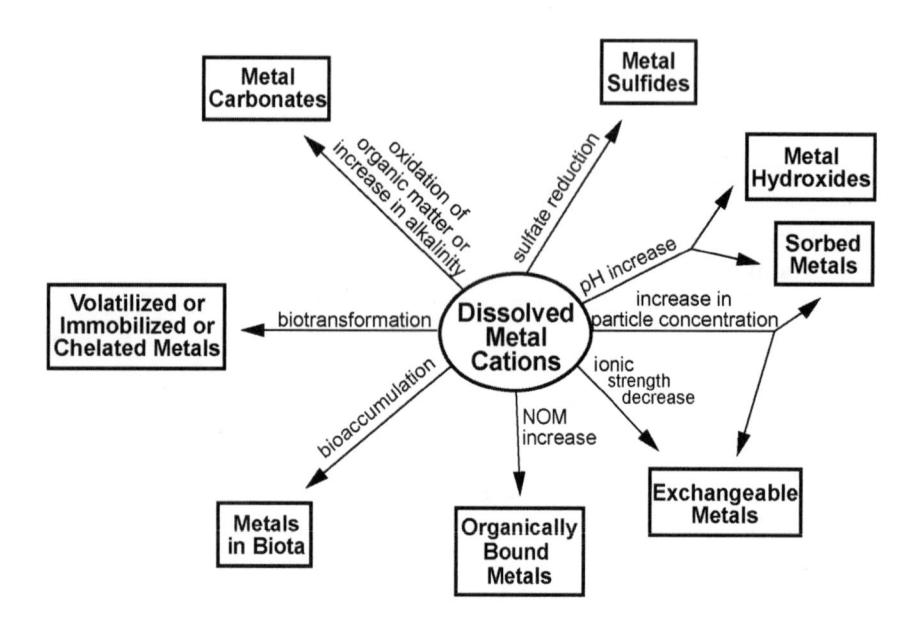

Figure 1. Diagram of some processes and geochemical conditions that can redistribute cationic dissolved metals in oxidizing, circumneutral-pH systems. Metals in each of the reservoirs (boxes) also can be redistributed by geochemical or biological processes or by changing geochemical conditions. NOM refers to natural organic matter (reprinted from Smith and Huyck, 1999, with permission).

metal mobility, and many reactions involving metals are kinetically controlled or biologically mediated. This rate dependence makes reactions extremely difficult to model (e.g., Langmuir and Mahoney, 1984). Some of the chemical reactions take place quickly (e.g., solute-water reactions), and other reactions take place more slowly (e.g., mineral recrystallization). In mining-influenced waters where many initial precipitates are amorphous or metastable, it is not likely that the residence time of the precipitates is long enough for the system to have reached equilibrium. Therefore, when modeling these systems, it is important to consider the solid phases that are actually present in the system instead of relying upon thermodynamically stable phases (Nordstrom and Alpers, 1999). Ritchie (1994) discusses some of the rates of processes in mine-waste systems.

Terminology and Scope

The term *mining-influenced waters* (MIW), introduced by Schmiermund and Drozd (1997), will be used in this report. MIW are affected by the weathering of rocks and minerals exposed by mining activities and may exhibit one or more of the characteristics of low pH, high sulfate, high Fe and Al, high noniron metals, and high turbidity (Schmiermund and Drozd, 1997). This report focuses on the behavior of metals, metalloids, and their inorganic compounds in areas that have been subjected to mining activities. The term *metal* is used in a general sense to mean a chemical element that, in aqueous solution, displays cationic behavior or that has an oxide that is soluble in acids (Parish, 1977). By this definition, elements that are nonmetals include H, the rare gases, B, C, Si, N, P, As, O, S, Se, Te, Po, F, Cl, Br, I, and At. A *metalloid* is an element with properties intermediate between those of metals and nonmetals. Metalloids include As, B, Ge, Po, Sb, Si, and Te. A *cation* is a positively charged ion; an *anion* is a negatively charged ion; and an *oxyanion* is an element that combines with oxygen to form an anionic species in aquatic systems (e.g., SO_4^{2-}, MoO_4^{2-}). A *ligand* is an anion or neutral molecule that can combine with a cation to form a complex. Common ligands in aquatic systems include hydroxyl (OH^-), carbonate (CO_3^{2-}), bicarbonate (HCO_3^-), phosphate (PO_4^{3-}), sulfate (SO_4^{2-}), sulfide (S^{2-}), hydrogen sulfide ion (HS^-), carboxyl (COOH), and dissolved organic carbon (DOC).

Many metals can be both essential and toxic, and their effects on organisms depend on concentration, speciation, and bioavailability. Some metals essential to plants or animals include Co, Cr, Cu, Fe, Mn, Mo, Ni, and Zn. Essential metals can exert toxic effects by being either too high in concentration or too low in concentration (deficient). Metals that are nonessential to biological functioning (e.g., Cd, Hg, and Pb) can be toxic at relatively low concentrations.

STRATEGIES TO PREDICT METAL MOBILITY AT MINING SITES

This section presents overall strategies to predict metal mobility at mining sites. The first topic is developing a conceptual model of a mining site, and the second topic is ways to estimate metal mobility. Guidelines to determine generalized relative mobility of metals under different environmental conditions are presented. Detailed information used to develop these guidelines is discussed in later sections of this report.

Conceptual Models

It is important to have an accurate conceptual model when assessing the potential effects of metals at a mining site. Metal speciation and transformations in the environment can be very

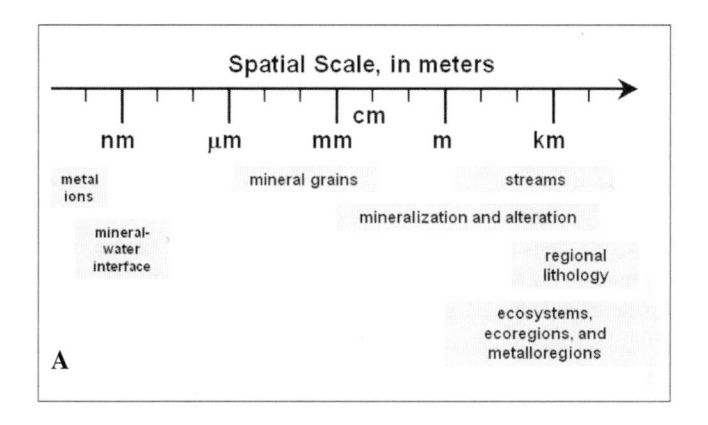

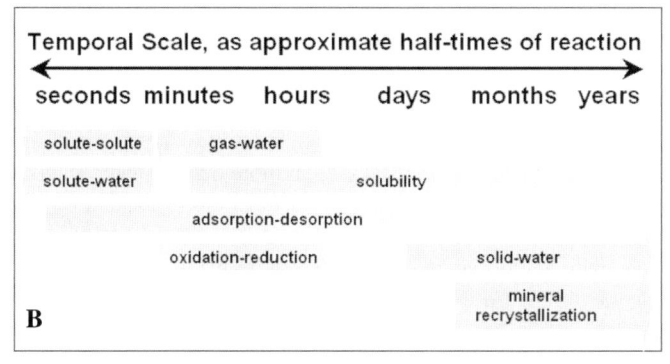

Figure 2. (A) Examples of differences in spatial scales of some factors that are influenced by geochemical processes. Note the wide spatial range of these factors. (B) Examples of differences in rates of some types of reactions that influence environmental geochemical conditions.

complicated, and a conceptual model will help focus on the most important factors to consider. Such a model should include relationships between the source, transport, and fate of the metals and should incorporate the mineralogical, hydrological, geological, and geochemical conditions at the site that might affect these factors. Conditions generally are site specific and are tied to receptor organisms of concern. Lefebvre et al. (2001) present an example of a conceptual model for mine-waste-rock piles. In the subsections that follow, some tools that aid in developing a conceptual model are briefly discussed.

Metal Source Characterization

When investigating metal mobility and bioavailability at a mining site, it is crucial to first consider and characterize the possible sources of metals in the environment. This is an important first step in developing a conceptual model. Numerous manuals and articles describe how to sample and characterize various potential metal sources (e.g., Al-Abed et al., 2006; Chao, 1984; Church et al., 2007; Cravotta and Kirby, 2004; Crock et al., 1999; Davis et al., 1993; Diehl et al., 2006; Ficklin and Mosier, 1999; Hageman, 2005; Hammarstrom and Smith, 2002; Jambor and Blowes, 1994; Jambor et al., 2003; Jenne and Luoma, 1977; Jordan and D'Alessandro, 2004; Macalady and Ranville, 1998; McLemore et al., 2007; MEND Manual, 2001; Mills and Robertson, 2006; Nimick et al., 2005; Plumlee and Ziegler, 2003; Plumlee et al., 2006; Price, 1997, 2005; Ranville et al., 2006; Sauvé, 2002; Smith et al., 2002, 2003; Tessier et al., 1979; U.S. EPA, 2001, 2004; U.S. Geological Survey; Wildeman et al., 2007). Maest et al. (2005a, b) provide a useful geochemical characterization toolbox that lists methods, references, and advantages and disadvantages of tools applicable to mining sites.

Geoenvironmental Models of Mineral Deposits

When developing a conceptual model of a mining site, it is important to consider geological aspects that resulted in the mineral deposit. Plumlee (1999), Plumlee et al. (1999), and Seal and Hammarstrom (2003) discuss how mineral deposits are classified according to similarities in their geologic characteristics and geologic setting and how this classification system may be extended to incorporate potential environmental effects of mineral deposits. This extended classification is termed *geoenvironmental models of mineral deposits* (du Bray, 1995). Geoenvironmental models can distinguish characteristics of various mineral deposits that may affect the geochemistry of aquatic systems. Geoenvironmental models provide information about natural geochemical variations associated with a particular type of mineral deposit and geochemical variations associated with its effluents, wastes, and mineral-processing facilities (Seal and Foley, 2002). Based on geoenvironmental models, potential metal sources and their likely concentration ranges can be determined for a given deposit type to provide an estimation of metal sources at a mining site.

Geoavailability

Once the important metal sources are identified, then considerations need to be turned to factors that will influence the mobility and bioavailability of metals at the site and away from the site. These factors are discussed in detail in later sections of this report.

The release of metals from solid phases is related to *geoavailability*. Geoavailability is that portion of a chemical element's or a compound's total content in an earth material that can be liberated to the surficial or near-surface environment (or biosphere) through mechanical, chemical, or biological processes. The geoavailability of a chemical element or a compound is related to the susceptibility and availability of its resident mineral phase(s) to these mechanical, chemical, or biological processes (Smith and Huyck, 1999; Smith, 1999a).

Figure 3 illustrates pathways and relationships between total metal content in an earth material and potential toxicity to an organism. Total metal is the abundance of a given metal in an earth material, and geoavailability is a function of the total metal content, access to weathering, and susceptibility to weathering. The definition for *bioavailability*, which is based upon Newman and Jagoe (1994), is the degree to which a contaminant in a potential source is free for uptake (movement into or onto an

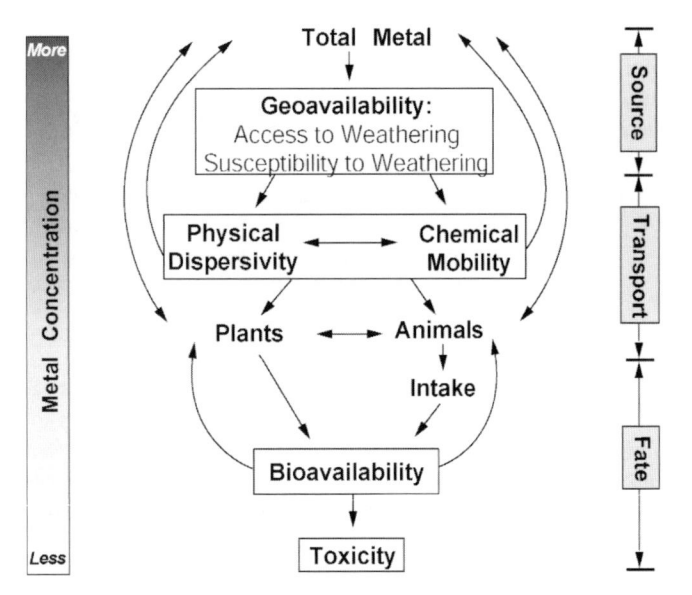

Figure 3. Diagram showing pathways and relationships between total metal in an earth material and toxicity. Geoavailability is that portion of a chemical element's or a compound's total content in an earth material that can be liberated to the surficial or near-surface environment (or biosphere) through mechanical, chemical, or biological processes. The gray scale on the left depicts that as a metal moves from one stage to another, generally less than 100% is transferred (modified from Smith and Huyck, 1999).

organism). In Figure 3, a distinction is made between plants and animals because bioavailability is generally a prerequisite for uptake in plants, whereas animals may intake (ingest, inhale, etc.) toxicants that subsequently pass through their bodies without any systemic uptake.

Each stage from total metal content in an earth material through toxicity in the surficial environment in Figure 3 is a reservoir with a distinct half-life. As a metal moves from one stage to another, generally less than 100% is transferred. Therefore, not all of the total metal content in an earth material is usually geoavailable, bioavailable, or toxic, and the gray scale in Figure 3 portrays this concept. Total metal content and geoavailability constitute the source factors; dispersivity and mobility comprise the transport factors; and uptake/intake, bioavailability, and toxicity represent the fate factors. Bioaccumulation links the fate and transport segments of the diagram.

Box Models

Box models can be a useful way to present a conceptual model of a mining site. Figure 4 shows an example of a simple box model developed to describe dissolved zinc concentrations in Lake Coeur d'Alene, Idaho (Balistrieri et al., 2002). Box models consist of reservoirs with inputs and outputs. Mass transfer between boxes is a function of residence time in the reservoir within each box. Box models use a mass-balance approach to represent processes that influence the element of interest. It is likely that not all of the processes considered in a box model are significant, but the action of creating and testing a box model helps develop an understanding of the important processes at the site.

Estimating Metal Mobility

Mobility refers to the capacity of an element to move within fluids after dissolution. It is difficult to predict element mobility quantitatively in surficial environments. Rather, mobility should be considered in a relative sense by empirically comparing the behavior of elements under changing environmental conditions, such as at geochemical barriers (which are discussed in a later section of this report). Table 1 lists the generalized relative mobility of elements expected under a variety of geochemical conditions. Factors controlling mobility include pH, solubility reactions, sorption reactions, and redox conditions (these factors are discussed in more detail in later sections of this report). Table 1 takes into account the tendency of the elements to sorb onto hydrous oxides or to precipitate (these processes are discussed in more detail in later sections of this report). Criteria for mobility distinctions are scaled by element abundance rather than being based on absolute solubility; no quantitative information can be inferred from Table 1. By comparing the different rows of Table 1, it is possible to make qualitative statements about the behavior of a given element under changing geochemical conditions. Data for Table 1 are derived from the author's personal experience with mine-drainage systems as well as from Vlasov (1966), Fuller (1977), Parish (1977), Perel'man (1977, 1986), Callahan et al. (1979), Lindsay (1979), Rose et al. (1979), Levinson (1980), Greenwood and Earnshaw (1984), Luka-

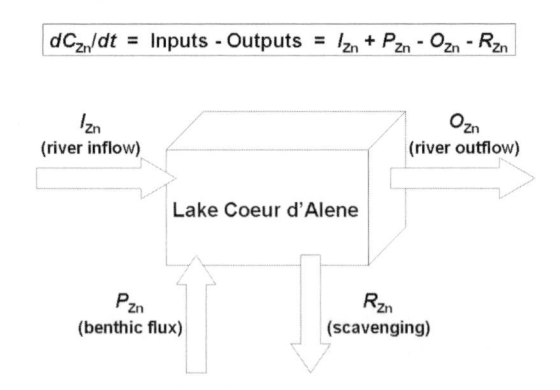

Figure 4. Simple box model to describe dissolved Zn concentrations in Lake Coeur d'Alene, Idaho. The lake is treated as a completely mixed system; dC_{Zn}/dt represents changes in dissolved Zn concentrations in the lake as a function of time; I_{Zn} represents the external inputs of dissolved Zn to the lake; P_{Zn} represents the internal sources of dissolved Zn to the lake water; O_{Zn} represents fluxes of dissolved Zn out of the lake; and R_{Zn} represents internal removal of dissolved Zn from the lake water (all as mg/L per day; modified from Balistrieri et al., 2002).

TABLE 1. GENERALIZED RELATIVE MOBILITY OF CHEMICAL ELEMENTS UNDER DIFFERENT ENVIRONMENTAL CONDITIONS

Environmental Conditions	Very Mobile	Mobile	Somewhat Mobile	Scarcely Mobile to Immobile
Oxidizing with pH < 3	Br, Cd, Cl, Co, Cu, F, I, Ni, Rn, S, Zn	Al, As, Ca, Fe, Hg, K, Mg, Mn, Na, P, Ra, REE, Se, Si, Sr, U, V	Ag, Ba, Be, Bi, Cr, Cs, Ga, Ge, Li, Mo, Pb, Rb, Sb, Th, Ti, Tl, W	Sc, Sn, Y, Zr
Oxidizing with pH > 5 to circumneutral, no iron substrates	Br, Cd, Cl, F, I, Rn, S, Zn	Ca, Mg, Mo, Na, Se, Sr, U, V	As, Ba, Bi, Co, Cr, Cs, Cu, Ge, Hg, K, Li, Mn, Ni, P, Ra, Rb, REE, Sb, Si, Tl	Ag, Al, Be, Fe, Ga, Sc, Sn, Th, Ti, W, Y, Zr
Oxidizing with pH > 5 to circumneutral, with abundant iron substrates	Br, Cl, F, I, Rn, S	Ca, Cd, Mg, Na, Sr, Zn	Ba, Bi, Co, Cs, Ge, Hg, K, Li, Mn, Ni, Rb, Sb, Se, Si, Tl	Ag, Al, As, Be, Cr, Cu, Fe, Ga, Mo, P, Pb, Ra, REE, Sc, Sn, Th, Ti, U, V, W, Y, Zr
Reducing with pH > 5 to circumneutral, no hydrogen sulfide	Br, Cl, F, I, Rn	Ca, Cd, Cu, Fe, Mg, Mn, Na, Ni, Pb, S, Sr, Zn	As, Ba, Co, Cr, Cs, Hg, K, Li, P, Ra, Rb, Si, Tl	Ag, Al, Be, Bi, Ga, Ge, Mo, REE, Sb, Sc, Se, Sn, Th, Ti, U, V, W, Y, Zr
Reducing with pH > 5 to circumneutral, with hydrogen sulfide	Br, Cl, F, I, Rn	Ca, Mg, Mn, Na, Sr	Ba, Cs, K, Li, P, Ra, Rb, Si, Tl	Ag, Al, As, Be, Bi, Cd, Co, Cr, Cu, Fe, Ga, Ge, Hg, Mo, Ni, Pb, REE, S, Sb, Sc, Se, Sn, Th, Ti, U, V, W, Y, Zn, Zr

Note: See text for details; information from Smith and Hyuck, 1999. REE = Rare-Earth Elements (which are treated here as a group, but individually can have somewhat different mobility behaviors).

shev (1984, 1986), Adriano (1986), Cotton and Wilkinson (1988), Hem (1989), and Kabata-Pendias and Pendias (1992).

Table 1 provides a general guide to predict metal behavior in surficial environments. This approach does not substitute for in-depth field studies and topical research; there is no reliable "cookbook" approach. The information in Table 1 may help to determine which elements could be mobile in a given environment and to anticipate the effects of various geochemical barriers. To use this approach in a natural setting, it is necessary to know something about the geochemical conditions. It is also necessary to have a good grasp of underlying chemical and geochemical principles (e.g., Garrels and Christ, 1965; Nordstrom and Munoz, 1994; Stumm and Morgan, 1996). Table 1 should be used only in a relative sense and does not provide any information about absolute concentrations or quantitative data.

CHEMICAL AND PHYSICAL PROPERTIES OF ELEMENTS

In order to understand the reasoning behind Table 1, it is necessary to drill down to the atomic scale to discuss some chemical and physical properties of chemical elements. This is because the behavior of a metal is determined largely by the chemical and physical characteristics of the metal. In order to generalize about the behavior of metals in the environment, it is important to understand the properties of a particular metal in addition to its geochemical environment.

Oxidation State

The oxidation state (also referred to as oxidation number or valence) of an element is important because it can have a significant effect on the mobility and interactions of the element. The oxidation state represents the charge that an atom appears to have when electrons are counted and may be either positive or negative. Oxidation states are used to track electrons in oxidation-reduction (redox) reactions. Table 2 lists the oxidation states of some chemical elements in aquatic systems.

Many elements can occur in more than one oxidation state in natural environments. Redox-sensitive elements include C, S, N, Fe, Mn, As, Cu, Cr, Hg, Mo, Sb, Se, U, V, and W.

Size

The size of an ion primarily depends on its oxidation state. Table 2 lists the oxidation states and effective ionic radii of some elements in aquatic systems. Note that an increase in the oxidation state results in a shrinkage in size. The ionic radius of an element is important in determining if it can take part in particular chemical and biochemical reactions. Also, elements with similar ionic radii and oxidation states can sometimes substitute for one another. For example, Cd^{2+} can substitute for Ca^{2+} in many geochemical and biological systems.

Electronegativity and Bonding

Electronegativity (EN) is the power of an atom in a molecule to attract electrons to itself (Pauling, 1960). Hence, EN is indicative of the types of compounds and the types of chemical bonds that a given element will form. In the periodic table of the elements, EN values increase in the direction of fluorine, which is the most electronegative element and is located in the upper right-hand corner. The Pauling scale is commonly used to quantify EN. In this scale fluorine is assigned a value of 4, and other elements have values ranging down to 0.7, which is the value for cesium, the least electronegative element. EN values for a variety of elements and complex ligands are listed in Table 3. Elements with high EN values (~2 or greater) are mainly nonmetals or potential ligands, whereas elements with low EN values (less than ~2) generally are metal cations.

EN relates to the type and strength of bonding between a metal and a ligand. A *covalent bond* is a type of chemical bond formed when an electron pair is shared between a metal and a ligand. Covalent bonds are relatively strong and tend to occur between elements with similar EN values. An *ionic bond* is a type of chemical bond based on electrostatic forces between two oppositely charged ions. Ionic bonds are commonly formed between metals and nonmetals, because metals tend to have low EN values and nonmetals tend to have high EN values. In reality, most metal-ligand bonds exhibit properties of both covalent and ionic bonding. However, two generalizations can be made:

1. A small difference in metal and ligand EN values leads to a predominantly covalent bond.
2. A large difference in metal and ligand EN values leads to a predominantly ionic bond.

Ionic Potential

Ionic potential (the ratio of oxidation number to ionic radius) of elements has been related to their mobility (Rose et al., 1979). As illustrated in Figure 5, elements with low ionic potential are generally mobile in the aquatic environment as simple cations (e.g., Na^+, Ca^{2+}), and elements with high ionic potential are generally mobile as oxyanions (e.g., sulfur in SO_4^{2-}, molybdenum in MoO_4^{2-}). Elements with high ionic potential tend to form covalent bonds rather than ionic bonds. Elements with intermediate ionic potential have a tendency to strongly sorb or hydrolyze and exhibit low solubility (see discussion in later sections); therefore, these elements are fairly immobile (Rose et al., 1979). The concept of ionic potential is useful in explaining how elements with apparently different chemical properties behave similarly during migration in the environment.

Classification of Chemical Elements

Inorganic-chemistry fundamentals can be useful in understanding metal behavior in the environment. Metals can be classified into groups based on their capacity for binding to different ligands. Several classification systems have developed through the years (e.g., Whitfield and Turner, 1983). One of the most useful classification systems was developed by Pearson (1963, 1968a, 1968b), who introduced hard and soft acid and base (HSAB) concepts to describe metals and ligands. Table 4 lists hard and soft acids and bases and their characteristics. Complexes are formed between metals (acids) and ligands (bases) in aqueous solutions and at interfaces (such as mineral or biological sur-

TABLE 2. OXIDATION STATES, EFFECTIVE IONIC RADII, AND THERMOCHEMICAL RADII OF SOME IONS IN AQUATIC SYSTEMS

Element	Chemical Symbol	Oxidation State	Radius (pm)[1][2]
Aluminum	Al	+3	67.5
Antimony	Sb	+3	90
		+5	74
Arsenic	As	+3	72
		+5	60
Barium	Ba	+2	149
Beryllium	Be	+2	59
Bismuth	Bi	+3	117
Cadmium	Cd	+2	109
Calcium	Ca	+2	114
Cerium	Ce	+3	115
Cesium	Cs	+1	181
Chromium	Cr	+3	75.5
		+6	58
Cobalt	Co	+2	79
Copper	Cu	+1	91
		+2	87
Iron	Fe	+2	92
		+3	78.5
Lanthanum	La	+3	117.2
Lead	Pb	+2	133
Lithium	Li	+1	90
Lutetium	Lu	+3	100.1
Magnesium	Mg	+2	86
Manganese	Mn	+2	97
Mercury	Hg	+1	133
		+2	116
Molybdenum	Mo	+4	79
		+6	73
Nickel	Ni	+2	83
Phosphorus	P	+5	52
Potassium	K	+1	152
Radium	Ra	+2	162[3]
Selenium	Se	+4	64
		+6	56
Silicon	Si	+4	54
Silver	Ag	+1	129
Sodium	Na	+1	116
Strontium	Sr	+2	132
Thallium	Tl	+1	164
		+3	102.5
Thorium	Th	+4	108
Tin	Sn	+4	83
Titanium	Ti	+4	74.5
Tungsten	W	+6	74
Uranium	U	+4	103
		+6	87
Vanadium	V	+3	78
		+4	72
		+5	68
Zinc	Zn	+2	88

Polyatomic Ion	Radius (pm)[4][2]
OH^-	119
NO_3^-	165
CO_3^{2-}	164
HCO_3^-	142
HS^-	193
SO_4^{2-}	244
NH_4^+	151

Note: Data from Huheey et al. (1993); pm = picometer (1 × 10⁻¹² m).
(1) Effective ionic radius for six-fold coordination
(2) 100 pm = 1 angstrom (Å) = 0.1 nm
(3) For eight-fold coordination
(4) Thermochemical radius of polyatomic ions.

TABLE 3. PAULING ELECTRONEGATIVITY (EN) VALUES FOR A VARIETY OF CHEMICAL SPECIES IN AQUATIC SYSTEMS

Element	Species	EN
Aluminum	Al^{3+}	1.61
Antimony	Sb^{3+}	**2.05**
Arsenic	As^{3+}	**2.0**
	As^{5+}	**2.2**
Barium	Ba^{2+}	0.89
Beryllium	Be^{2+}	1.57
Bismuth	Bi^{3+}	**2.02**
Bromine	Br^-	**2.96**
Cadmium	Cd^{2+}	1.69
Calcium	Ca^{2+}	1.00
Carbon	C^{4+}	**2.55**
Cerium	Ce^{3+}	1.12
Cesium	Cs^+	0.79
Chlorine	Cl^-	**3.16**
Chromium	Cr^{3+}	1.6
	Cr^{6+}	**2.1**
Cobalt	Co^{2+}	1.88
Copper	Cu^+	1.8
	Cu^{2+}	**2.0**
Fluorine	F^-	**3.98**
Gold	Au^+	**2.54**
	Au^{3+}	**2.9**
Hydrogen	H^+	**2.20**
Iron	Fe^{2+}	1.7
	Fe^{3+}	1.8
Lanthanum	La^{3+}	1.10
Lead	Pb^{2+}	1.6
Lithium	Li^+	0.98
Lutetium	Lu^{3+}	1.27
Magnesium	Mg^{2+}	1.31
Manganese	Mn^{2+}	1.4
Mercury	Hg^+	1.8
	Hg^{2+}	**2.00**
Molybdenum	Mo^{4+}	1.6
	Mo^{6+}	**2.1**
Nickel	Ni^{2+}	1.91
Oxygen	O^{2-}	**3.44**
Potassium	K^+	0.82
Radium	Ra^{2+}	(0.83)
Selenium		**2.55**
Silicon	Si^{4+}	1.90
Silver	Ag^+	1.93
Sodium	Na^+	0.93
Strontium	Sr^{2+}	0.95
Thallium	Tl^+	1.5
	Tl^{3+}	**2.04**
Thorium	Th^{4+}	1.1
Tin	Sn^{4+}	1.96
Titanium	Ti^{4+}	1.54
Tungsten	$W(II)$	**2.36**
Uranium	U^{4+}	1.3
	U^{6+}	1.9
Vanadium	V^{3+}	1.35
	V^{4+}	1.6
	V^{5+}	1.8
Zinc	Zn^{2+}	1.65

Complex Ligand	Coord. Number	EN
OH^-	1	**3.1**
	2	**2.75**
	3 and 4	**2.15**
NO_3^-		**3.5**
$H_2PO_4^-$		**3.15**
HPO_4^{2-}		**2.8**
CO_3^{2-}		(**2.5**)
HCO_3^-		(~**4**)
HS^-		(**2.33**)
SO_4^{2-}		**3.7**

Note: Data from Huheey et al. (1993); Langmuir (1997); EN = electronegativity (in Pauling Scale units); values in parentheses are estimates; bold values are ≥2 (see text for details).

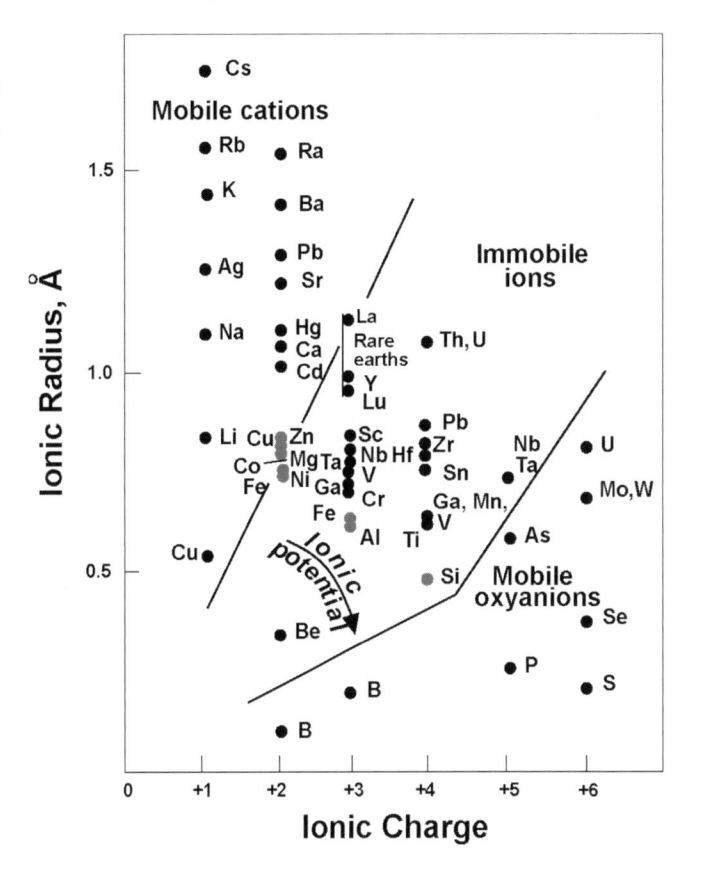

Figure 5. Mobility of elements in the surficial environment as a function of ionic potential (reprinted from Rose et al., 1979, with permission of the author).

faces). Hard-metal cations preferentially form complexes with F (the most electronegative element) and with ligands having O as the electron donor (e.g., COOH and PO_4^{3-}). Water is strongly attracted to these metals, and they do not form sulfides (complexes or precipitates). Hard-metal cations tend to form relatively insoluble precipitates with OH^-, CO_3^{2-}, and PO_4^{3-} (Stumm and Morgan, 1996). Soft-metal cations preferentially form complexes with ligands containing I, S, or N. Soft-metal cations form insoluble sulfides and soluble complexes with S^{2-} and HS^- (Stumm and Morgan, 1996).

The HSAB classification is a useful concept to help explain the strength of metal complexing and metal toxicity. According to this concept, cations are Lewis acids and act as an electron acceptor, and anions are Lewis bases and act as an electron donor. The term *soft* refers to an electron cloud that is readily deformable so that the electrons are relatively mobile (i.e., polarizable). The term *hard* refers to an electron cloud that is relatively rigid so that the electrons are relatively immobile (i.e., nonpolarizable). Soft species prefer to participate in covalent bonds, and hard species prefer to participate in ionic bonds (Langmuir, 1997). Hard acids tend to bind to hard bases, and soft acids tend to bind to soft bases. The terms *hard* and *soft* are relative, and there are borderline cases between hard and soft for both acids and bases. Generalizations about the speciation, behavior, and mobility of elements in aquatic

TABLE 4. HARD AND SOFT ACIDS AND BASES, AND THEIR CHARACTERISTICS

Hard (Class A)		Soft (Class B)
Often macronutrients		Often toxic
Ionic bonding		Covalent bonding (more irreversible)
Binding preference is with oxygen		Binding preference is with sulfur and nitrogen
Prefers F > O > N = Cl > Br > I > S		Prefers S > I > Br > Cl = N > O > F
$OH^- > RO^- > RCO_2^-$		
$CO_3^{2-} >> NO_3^-$		
$PO_4^{3-} >> SO_4^{2-} >> ClO_4^-$		
pH sensitive		

Hard Acids	Borderline Acids	Soft Acids
Al^{3+}, As^{3+}, Be^{2+}, Ca^{2+}, Ce^{4+}, Co^{3+}, CO_2, Cr^{3+}, Cr^{6+}, Fe^{3+}, H^+, K^+, La^{3+}, Li^+, Mg^{2+}, Mn^{2+}, Na^+, Sc^{3+}, Si^{4+}, Sn^{4+}, SO_3, Sr^{2+}, Th^{4+}, Ti^{4+}, U^{4+}, UO_2^{2+}, VO^{2+}, Zr^{4+}	Bi^{3+}, Co^{2+}, Cu^{2+}, Fe^{2+}, Ni^{2+}, Pb^{2+}, Sb^{3+}, Sn^{2+}, SO_2, Zn^{2+}	Ag^+, Au^+, Cd^{2+}, Cu^+, Hg^{2+}, Hg_2^{2+}, Tl^+

Hard Bases	Borderline Bases	Soft Bases
NH_3, RNH_2, N_2H_4 H_2O, OH^-, O^{2-}, ROH, RO^-, R_2O CH_3COO^-, CO_3^{2-}, NO_3^-, PO_4^{3-}, SO_4^{2-}, ClO_4^- F^-, (Cl^-)	$C_6H_5NH_2$, C_5H_5N, N_3^-, N_2 NO_2^-, SO_3^{2-} Br^-	H^- R^-, C_2H_4, C_6H_6, CN^-, RNC, CO SCN^-, R_3P, $(RO)_3P$, R_3As R_2S, RSH, RS^-, $S_2O_3^{2-}$ I^-

Note: Information from Huheey et al. (1993); Langmuir (1997); Stumm and Morgan (1996); R = organic molecule.

systems can be made based on this type of HSAB classification system.

Using HSAB to Estimate Bioavailability

Nieboer and Richardson (1980) modified existing metal-classification systems to make them more applicable to biological systems. According to Nieboer and Richardson's classification, Class A (hard) metals, which tend to seek oxygen-containing ligands, comprise all the macronutrient metals (such as K and Ca). Class B (soft) metals, which tend to seek nitrogen- and sulfur-containing groups, comprise many of the more toxic metals. Borderline metals, which have intermediate properties, include most of the common metals. As shown in Figure 6, there is a distinct break between Class A metals and the borderline group, but there is little distinction between the borderline group and Class B metals. This type of approach can provide a general set of criteria by which the actions of different metals can be compared. For example, Class B metals may displace borderline metals, such as Zn or Cu, from enzymes. The toxicity of a borderline metal depends on its Class B character; it will be able to displace many Class A metals and, depending upon their relative affinities, other borderline metals. Nieboer and Fletcher (1996) discuss several chemical and physical factors of metals that relate to their reactivity and toxicity.

Walker et al. (2003) provide a review of reported correlations between physical and chemical properties of cations and toxicity to mammalian and nonmammalian species using in vitro and in vivo assays. They conclude that certain useful cor-

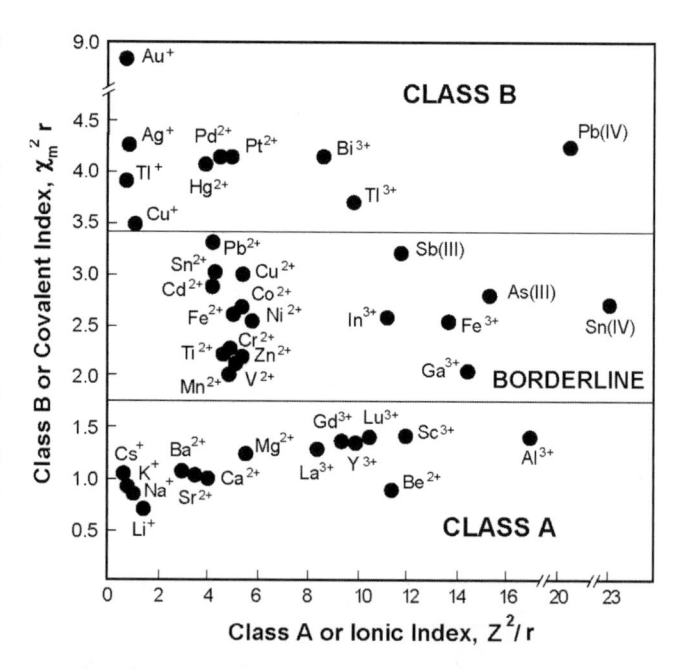

Figure 6. Chemical classification of metal ions according to Nieboer and Richardson (1980). χ_m is the metal-ion electronegativity, r is the metal ionic radius, and Z is the formal charge of the metal ion. Oxidation states given by Roman numerals imply that simple cations do not exist (modified from Nieboer and Richardson, 1980).

relations can be made between several physical and chemical properties of ions (mostly cations) and toxicity of metals. McKinney et al. (2000) provide a review of qualitative and quantitative modeling methods that relate chemical structure to biological activity. These structure-activity relationships (SARs) are being applied to the prediction and characterization of chemical toxicity. Quantitative ion character-activity relationships (QICARs) that use metal-ligand binding characteristics to predict metal toxicity are currently under development (Newman et al., 1998; Ownby and Newman, 2003). The QICAR work demonstrates the feasibility of predicting metal toxicity from metal-ion characteristics.

CHEMICAL SPECIATION

An understanding of metal speciation is key to understanding metal mobility, bioavailability, and toxicity. Different chemical species of a given metal often have different mobility behavior and toxicological effects. The terms *species* and *speciation* are used in the literature in a variety of ways. For the purposes of this report, distinct chemical species are "chemical compounds that differ in isotopic composition, conformation, oxidation or electronic state, or in the nature of their complexed or covalently bound substituents," and speciation is the "distribution of an element amongst defined chemical species in a system" (Templeton et al., 2000).

The general formula for metal cations is usually written as M^{n+}. This is a bit misleading because metal ions dissolved in water are not present as bare cations but, rather, are complexed with water molecules. When a metal binds to a ligand other than water, the ligand must substitute for the complexed water molecules. This substitution is faster for some metals than for others. Rules of thumb for this rate of substitution are:

1. For metal ions of the same charge, substitution rates for the same ligand will increase with increasing metal-ion size (e.g., $Be^{2+} < Mg^{2+} < Ca^{2+} < Sr^{2+} < Ba^{2+}$).
2. For metal ions of about the same size, substitution rates will increase with decreasing metal-cation charge (e.g., $Mg^{2+} < Li^{+}$).

These rules of thumb can help determine why some metals readily bind with various ligands whereas other metals do not. This behavior is related to some of the same principles behind HSAB behavior.

Aqueous metal species occur as free ions (which are actually complexed with water molecules as discussed above) or as metal complexes. The formation of metal complexes in solution tends to increase metal mobility. Total metal concentration does not distinguish between the various species. For many metals, the free ion is thought to be the primary species that causes toxicity to aquatic organisms. Therefore, to achieve a reliable estimate of metal bioavailability, it is necessary to determine metal species. However, it should be noted that total metal concentration can provide an upper limit for estimation of metal bioavailability and toxicity.

There is ongoing research to find analytical and computational approaches to determine metal speciation. D'Amore et al. (2005) provide a recent review of speciation methods for metals in soils, and Buffle and Horvai (2000) and Ure and Davidson (2002) provide information on speciation methods for a variety of applications.

Factors that can influence metal speciation include pH, redox conditions, inorganic ligands, organic ligands (DOC), and competition from other ions. Several books provide detailed explanations about factors that influence metal speciation (e.g., Cotton and Wilkinson, 1988; Drever, 1997; Garrels and Christ, 1965; Greenwood and Earnshaw, 1984; Hem, 1989; Huheey et al., 1993; Langmuir, 1997; Morel and Herring, 1993; Nordstrom and Munoz, 1994; Parish, 1977; Stumm and Morgan, 1996). The relationship between metal speciation, mobility, and bioavailability has been reviewed by several authors (e.g., Allen, 2002; Allen et al., 1980; Bourg, 1988; Forstner, 1987; Luoma, 1983; Luoma and Carter, 1993; Nieboer and Fletcher, 1996; Pagenkopf, 1983; Tessier and Turner, 1995). Smith and Huyck (1999) provide a discussion of the links between metal abundance, mobility, bioavailability, and toxicity in mining environments.

Transition Metals

The transition metals (Ti, V, Cr, Mn, Fe, Co, Ni, and Cu) are in the center of the periodic table of the elements. Transition metals have unusual properties that affect their environmental chemistry. One important property is that they tend to have multiple oxidation states. Also, transition metals tend to form a variety of complexes and have a reasonably well-established rule for the sequence of complex stability based on empirical observation. According to this rule (the Irving-Williams order), the stability of complexes follows the order:

$$Mn^{2+} < Fe^{2+} < Co^{2+} < Ni^{2+} < Cu^{2+} > Zn^{2+}$$

Organometallic Transformations

Some metals can be transformed, either biotically or abiotically, into organometallic compounds, which are compounds that have a metal-carbon bond. Methylation, when a methyl group (CH_3) combines with a metal, is an example of the formation of an organometallic compound and is favored by anoxic, high-temperature environments. Mercury is perhaps the best-known example of a metal that undergoes organometallic transformations, but As, Pb, Se, and Sn may also be transformed into organometallic compounds (Craig, 2003). Organometallic transformations can affect metal mobility and toxicity. For the example of Hg, methylmercury (CH_3Hg) is the most bioavailable and toxic form of Hg (Gerould, 2000).

The Role of pH

Most metals found in MIW are cations (e.g., Cd, Cu, Ni, Pb, Zn), and the predominant charge on most of the cations is +2. For this reason, mobility and bioavailability are determined primarily by pH and are enhanced under acidic conditions. Figure 7 is an example of a Ficklin diagram, in which total concentrations of base metals are plotted against pH. This diagram clearly demonstrates the inverse relationship between high cationic metal concentrations and low-pH values. Zinc tends to dominate the base-metal concentrations in these types of plots, but individual-element plots exhibit similar behavior. Therefore, metal cations tend to be more mobile under low-pH conditions.

Some metals and metalloids (e.g., As, Cr, Mo, Se, Te, V) can combine with oxygen to form a stable, negatively charged (anionic) species called an oxyanion. As discussed above, elements with high ionic potential have a tendency to form oxyanions (e.g., MoO_4^{2-}). Because of their negative charge, oxyanions have very different mobility characteristics than do cationic species. Table 5 identifies elements that are anionic or cationic in aquatic systems. Cationic species tend to be more mobile under low-pH conditions and less mobile under high-pH conditions, whereas the opposite is true for anionic species. There are also distinct differences in bioavailability characteristics between cationic and anionic species. For example, oxyanions can be transported through living-cell membranes by diffusion-controlled processes (Wood, 1988). Arsenate may replace phosphate, and Se may replace S in many biological systems.

The pH of a system can also control which ligands are available for binding. Aqueous species of the carbonate system change with changing pH. At low pH, minimal CO_3^{2-} is available for metal binding, but at pH > 8, CO_3^{2-} can become a predominant ligand for many metals. The same is true for OH^-, which becomes more abundant with increasing pH.

The Role of Redox Chemistry

Table 5 lists some redox-sensitive elements in aquatic systems. A redox-sensitive element will generally undergo a change in mobility under different oxidizing or reducing conditions. For example, chromium dissolves as it is oxidized to chromium (VI) and precipitates upon reduction to chromium (III); this is important because chromium (VI) is much more toxic than is chromium (III). Similarly, uranium is immobile under reducing conditions but can be mobile under oxidizing conditions. Conversely, iron and manganese may be soluble under reducing conditions; consequently, metals sorbed onto iron oxides and manganese oxides can be released under reducing conditions.

It is very difficult to measure redox conditions in natural environments. Also, disequilibrium between redox couples (e.g., Fe^{2+} and Fe^{3+}) is common. Therefore, it is good practice to directly measure redox-sensitive species of interest whenever possible (Nordstrom, 2002).

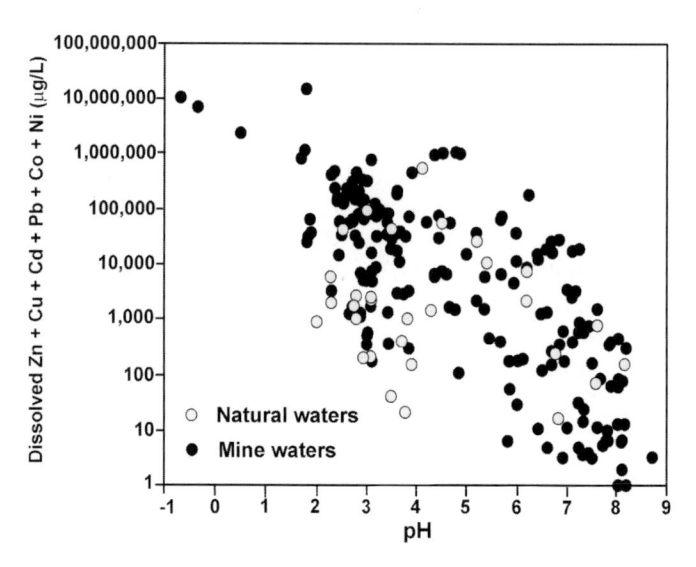

Figure 7. Ficklin diagram showing how the sum of dissolved base metals (Zn, Cu, Cd, Pb, Co, and Ni) varies with pH in natural (gray circles) and mine (black circles) waters draining diverse mineral-deposit types. Note that the trend is for lower-pH waters to contain higher concentrations of metals; however, higher-pH waters may still contain significant metal concentrations. For the diagram, ppb and µg/L are assumed to be equivalent (modified from Plumlee et al., 1999).

CHARACTERISTICS OF MINING-INFLUENCED WATERS AND IMPORTANT GEOCHEMICAL PROCESSES

Mining-influenced waters (MIW) generally have low-pH values, high concentrations of SO_4^{2-}, and high concentrations of Fe, Al, Mn, and several other metals. This unusual composition makes these systems somewhat unique when defining geochemical processes, reactants, and phases that control metal mobility. Discussion of the formation and composition of MIW can be found in Alpers and Blowes (1994), Ficklin et al. (1992), Nordstrom and Alpers (1999), Plumlee et al. (1999), Schmiermund and Drozd (1997), and Smith (2005a).

Solubility Reactions

In this report, the term *solubility* refers to the amount of a substance that can be dissolved in water at a given temperature and pressure. This parameter is used in environmental studies to help determine the fate of substances. Solubility in water is described by a solubility product (K_{sp}), which is the equilibrium constant for a solubility reaction. The tendency for a metal to form a solid compound is related to the chemical and physical properties previously discussed. Some metals can make extremely insoluble compounds (very low K_{sp} values; e.g., Pb), so these metals tend to precipitate as solids and have limited mobility. Other metals (e.g., Zn) tend to be relatively mobile

TABLE 5. GENERAL CHARACTERISTICS OF SOME CHEMICAL ELEMENTS IN SIMPLE SURFACE OR NEAR-SURFACE AQUATIC SYSTEMS[1][2]

Element	Chemical Symbol	Anionic[3]	Cationic	Redox-Sensitive[4]	Commonly Forms Sulfides
Aluminum	Al		X		
Antimony	Sb	X		X	X
Arsenic	As	X		X	X
Barium	Ba		X		
Beryllium	Be		X		
Cadmium	Cd		X		X
Chromium	Cr	X	X	X	
Cobalt	Co		X		X
Copper	Cu		X	X	X
Iron	Fe		X	X	X
Lead	Pb		X	(X[6])	X
Lithium	Li		X		
Manganese	Mn		X	X	
Mercury	Hg		X	X	X
Molybdenum	Mo	X	X[5]	X	X
Nickel	Ni		X		X
Selenium	Se	X		X	
Silver	Ag		X		X
Thallium	Tl		X	X	X
Thorium	Th		X	(X[6])	
Uranium	U	X	X	X	
Vanadium	V	X	X	X	
Zinc	Zn		X		X

Note: Modified from Smith and Huyck (1999).
(1) This table is meant as a simple guide for element behavior under normal surface or near-surface aqueous conditions.
(2) This table does not include complexes with other elements.
(3) Anionic species exist as oxyanions.
(4) Elements that change oxidation state and oftentimes behavior under different redox conditions.
(5) Cationic species exist for Mo but are rare in aquatic systems.
(6) Some of the elements, such as Pb and Th, are redox-sensitive only under extreme conditions.

because they don't readily form insoluble solids. Due to the high SO_4^{2-} concentrations in MIW, metals that form strong bonds (and relatively insoluble precipitates) with SO_4^{2-} would be expected to precipitate and be relatively immobile. Examples of relatively insoluble sulfate minerals are anglesite ($PbSO_4$) and barite ($BaSO_4$). Table 5 lists chemical elements that form sulfide precipitates under reducing conditions. Nordstrom and Alpers (1999) provide a detailed discussion of the solid phases that may form from MIW.

Metal complexation can affect the concentration and transport of metal ions. Anions, such as SO_4^{2-}, are commonly elevated in MIW. Metal complexation with these anions can increase dissolved metal concentrations above what is usually observed for solubility reactions. Cravotta (2006) examined data for 140 water samples from abandoned Pennsylvania coal mines, with pH values ranging from 2.7 to 7.3. He found that formation of aluminum-sulfate complexes greatly increased total dissolved Al concentrations at equilibrium with aluminum hydroxide and hydroxysulfate minerals. Similarly, ferric-iron-sulfate complexes increased dissolved Fe^{3+} concentrations in equilibrium with iron hydroxide or hydroxysulfate minerals.

The pH is a major control on the solubility of most metal compounds. The solubility of many metals is amphoteric, which means that the metals have a tendency to dissolve and form cations at low pH and anions at high pH, with minimal solubility at intermediate pH. For example, $Al(OH)_3$ has its minimal solubility between pH 6 and 7 (Bigham and Nordstrom, 2000; Nordstrom and Ball, 1986). The pH value of minimal solubility is different for different metals. This concept is illustrated in Figure 8. Note that gibbsite (aluminum hydroxide) and ferrihydrite (iron oxyhydroxide) have minimal solubility between pH 6 and 8, and that hydroxides of Cd, Fe(II), Zn, and Cu have minimal

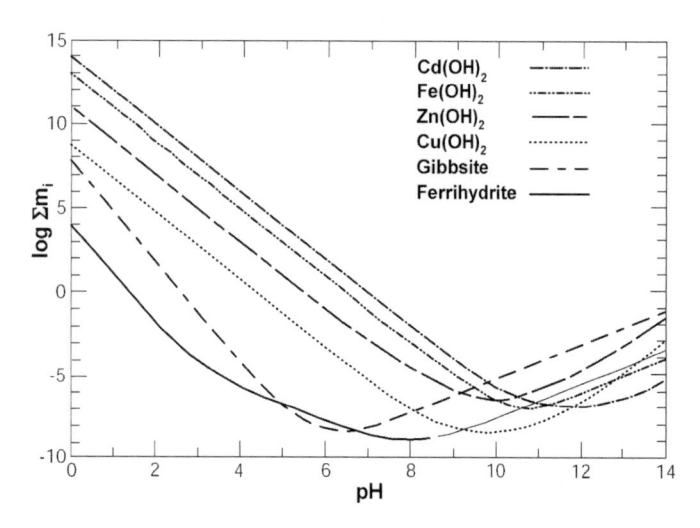

Figure 8. Solubility curves for hydroxides of Al, Fe^{3+}, Fe^{2+}, Cu, Zn, and Cd (reprinted from Nordstrom and Alpers, 1999, with permission).

solubility at much higher pH values (above pH 9). Also note that for a given pH value in Figure 8 the solubility is different for the different metals.

Aluminum mobility at mining sites is related to pH and SO$_4^{2-}$ concentration. Nordstrom (1982) found that in MIW, aluminum-sulfate and aluminum-hydroxysulfate minerals are more stable than are more common aluminum minerals found in soil. In water with pH < 4.5 or 5, dissolved Al tends to remain in solution, but in water with pH > 5, Al tends to precipitate as a solid (Nordstrom and Ball, 1986). It is common to observe white aluminum-hydroxysulfate precipitates at mining sites where the water pH has risen to a value above pH 5. Also, Al solubility influences its bioavailability. Aluminum can be fairly toxic to aquatic life, but at circumneutral pH values Al is relatively insoluble and, hence, not very bioavailable to aquatic life.

Iron mobility at mining sites is also related to pH and SO$_4^{2-}$ concentration. Saturation of iron-hydroxysulfate minerals generally occurs around pH 4. So, at pH > 4, Fe precipitates as a solid and is no longer mobile. These iron precipitates form the yellow-orange-red precipitates that form on streambeds at many mining sites. Photoreduction of Fe in acidic streams also can play a role in the mobility of iron (McKnight et al., 2001). Solid-iron phases are known to strongly sorb many metals. The role of iron precipitates in controlling aqueous metal concentrations by sorption processes is discussed in the following section.

Nordstrom and Alpers (1999) compiled a list of minerals that likely would control metal concentrations in MIW. Generally, these minerals are either relatively insoluble or have components that are common in MIW (such as SO$_4^{2-}$). The minerals include alunogen, anglesite, barite, basaluminite, calcite, cerussite, chalcanthite, epsomite, ferrihydrite, gibbsite, goslarite, gypsum, halotrichite-pickeringite, manganese oxides, melanterite, otavite, rhodochrosite, schwertmannite, scorodite, siderite, microcrystalline silica, smithsonite, and witherite.

Sorption Reactions

Sorption reactions, involving both inorganic and organic particulates, largely control the fate of many trace elements in natural systems. Metal sorption is strongly pH-dependent and a function of metal-complex formation and ionic strength (Dzombak and Morel, 1987). At many mining sites there are abundant iron- and aluminum-oxide precipitates. These precipitates can act as effective sorbents for a variety of metals (Smith, 1999b; Smith et al., 1998).

The term *sorption* is a general term that describes removal of a solute from solution to a contiguous solid phase and is used when the specific removal mechanism is not known. *Sorbate*, or *adsorbate*, refers to the solute that sorbs on the solid phase. *Sorbent*, or *adsorbent*, is the solid phase on which the sorbate sorbs. *Adsorption* refers to the two-dimensional accumulation of an adsorbate at a solid surface. The term *absorption* is used when there is diffusion of the sorbate into the solid phase. Absorption processes usually show a significant time dependency. Sposito (1986) provides a more detailed description of these terms.

The formation and dissolution of iron-hydroxysulfate minerals such as jarosite and schwertmannite can influence the mobility of metals in the environment. Jarosite [KFe$_3$(SO$_4$)$_2$(OH)$_6$] is a ferric sulfate mineral that forms under acidic conditions. It can incorporate Pb, Hg, Cu, Zn, Ag, and Ra by substitution for structural K or Fe, and it can incorporate anions such as chromate, arsenate, and selenate by substitution for SO$_4^{2-}$ (Dutrizac and Jambor, 1987). Schwertmannite [Fe$_8$O$_8$(OH)$_6$(SO$_4$)], informally known as yellowboy, is a poorly crystalline mineral with high specific surface area. It occurs as a precipitate from acidic, sulfate-rich waters (Bigham et al., 1996), such as acidic mine-drainage environments. Schwertmannite may accumulate metals, such as Cu, Zn, Ni, Se, and As, by substitution into the crystalline structure or sorption (Smith et al., 1998; Smith, 1999b). Although metals may be immobilized by coprecipitation or sorption with iron hydroxysulfate minerals, transport or burial of the materials or changes in the local redox environment could lead to conditions favoring remobilization by dissolution or desorption.

Trace elements partition between dissolved and particulate phases, and this partitioning can influence their transport and bioavailability (Luoma and Davis, 1983; Jenne and Zachara, 1987). In fact, sorption processes appear to control metal partitioning in most natural aquatic systems (Jenne, 1968; Hem, 1989). Partitioning of a metal between solid and solution phases is influenced by several factors. Generally, conditions that cause metals to be present in the solution phase include low-pH conditions, reducing conditions, low particulate loads, and/or high dissolved concentrations of a strong complexing agent.

For sorption of metals on oxide minerals, solution pH is the primary variable. Typically, cation adsorption increases with increasing pH from near zero to nearly 100% over a critical-pH range of 1–2 units (James and Healy, 1972; Kinniburgh and Jackson, 1981; Davis and Hayes, 1986). This critical-pH range is termed the *adsorption edge*, and its placement seems to be char-

TABLE 6. CRITICAL pH RANGES FOR SORPTION OF DIVALENT METAL CATIONS ON HYDROUS IRON AND ALUMINUM OXIDES

Cation	Critical pH Range
Cu, Pb, Hg	3–5
Zn, Co, Ni, Cd	5–6.5
Mn	6.5–7.5
Mg, Ca, Sr	6.5–9

Note: After Kinniburgh and Jackson, 1981; generally, the critical pH range for a given cation is higher for silica and lower for manganese oxides.

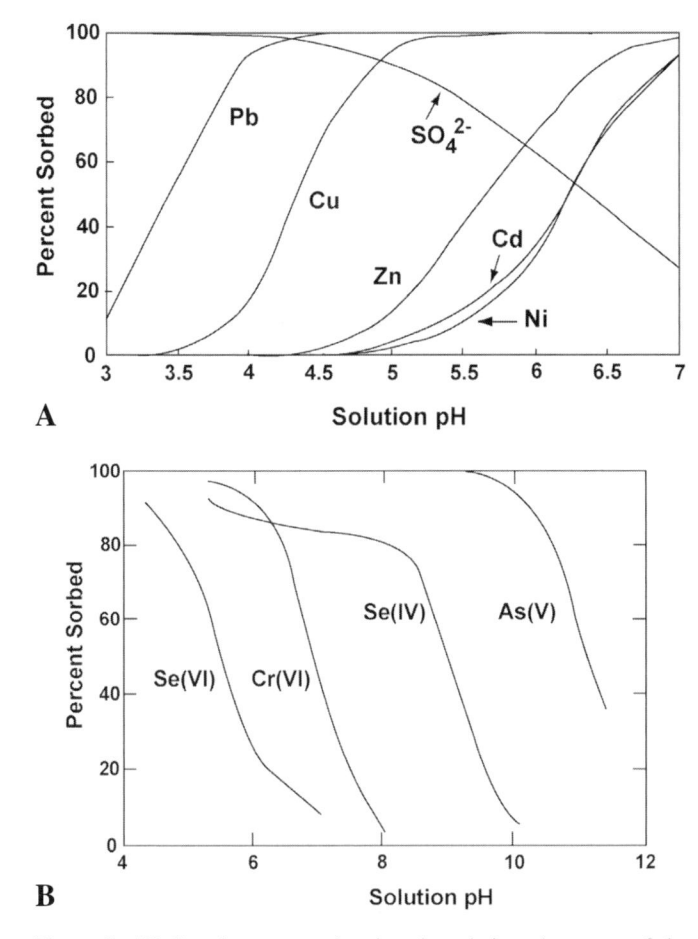

Figure 9. (A) Sorption curves showing the relative placement of the critical-pH range of metals and sulfate on hydrous iron oxide (modified from Smith and Macalady, 1991). (B) Sorption curves showing the relative placement of the critical-pH range of selected oxyanions on hydrous ferric oxide (modified from Davis and Kent, 1990).

acteristic of the particular adsorbate and, to a lesser extent, to the particular adsorbent (Spark et al., 1995). The critical-pH range (adsorption edge) is illustrated in Figure 9. Anion adsorption (Fig. 9B) is the mirror image of cation adsorption (Fig. 9A) in that anion adsorption tends to decrease with increasing pH. For a given sorbate concentration, increasing the amount of adsorbent material will shift down the pH of the adsorption edge for cations and shift up the pH of the adsorption edge for anions. Table 6 lists the critical pH ranges for metal sorption onto oxide sorbent materials.

The distribution of a metal between aqueous and solid phases can be described by a partition coefficient (K_d), which is the ratio between the metal on the solid phase and the metal in solution. Partition coefficients are commonly used in computer transport models. It is important to keep in mind that K_d values are not constants and that they vary across different conditions, such as type of solid material, pH, and oxidation state of the metal. Soil and sediment with high K_d values have a high sorption or buffering capacity for added metals.

Another approach that incorporates sorption into computer models is surface-complexation (Stumm et al., 1970, 1976; Schindler and Gamsjager, 1972; Schindler et al., 1976). In this approach, sorption of ions on surfaces of oxide minerals is treated as analogous to the formation of aqueous complexes. Unlike partition coefficients, surface-complexation models have predictive capabilities beyond the measured conditions.

IMPORTANT ENVIRONMENTAL CONDITIONS

When metals are introduced into a stream, interactions such as dilution, chemical transformations, degradation, settling, resuspension, and other processes take place. Conditions in the stream are also a factor (e.g., pH, organic content, suspended solids, and numerous other factors) and can significantly affect how a metal will behave. Factors affecting the chemical composition of most surface waters are climate, lithology, geoavailability of elements, vegetation, topography, flow rates, biological activity, and time. The composition of water is controlled by interactions with earth materials through which the water flows. It is possible to make generalizations about some of the environmental factors that control metal chemistry, which allows for estimates of metal behavior, mobility, and fate.

Geochemical Gradients and Barriers

Perel'man (1977) discusses the importance of geochemical gradients, which describe gradual changes of a landscape, and of geochemical barriers, which describe abrupt changes. An example of a geochemical gradient might be the vertical and horizontal distribution of certain elements away from a mineral deposit within a constant lithology; for a given element, an anomalous concentration eventually declines to a background concentration at some distance away from the deposit. Another example of a geochemical gradient is the concentration plume for some elements downwind from a smelter. Perel'man (1986) defines geochemical barriers as zones of the Earth's crust with sharp physical or chemical gradients that are commonly associated with accumulation of elements. Geochemical barriers comprise abrupt changes in physical or chemical environments in the path of migration of elements causing the precipitation of certain elements from solution. Geochemical barriers include mechanical, physicochemical, biochemical, and anthropogenic (or technogenic) types. Complex barriers may be created when two or more barrier types are superimposed. Complex barriers are a common occurrence because geochemical processes are often linked and result in changing geochemical conditions. Perel'man (1977, 1986) gives a more in-depth discussion of geochemical barriers. This concept can help to forecast element distributions in the surficial environment and explain metal transport and mobility.

pH Barriers

Acidic barriers develop when pH values decrease. Under these conditions, elements that form oxyanions, such as Mo, as well as certain complexes, generally become less mobile, whereas many cationic metals, such as Cu, generally become more mobile. Solubility relationships can play an important role. For example, Al is usually fairly mobile below a pH of ~4, but will gradually precipitate between a pH of ~5 and 9. On the other hand, Si (as SiO_2) is relatively insoluble at low pH and becomes more soluble at high pH. One of the most important effects of developing low-pH environments is the destruction of the carbonate-bicarbonate buffering system, a feedback mechanism that controls the extent of pH change in an aquatic system. Below a pH of ~4.5, carbonate and bicarbonate are converted to carbonic acid. Upon such acidification, the water loses its capacity to buffer changes in pH, and many photosynthetic organisms that use bicarbonate as their inorganic carbon source become stressed or die. Once damaged, the alkalinity of a natural system may take significant time to recover, even if no further acid is added to the system. The carbonate-bicarbonate system may have both a direct and an indirect effect on the mobility of several elements.

Alkaline barriers develop where acidic waters encounter alkaline conditions over a short distance (e.g., oxidation zones of pyrite in limestone host rock). This type of barrier mostly retains those elements that migrate easily under acidic conditions and precipitate as hydroxides or carbonates under alkaline conditions (such as Fe, Al, Cu, Ni, and Co). During the shift to alkaline conditions, hydrous iron, aluminum, and manganese oxides may sorb trace metals and create an alkaline/adsorption complex barrier.

Redox Barriers

In most surficial aquatic systems, atmospheric oxygen is the primary oxidant, and organic matter is the primary reductant. At mining sites, other reductants may include FeS, FeS_2, Fe^{2+}, Mn^{2+}, or H_2S. There is a redox balance that depends upon the rate of oxygen depletion versus the rate of oxygen replenishment. If the rate of oxygen depletion is greater, redox-sensitive elements may undergo transformations from one chemical species to another. At circumneutral pH values, some redox reactions that may take place are listed below in order of increasingly reducing conditions:

$NO_3^- \rightarrow N_{2\,(g)}$

$MnO_{2\,(s)} \rightarrow Mn^{2+}$

$NO_3^- \rightarrow NO_2^-$

$NO_2^- \rightarrow NH_4^+$

$Fe(OH)_{3\,(s)} \rightarrow Fe^{2+}$

$SO_4^{2-} \rightarrow H_2S$

$HCO_3^- \rightarrow CH_4$

$HCO_3^- \rightarrow CH_2O$

Reducing barriers can be divided into those that contain hydrogen sulfide and those that do not (referred to as reducing gley environments; Perel'man, 1986). Berner (1981) proposed a simple redox classification based on the presence or absence of dissolved oxygen (DO) and sulfide. Under this scheme, oxic conditions exist where DO > 30 μM. Anoxic systems are divided into those with and without measurable sulfide (sulfidic and nonsulfidic, respectively). In sulfidic systems, many metals may precipitate as sulfide minerals.

Reducing hydrogen sulfide barriers develop where oxidizing waters come into contact with a reducing hydrogen sulfide environment or with sulfide minerals, or where deoxygenated sulfate-rich water encounters an accumulation of organic matter. Insoluble sulfides of elements such as Fe, Cu, Zn, Pb, Co, Ni, and Ag may precipitate at reducing barriers that contain hydrogen sulfide. Reducing gley barriers can form where water infiltrates soil and the weathering crust, and where free oxygen is lost or consumed. Depending on the pH, reducing gley waters are usually favorable for the transport of many ore-forming elements; additionally, elements such as Se, Cu, U, Mo, V, Cr, Ag, and As are known to accumulate at some reducing gley barriers (Perel'man, 1986). For example, roll-front-type uranium deposits may form under such conditions. Both manganese and iron oxides may undergo dissolution in reducing environments. These oxides provide important substrates for metal sorption and coprecipitation, so dissolution of these oxides may result in release of associated metals in reducing environments.

Oxidizing barriers occur where oxygen is introduced into anoxic waters or where anoxic ground water is discharged to the surficial environment. Iron, and possibly manganese, may precipitate at these barriers. Because hydrous iron and manganese oxides are good sorbents for metals (such as Cu and Co), a complex barrier may form by combining an oxidizing barrier with an adsorption barrier.

Evaporation Barriers

Evaporation barriers are often indicated by the presence of salt crusts or efflorescent salts, and Na, Mg, Ca, Cl, S, and CO_3^{2-}

salts may precipitate at these barriers. Evaporation barriers may be temporary and related to changing climatic conditions. For example, in some mine-waste-rock piles, efflorescent salts, enriched in elements such as Fe, Al, Cu, and S, may form during the dry season. These salts will be flushed from the system during a subsequent wet period and may cause a brief spike in metal content and acidity of the storm-water runoff (Nordstrom and Alpers, 1999).

Adsorption Barriers

Adsorption barriers are typically part of complex barriers. The most common sorbents (e.g., hydrous iron, aluminum, and manganese oxides, organic matter, and clay minerals) have different affinities for elements under different geochemical conditions. Adsorption reactions are known to control trace-metal concentrations in many natural systems.

Temperature/Pressure Barriers

Temperature/pressure (thermodynamic) barriers are formed in areas with temperature and pressure variations. One example of such a barrier is the degassing of carbon-dioxide-rich ground water as pressure drops and the subsequent deposition of carbonate minerals. Trace elements, such as Pb and Cd, can precipitate as carbonate minerals or coprecipitate with $CaCO_3$.

CONSIDERATIONS AND POSSIBLE APPLICATIONS

Understanding the chemical and physical factors of metals and geochemical processes that can influence metal mobility, distribution, and bioavailability can aid in forecasting the potential ecological effects of metals in surficial mining environments. The following sections briefly describe some of the applications where incorporating information about metal mobility is useful.

Risk Assessment

Ecological risk assessments are becoming increasingly important in evaluating the effects of historical mining as well as in predicting the potential effects of present and future mining. The U.S. Environmental Protection Agency (U.S. EPA) adopted a framework for ecological risk assessment (U.S. EPA, 1998) that includes planning, problem formulation, analysis, interpretation and risk characterization, communicating results, and risk management. This approach has been applied to the U.S. EPA Framework for Metals Assessment (U.S. EPA, 2007b). Most risk-assessment approaches have been developed for synthetic organic compounds. Because of the differences in environmental behavior between metals and organic compounds, it is generally becoming accepted that risk assessments for metals should be designed differently than those for organic compounds (Lee and Allen, 1998).

Environmental chemistry can be used to determine metal speciation for use in risk assessments and to assess the mobility of metals in the environment. In risk assessments involving metals, it is essential to identify factors that control metal transformations between bioavailable and nonbioavailable forms (Campbell et al., 2006; U.S. EPA, 2007b; Waeterschoot et al., 2003). The unit-world model, which is under development, will provide a quantitative method to assess risks posed by metals from their source of contamination, through transport in the aquatic environment, to uptake by biological receptors (Allen et al., 2000). Risk assessments can be performed at various scales from site-specific to watershed to regional. Approaches for regional risk assessments are still in development (e.g., Landis and Wiegers, 1997).

Water-Quality Criteria: Approaches to Determine the Bioavailable Metal Fraction

Ambient water-quality criteria have been developed by the U.S. EPA in support of the Clean Water Act. The goal of the criteria is to protect the physical, chemical, and biological integrity of waters of the United States. Several accommodations have evolved in the criteria to account for site-specific differences in metal bioavailability. For example, dissolved concentrations replaced total concentrations in some of the criteria. Also, criteria for several metals have been expressed as a function of water hardness to account for the protective effects of Ca and Mg on aquatic metal toxicity. The water effect ratio (WER) is an empirical approach that was developed to account for site-specific conditions where the water chemistry can alter metal bioavailability and toxicity (U.S. EPA, 1994). In this approach, toxicity tests are performed with site-specific water and compared with results from tests performed with laboratory water. The water-quality criterion is then adjusted to reflect the influence of the site-specific water.

Recently, the U.S. EPA has adopted a computational approach to address aquatic metal toxicity. The biotic ligand model (BLM) is a computer model that mathematically estimates the effects of water chemistry on the speciation and bioavailability of metals and on their acute toxicity to aquatic biota (Di Toro et al., 2001; Gorsuch et al., 2002; Niyogi and Wood, 2004; Paquin et al., 2002; Santore et al., 2001; Slaveykova and Wilkinson, 2005; Villavicencio et al., 2005). It is being used to develop site-specific water-quality criteria and to assess aquatic risk for metal exposure. The BLM has been incorporated into the 2007 update of the ambient water-quality criteria for Cu (U.S. EPA, 2007a) and is being used to determine regulatory site-specific concentration criteria for Cu. Given site-specific water chemistry, a chosen metal, and a chosen organism, the BLM predicts the LC_{50} (lethal concentration 50, which is the metal concentration that results in the death of 50% of a group of test organisms) for the chosen metal and organism. The BLM also can be used for predictive ecological risk assessments (Smith, 2005b; Smith et al., 2006). Collection of dissolved organic carbon data is imperative for use of the BLM.

Baseline Metal Concentrations

The term *baseline* has various meanings, especially when used in an environmental context. An environmental baseline is a summary of existing conditions over some time frame for some environmental system or material of interest (Lee and Helsel, 2005). Baselines can take human influences into account. In contrast, an environmental background describes the natural tendency of an environmental system or material in the absence of human influences (Lee and Helsel, 2005). Determination of an environmental background is very difficult, if not impossible. Determination of an environmental baseline also can present some difficulties because baseline concentrations may be below or near analytical detection limits. Methods to handle data with less-than values are available in the literature (e.g., Helsel, 2005).

Metals in the environment may result from natural geologic processes as well as from mining activities. Therefore, both the environmental baseline and background of metals can be elevated in mineralized areas. Baseline concentrations of elements are important determinants for risk assessment because organisms may adapt to elevated concentrations (Chapman et al., 1998). Consequently, metal-toxicity thresholds for organisms in environments with elevated metal concentrations may be greater than those for organisms adapted to low-concentration conditions (McLaughlin and Smolders, 2001).

Because metals are naturally occurring substances, biota have evolved in the presence of metals. It is important to identify environmental controlling factors, such as pH, organic matter, iron, and redox conditions, and their effect on metal concentration and exposure to biota. For site-specific assessments, the key aspect is to identify the biota most likely to be susceptible to metals. For regional assessments, the approach is often taken to protect highly sensitive species with limited distributions, which results in criteria that are overprotective for the larger area. One interesting approach is the concept of metalloregions, or metal-related ecoregions (McLaughlin and Smolders, 2001). This approach considers conditions within each type of metal-controlling environment. A series of fact sheets published by the International Council on Mining and Metals (ICMM; available at http://www.icmm.com) discusses relationships between metals, baseline conditions, bioavailability, and risk assessment.

Landscape Geochemistry

Landscape geochemistry can help define metal distributions in the environment. Landscape geochemistry focuses on the interaction of the lithosphere with the hydrosphere, biosphere, and atmosphere, and links exploration geochemistry with environmental science (Fortescue, 1980). Landscape geochemistry is a holistic approach to the study of the geochemistry of the environment in that it involves element cycles and may involve local, regional, and global studies. Fortescue (1992) reviews the development of landscape geochemistry and provides the foundation of how it relates to environmental science.

Fortescue (1992) proposes the establishment of a discipline of global landscape geochemistry (GLG), which may provide the foundation for future developments in applied and environmental geochemistry and which is necessary to adequately address current geoenvironmental problems. GLG regional geochemical mapping can be used to delineate geochemical provinces, identify local geochemical enrichments in mineral deposits, determine baseline environmental geochemistry, monitor environmental changes in soil and water geochemistry in response to human activities, evaluate the nutritional status of plants and animals, and study human health. Fortescue (1992) notes that there is a need to map geochemical landscapes as an essential preliminary step to the study of environmental geochemistry. Geochemical maps based on the analysis of rocks, soils, sediments, waters, and vegetation, originally compiled for mineral exploration purposes, may be extended to multipurpose geochemical surveys that have applications in agriculture, pollution studies, and human health (Webb, 1964). However, geochemical analyses for mineral exploration purposes have generally been designed to be cost effective; consequently, the quality of the geochemical data often is inadequate for many environmental applications.

SUMMARY

The purpose of this report is to present some strategies to predict metal mobility at mining sites. Metals occur naturally in the environment. Some metals are essential or beneficial to living organisms, but many metals are potentially toxic. In mineralized environments, metal concentrations tend to be elevated compared with natural abundance. Mining activities can increase metal concentrations in the environment. Once introduced, metals persist in the environment and can only be reduced by physical removal. However, metal speciation can influence metal distribution, transport, and bioavailability within the environment. Consequently, the forms, transformations, and geochemical environments of metals need to be considered when evaluating potential metal mobility at mining sites. Understanding the factors that influence metal mobility can aid in forecasting potential ecological effects of metals in mining environments.

Physical and chemical properties of metals at the atomic level are responsible for differences in their environmental behavior. Some important properties include oxidation state, size, and electronegativity. Metals can be classified into groups based on their capacity for binding to different ligands. The hard and soft acid and base classification is a useful concept to help explain the strength of metal complexation, metal behavior, and bioavailability. Qualitative and quantitative modeling methods are being developed that relate the chemistry of metals to biological activity.

Different chemical species of a given metal generally have different mobility and bioavailability behavior and toxicological effects. Factors that can influence speciation include pH, redox conditions, availability of inorganic and organic ligands, and competition from other ions. Most metals exist as cations, although some metals exist as oxyanions. Cationic species tend to be more mobile under low-pH conditions and less mobile under high-pH conditions, whereas the opposite is true for anionic species. The pH of a system also can control which ligands are available for binding. For redox-sensitive elements, such as Cr, As, and Se, a change in mobility will occur under different redox conditions.

Mining-influenced waters generally have low-pH values, high concentrations of sulfate, and high concentrations of Fe, Al, Mn, and several other metals. This unusual composition makes these systems somewhat unique when defining the geochemical processes, reactants, and phases that control metal mobility. Solubility reactions generally control Fe and Al concentrations, as well as some other metal concentrations. The pH is important in controlling the solubility of most metal compounds. Sorption reactions, involving both inorganic and organic particulates, largely control the fate of many trace elements in natural systems. Metal sorption is strongly pH-dependent and is influenced by metal-complex formation and ionic strength. At many mining sites there are abundant iron- and aluminum-oxide precipitates that can act as effective sorbents for a variety of metals.

A conceptual model can help integrate and prioritize the importance of metal speciation and changing geochemical environments at a mining site. Such a model should include relationships and pathways between the sources, transport mechanisms, and fate of the metals, and should incorporate the mineralogical, hydrological, geological, and geochemical conditions at the site that might affect these pathways. An important aspect of a conceptual model is source characterization.

It is difficult to quantitatively predict metal mobility in surficial environments; however, mobility can be considered in a relative sense by comparing metal behavior under changing environmental conditions that control metal mobility (e.g., pH conditions, redox conditions, solubility reactions, and sorption reactions). Knowledge of how metal speciation responds to changing environmental conditions can be used to predict metal mobility. Hence, it is possible to make generalizations about relative metal mobility under different conditions.

There are several applications where incorporating information about metal mobility is useful. The U.S. EPA has developed an ecological risk assessment framework for metals to specifically address the speciation characteristics of metals. Also, water-quality criteria are being rewritten to address metal speciation using the biotic ligand model. Finally, considerations of metal-baseline conditions are being used to understand the potential effects of metals introduced to the environment.

ACKNOWLEDGMENTS

The author would like to thank Andrew Archuleta, Lopaka (Rob) Lee, and Carol Russell for their helpful reviews. Preparation of this report was funded by the U.S. Geological Survey Mineral Resources Program and the U.S. Geological Survey Toxic Substances Hydrology Program.

REFERENCES CITED

Adriano, D.C., 1986, Trace elements in the terrestrial environment: Springer-Verlag, New York, 533 p.

Al-Abed, S.R., Hageman, P.L., Jegadeesan, G., Madhavan, N., and Allen, D., 2006, Comparative evaluation of short-term leach tests for heavy metal release from mineral processing waste: Science of the Total Environment, v. 364, p. 14–23.

Allen, H.E., ed., 2002, Bioavailability of metals in terrestrial ecosystems: Importance of partitioning for bioavailability to invertebrates, microbes, and plants: Pensacola, Florida, SETAC Press.

Allen, H.E., Hall, R.H., and Brisbin, T.D., 1980, Metal speciation effects on aquatic toxicity: Environmental Science & Technology, v. 14, p. 441–443, doi: 10.1021/es60164a002.

Allen, H.E., Di Toro, D.M., Paquin, P.R., and Santore, R.C., 2000, The "unit world" model: A proposed uniform model to prioritize environmental hazards of metals and inorganic metal compounds: ICME Newsletter (Canada), v. 8, no. 3, p. 5–6.

Alpers, C.N. and Blowes, D.W., eds., 1994, Environmental geochemistry of sulfide oxidation: Washington, D.C., American Chemical Society Symposium Series 550, 681 p.

Balistrieri, L.S., Box, S.E., Bookstrom, A.A., Hooper, R.L., and Mahoney, J.B., 2002, Impacts of historical mining in the Coeur d'Alene River Basin, *in* Balistrieri, L.S., and Stillings, L.L., eds., Pathways of metal transfer from mineralized sources to bioreceptors: A synthesis of the Mineral Resources Program's past environmental studies in the western United States and future research directions, Chapter 6: U.S. Geological Survey Bulletin 2191. (Available online at http://geopubs.wr.usgs.gov/bulletin/b2191/.)

Berner, R.A., 1981, A new geochemical classification of sedimentary environments: Journal of Sedimentary Petrology, v. 51, p. 359–365.

Bigham, J.M. and Nordstrom, D.K., 2000, Iron and aluminum hydroxysulfates from acid sulfate waters, *in* Alpers, C.N., Jambor, J.L., and Nordstrom, D.K., eds., Sulfate minerals: Crystallography, geochemistry, and environmental significance, Reviews in Mineralogy and Geochemistry, vol. 40: Washington, D.C., Mineralogical Society of America and The Geochemical Society, p. 351–403.

Bigham, J.M., Schwertmann, U., Traina, S.J., Winland, R.L., and Wolf, M., 1996, Schwertmannite and the chemical modeling of iron in acid sulfate waters: Geochimica et Cosmochimica Acta, v. 60, p. 2111–2121, doi: 10.1016/0016-7037(96)00091-9.

Bourg, A.C.M., 1988, Metals in aquatic and terrestrial systems: Sorption, speciation, and mobilization, *in* Salomons, W., and Forstner, U., eds., Chemistry and biology of solid wastes: Berlin, Springer-Verlag, p. 3–32.

Buffle, J., and Horvai, G., eds., 2000, In situ monitoring of aquatic systems-chemical analysis and speciation, IUPAC Series on Analytical and Physical Chemistry of Environmental Systems, vol. 6: Chichester, John Wiley & Sons, 632 p.

Callahan, M.A., Slimak, M.W., Gabel, N.W., May, I.P., Fowler, C.F., Freed, J.R., Jennings, P., Durfee, R.L., Whitmore, F.C., Maestri, B., Mabey, W.R., Holt, B.R., and Gould, C., 1979, Water-related environmental fate of 129 priority pollutants, vol. 1, Introduction and technical background, metals and inorganics, pesticides and PCBs: U.S. Environmental Protection Agency, EPA-440/4-79-029a.

Campbell, P.G.C., Chapman, P.M., and Hale, B.A., 2006, Risk assessment of metals in the environment, *in* Issues in Environmental Science and Technology, No. 22: The Royal Society of Chemistry, London.

Chao, T.T., 1984, Use of partial dissolution techniques in geochemical exploration: Journal Geochemical Exploration, v. 20, p. 101–135, doi: 10.1016/0375-6742(84)90078-5.

Chapman, P.M., Wang, F., Janssen, C., Persoone, G., and Allen, H.E., 1998, Ecotoxicology of metals in aquatic sediments: Binding and release, bioavailability, risk assessment and remediation: Canadian Journal of Fisheries and Aquatic Sciences, v. 55, p. 2221–2243, doi: 10.1139/cjfas-55-10-2221.

Church, S.E., von Guerard, P., and Finger, S.E., eds., 2007, Integrated investigations of environmental effects of historical mining in the Animas River watershed, San Juan County, Colorado: U.S. Geological Survey Professional Paper 1651, 1096 p.

Cotton, F.A., and Wilkinson, G., 1988, Advanced inorganic chemistry (5th edition): John Wiley & Sons, New York, 1455 p.

Craig, P.J., ed., 2003, Organometallic compounds in the environment (2nd edition): Chichester, England, John Wiley & Sons Ltd., 415 p.

Cravotta, C.A. III, 2006, Relations among pH, sulfate, and metals concentrations in anthracite and bituminous coal-mine discharges, Pennsylvania, in Proceedings of the Seventh International Conference on Acid Rock Drainage (7th ICARD), St. Louis, Missouri, March 26–30, 2006.

Cravotta, C.A. III and Kirby, C.S., 2004, Acidity and alkalinity in mine drainage: Practical considerations, in 2004 National Meeting of the American Society of Mining and Reclamation, April 18–24, 2004: ASMR, p. 334–365.

Crock, J.G., Arbogast, B.F., and Lamothe, P.J., 1999, Laboratory methods for the analysis of environmental samples, in Plumlee, G.S., and Logsdon, M.J., eds., The environmental geochemistry of mineral deposits, Part A: Processes, techniques, and health issues, Reviews in Economic Geology, vol. 6A: Littleton, Colorado, Society of Economic Geologists, Inc., p. 265–287.

D'Amore, J.J., Al-Abed, S.R., Scheckel, K.G., and Ryan, J.A., 2005, Methods for speciation of metals in soils: A review: Journal of Environmental Quality, v. 34, p. 1707–1745, doi: 10.2134/jeq2004.0014.

Davis, A., Drexler, J.W., Ruby, M.V., and Nicholson, A., 1993, Micromineralogy of mine wastes in relation to lead bioavailability, Butte, Montana: Environmental Science & Technology, v. 27, p. 1415–1425, doi: 10.1021/es00044a018.

Davis, J.A., and Hayes, K.F., 1986, Geochemical processes at mineral surfaces: An overview, in Davis, J.A., and Hayes, K.F., eds., Geochemical processes at mineral surfaces, American Chemical Society Symposium Series 323: Washington, D.C., American Chemical Society, p. 2–18.

Davis, J.A., and Kent, D.B., 1990, Surface complexation modeling in aqueous geochemistry, in Hochella, M.F., and White, A.F., eds., Mineral-water interface geochemistry, Reviews in Mineralogy, vol. 23: Washington, D.C., Mineralogical Society of America, p. 177–260.

Diehl, S.F., Hageman, P.L., and Smith, K.S., 2006, What's weathering in mine waste? Mineral dissolution and field leach studies, Leadville and Montezuma Mining Districts, Colorado, in Proceedings of the Seventh International Conference on Acid Rock Drainage (7th ICARD), St. Louis, Missouri, March 26–30, 2006.

Di Toro, D.M., Allen, H.E., Bergman, H.L., Meyer, J.S., Paquin, P.R., and Santore, R.C., 2001, Biotic ligand model of the acute toxicity of metals, 1. Technical basis: Environmental Toxicology and Chemistry, v. 20, no. 10, p. 2383–2396, doi: 10.1897/1551-5028(2001)020<2383:BLMOTA>2.0.CO;2.

Drever, J.I., 1997, The geochemistry of natural waters (3rd edition): Prentice-Hall, Inc., Upper Saddle River, New Jersey, 436 p.

du Bray, E.A., ed., 1995, Preliminary compilation of descriptive geoenvironmental mineral deposit models: U.S. Geological Survey Open-File Report 95-831, 272 p. (Available online at http://pubs.usgs.gov/of/1995/ofr-95-0831/; accessed January 12, 2005.)

Dutrizac, J.E. and Jambor, J.L., 1987, The behavior of arsenic during jarosite precipitation: Arsenic precipitation at 97 °C from sulfate or chloride media: Canadian Metallurgical Quarterly, v. 26, p. 91–101.

Dzombak, D.A. and Morel, F.M.M., 1987, Adsorption of inorganic pollutants in aquatic systems: Journal of Hydraulic Engineering, v. 113, p. 430–475.

Ficklin, W.H. and Mosier, E.L., 1999, Field methods for sampling and analysis of environmental samples for unstable and selected stable constituents, in Plumlee, G.S., and Logsdon, M.J., eds., The environmental geochemistry of mineral deposits, Part A: Processes, techniques, and health issues, Reviews in Economic Geology, vol. 6A: Littleton, Colorado, Society of Economic Geologists, Inc., p. 249–264.

Ficklin, W.H., Plumlee, G.S., Smith, K.S., and McHugh, J.B., 1992, Geochemical classification of mine drainages and natural drainages in mineralized areas, in Kharaka, Y.K., and Maest, A.S., eds., Water-rock interaction, vol. 1, Seventh International Symposium on Water-Rock Interaction: A.A. Balkema, Rotterdam, p. 381–384.

Forstner, U., 1987, Changes in metal mobilities in aquatic and terrestrial cycles, in Patterson, J.W., and Passino, R., eds., Metals speciation, separation, and recovery: Lewis Publishers, Chelsea, Michigan, p. 3–26.

Fortescue, J.A.C., 1980, Environmental geochemistry: A holistic approach: Springer-Verlag, New York, 347 p.

Fortescue, J.A.C., 1992, Landscape geochemistry: Retrospect and prospect, 1990: Applied Geochemistry, v. 7, p. 1–53, doi: 10.1016/0883-2927(92) 90012-R.

Fuller, W.H., 1977, Movement of selected metals, asbestos, and cyanide in soil: Applications to waste disposal problems: U.S. Environmental Protection Agency, EPA-600/2-77-020, 242 p.

Garrels, R.M., and Christ, C.L., 1965, Solutions, minerals, and equilibria: Freeman, Cooper & Company, San Francisco, 450 p.

Gerould, S., 2000, Mercury in the environment: U.S. Geological Survey Fact Sheet 146–00. (Available online at http://www.usgs.gov/themes/factsheet/146-00/.)

Gorsuch, J.W., Janssen, C.R., Lee, C.M., and Reiley, M.C., eds., 2002, Special issue: The biotic ligand model for metals: Current research, future directions, regulatory implications: Comparative Biochemistry and Physiology, Part C: Toxicology and Pharmacology, v. 133C, no. 1-2, September 2002, 343 p.

Greenwood, N.N. and Earnshaw, A., 1984, Chemistry of the elements: Pergamon Press, New York, 1542 p.

Hageman, P.L., 2005, A simple field leach test to assess potential leaching of soluble constituents from mine wastes, soils, and other geologic materials: U.S. Geological Survey Fact Sheet 2005-3100. (Available online at http://pubs.usgs.gov/fs/2005/3100/pdf/FS-3100_508.pdf.)

Hammarstrom, J.M. and Smith, K.S., 2002, Geochemical and mineralogic characterization of solids and their effects on waters in metal-mining environments, in Seal, R.R. II, and Foley, N.K., eds., Progress on geoenvironmental models for selected mineral deposit types: U.S. Geological Survey Open-File Report 02-0195, p. 8–54. (Available online at http://pubs.usgs.gov/of/2002/of02-195/.)

Helsel, D.R., 2005, Nondetects and data analysis: Statistics for censored environmental data: John Wiley & Sons, New York.

Hem, J.D., 1989, Study and interpretation of the chemical characteristics of natural water (3rd edition): U.S. Geological Survey Water-Supply Paper 2254, 263 p. (Available online at http://water.usgs.gov/pubs/wsp/wsp2254/; accessed January 12, 2005.)

Huheey, J.E., Keiter, E.A., and Keiter, R.L., 1993, Inorganic chemistry: Principles of structure and reactivity (4th edition): HarperCollins College Publishers, New York, 964 p.

Jambor, J.L. and Blowes, D.W., eds., 1994, Mineralogical Association of Canada short course on environmental geochemistry of sulfide mine wastes, vol. 22: Mineralogical Society of Canada.

Jambor, J.L., Blowes, D.W., and Ritchie, A., eds., 2003, Environmental aspects of mine wastes, Mineralogical Association of Canada Short Course Series, vol. 31: Mineralogical Association of Canada.

James, R.O. and Healy, T.W., 1972, Adsorption of hydrolyzable metal ions at the oxide-water interface, Part III: A thermodynamic model of adsorption: Journal of Colloid and Interface Science, v. 40, p. 65–79, doi: 10.1016/0021-9797(72)90174-9.

Jenne, E.A., 1968, Controls on Mn, Fe, Co, Ni, Cu, and Zn concentrations in soils and water: The significant role of hydrous Mn and Fe oxides, in Gould, R.F., ed., Trace inorganics in water, Advances in Chemistry Series No. 73: Washington, D.C., American Chemical Society, p. 337–387.

Jenne, E.A., and Luoma, S.N., 1977, Forms of trace elements in soils, sediments, and associated waters: An overview of their determination and biological availability, in Wildung, R.E., and Drucker, H., eds., Biological implications of metals in the environment, Technical Information Center, Energy Research and Development Administration Symposium Series 42: NTIS, Springfield, Virginia, CONF-750929, p. 110–143.

Jenne, E.A., and Zachara, J.M., 1987, Factors influencing the sorption of metals, in Dickson, K.L., Maki, A.W., and Brungs, W.A., eds., Fate and effects of sediment-bound chemicals in aquatic systems, SETAC Special Publication Series: Pergamon Press, New York, p. 83–98.

Jordan, G., and D'Alessandro, M. eds., 2004, Mining, mining waste and related environmental issues: Problems and solutions in Central and Eastern European Candidate Countries: Joint Research Centre of the European Commission, Ispra, EUR 20868 EN, 208 p.

Kabata-Pendias, A. and Pendias, H., 1992, Trace elements in soils and plants (2nd edition): CRC Press, Inc., Boca Raton, Florida, 342 p.

Kinniburgh, D.G. and Jackson, M.L., 1981, Cation adsorption by hydrous metal oxides and clay, in Anderson, M.A., and Rubin, A.J., eds., Adsorption of inorganics at solid-liquid interfaces: Ann Arbor Science, Ann Arbor, Michigan, p. 91–160.

Landis, W.G. and Wiegers, J.A., 1997, Design considerations and a suggested approach for regional and comparative ecological risk assessment: Human and Ecological Risk Assessment, v. 3, p. 287–297.

Langmuir, D., 1997, Aqueous environmental geochemistry: Prentice-Hall, 600 p.

Langmuir, D. and Mahoney, J.J., 1984, Chemical equilibrium and kinetics of geochemical processes in ground water studies, in B. Hitchon and E. Wallick, eds., Practical applications of ground water geochemistry: National Water Well Association, Dublin, Ohio, p. 69–95.

Lee, C.M. and Allen, H.E., 1998, The ecological risk assessment of copper differs from that of hydrophobic organic chemicals: Human and Ecological Risk Assessment, v. 4, p. 605–617, doi: 10.1080/10807039891284442.

Lee, L. and Helsel, D., 2005, Baseline models of trace elements in major aquifers of the United States: Applied Geochemistry, v. 20, p. 1560–1570, doi: 10.1016/j.apgeochem.2005.03.008.

Lefebvre, R., Hockley, D., Smolensky, J., and Gelinas, P., 2001, Multiphase transfer processes in waste rock piles producing acid mine drainage, Part 1: Conceptual model and system conceptualization: Journal of Contaminant Hydrology, v. 52, p. 137–164, doi: 10.1016/S0169-7722(01)00156-5.

Levinson, A.A., 1980, Introduction to exploration geochemistry (2nd edition): Applied Publishing, Ltd., Wilmette, Illinois, 924 p.

Lindsay, W.L., 1979, Chemical equilibria in soils: John Wiley and Sons, New York, 449 p.

Lukashev, V.K., 1984, Mode of occurrence of elements in secondary environments: Journal of Geochemical Exploration, v. 21, p. 73–87, doi: 10.1016/0375-6742(84)90035-9.

Lukashev, V.K., 1986, Some scientific and applied problems of supergene geochemistry in the U.S.S.R: Applied Geochemistry, v. 1, p. 441–449, doi: 10.1016/0883-2927(86)90049-1.

Luoma, S.N., 1983, Bioavailability of trace metals to aquatic organisms: A review: The Science of the Total Environment, v. 28, p. 1–22.

Luoma, S.N. and Carter, J.L., 1993, Understanding the toxicity of contaminants in sediments: Beyond the bioassay-based paradigm: Environmental Toxicology and Chemistry, v. 12, p. 793–796.

Luoma, S.N. and Davis, J.A., 1983, Requirements for modeling trace metal partitioning in oxidized estuarine sediments: Marine Chemistry, v. 12, p. 159–181, doi: 10.1016/0304-4203(83)90078-6.

Macalady, D.L. and Ranville, J.F., 1998, The chemistry and geochemistry of natural organic matter (NOM), in Macalady, D.L., ed., Perspectives in environmental chemistry: Oxford University Press, New York, p. 94–137.

Maest, A., Kuipers, J., Travers, C., and Atkins, D., 2005a, Evaluation of methods and models used to predict water quality at hardrock mine sites: Sources of uncertainty and recommendations for improvement, in Proceedings, Society for Mining, Metallurgy, and Exploration Annual Meeting and Exhibit, Salt Lake City, Utah, February 28–March 2, 2005, Preprint 05-90: Society for Mining, Metallurgy, and Exploration, Inc., Littleton, Colorado.

Maest, A.S., Kuipers, J.R., Travers, C.L., and Atkins, D.A., 2005b, Predicting water quality at hardrock mines: methods and models, uncertainties, and state-of-the-art: Buka Environmental and Kuipers & Associates, 77 p. (Available online at http://www.earthworksaction.org/pubs/PredictionsReportFinal.pdf.)

McKinney, J.D., Richard, A., Waller, C., Newman, M.C., and Gerberick, F., 2000, The practice of structure activity relationships (SAR) in toxicology: Toxicological Sciences, v. 56, p. 8–17, doi: 10.1093/toxsci/56.1.8.

McKnight, D.M., Kimball, B.A., and Runkel, R.L., 2001, pH dependence of iron photoreduction in a Rocky Mountain stream affected by acid mine drainage: Hydrological Processes, v. 15, p. 1979–1992, doi: 10.1002/hyp.251.

McLaughlin, M.J. and Smolders, E., 2001, Background zinc concentrations in soil affect the zinc sensitivity of soil microbial processes: A rationale for the metalloregion approach to risk assessments: Environmental Toxicology and Chemistry, v. 20, p. 2639–2643, doi: 10.1897/1551-5028(2001)020<2639:BZCISA>2.0.CO;2.

McLemore, V.T., Smith, K.S., Russell, C.C., and the Sampling and Monitoring Committee of the Acid Drainage Technology Initiative–Metal Mining Sector, 2007, this volume, Sampling and Monitoring for Closure, in DeGraff, J.V., ed., Understanding and Responding to Hazardous Substances at Mine Sites in the Western United States: Geological Society of America Reviews in Engineering Geology, v. XVII, doi: 10.1130/2007.4017(11).

MEND Manual, 2001, summarizes the work completed by Mine Environment Neutral Drainage (MEND) in Canada in a six-volume set. (See http://www.nrcan.gc.ca/mms/canmet-mtb/mmsl-lmsm/mend/mendmuale.htm.)

Mills, C. and Robertson, A., 2006, Acid rock drainage at Enviromine: Infomine Website, accessed on April 30, 2006, at http://technology.infomine.com/enviromine/ard/home.htm.

Morel, F.M.M. and Herring, J.G., 1993, Principles and applications of aquatic chemistry: John Wiley & Sons, Inc., New York, 588 p.

Newman, M.C. and Jagoe, C.H., 1994, Inorganic toxicants: Ligands and the bioavailability of metals in aquatic environments, in Hamelink, J.L., Landrum, P.F., Bergman, H.L., and Benson, W.H., eds., Bioavailability: Physical, chemical, and biological interactions, SETAC Special Publications Series: CRC Press, Inc., Boca Raton, Florida, p. 39–61.

Newman, M.C., McCloskey, J.T., and Tatara, C.P., 1998, Using metal–ligand binding characteristics to predict metal toxicity: Quantitative ion character-activity relationships (QICARs): Environmental Health Perspectives, v. 106, Suppl. 6, p. 1419–1425.

Nieboer, E. and Fletcher, G.G., 1996, Determinants of reactivity in metal toxicology, in Chang, L.W., ed., Toxicology of metals: Boca Raton, Florida, CRC Press, p. 113–132.

Nieboer, E. and Richardson, D.H.S., 1980, The replacement of the nondescript term "heavy metals" by a biologically and chemically significant classification of metal ions: Environmental Pollution (Series B), v. 1, p. 3–26.

Nimick, D.A., Church, S.E., and Finger, S.E., eds., 2005, Integrated investigations of environmental effects of historical mining in the Basin and Boulder Mining Districts, Boulder River watershed, Jefferson County, Montana: U.S. Geological Survey Professional Paper 1652. (Available online at http://pubs.usgs.gov/pp/2004/1652/.)

Niyogi, S. and Wood, C.M., 2004, Biotic ligand model: A flexible tool for developing site-specific water quality guidelines for metals: Environmental Science & Technology, v. 38, no. 23, p. 6177–6192, doi: 10.1021/es0496524.

Nordstrom, D.K., 1982, The effects of sulfate on aluminum concentrations in natural waters: Some stability relations in the system Al_2O_3-SO_3-H_2O at 298 K: Geochimica et Cosmochimica Acta, v. 46, p. 681–692, doi: 10.1016/0016-7037(82)90168-5.

Nordstrom, D.K., 2002, Aqueous redox chemistry and the behavior of iron in acid mine waters, in Wilkin, R.T., Ludwig, R.D., and Ford, R.G., eds, Proceedings of the Workshop on Monitoring Oxidation-Reduction Processes for Ground-water Restoration, Dallas, Texas, April 25-27, 2000: Cincinnati, Ohio, U.S. Environmental Protection Agency, EPA/600/R-02/002, p. 43–47. (Available online at http://water.usgs.gov/nrp/proj.bib/Publications/nordstrom_epa.pdf.)

Nordstrom, D.K. and Alpers, C.N., 1999, Geochemistry of acid mine waters, in Plumlee, G.S., and Logsdon, M.J., eds., The environmental geochemistry of mineral deposits, Part A: Processes, techniques, and health issues, Reviews in Economic Geology, vol. 6A: Littleton, Colorado, Society of Economic Geologists, p. 133–160.

Nordstrom, D.K. and Ball, J.W., 1986, The geochemical behavior of aluminum in acidified surface waters: Science, v. 232, p. 54–58, doi: 10.1126/science.232.4746.54.

Nordstrom, D.K. and Munoz, J.L., 1994, Geochemical thermodynamics (2nd edition): Boston, Blackwell.

Ownby, D.R., and Newman, M.C., 2003, Advances in quantitative ion characteractivity relationships (QICARs): Using metal-ligand binding characteristics to predict metal toxicity: QSAR & Combinatorial Science, v. 22, p. 241-246, doi: 10.1002/qsar.200390018.

Pagenkopf, G.K., 1983, Gill surface interaction model for trace metal toxicity to fishes: Role of complexation, pH, and water hardness: Environmental Science & Technology, v. 17, p. 342–347, doi: 10.1021/es00112a007.

Paquin, P.R., Gorsuch, J.W., Apte, S., Batley, G.E., Bowles, K.C., Campbell, P.G.C., Delos, C.G., Di Toro, D.M., Dwyer, R.L., Galvez, F., Gensemer, R.W., Goss, G.G., Hogstrand, C., Janssen, C.R., McGeer, J.C., Naddy, R.B., Playle, R.C., Santore, R.C., Schneider, U., Stubblefield, W.A., Wood, C.M., and Wu, K.B., 2002, The biotic ligand model: A historical overview: Comparative Biochemistry and Physiology, v. 133C, p. 3–35.

Parish, R.V., 1977, The metallic elements: Longman, Inc., New York, 254 p.

Pauling, L., 1960, The nature of the chemical bond (3rd edition): Cornell University Press, Ithaca, New York.

Pearson, R.G., 1963, Hard and soft acids and bases: Journal of the American Chemical Society, v. 85, p. 3533–3539, doi: 10.1021/ja00905a001.

Pearson, R.G., 1968a, Hard and soft acids and bases, HSAB, Part I: Fundamental principles: Journal of Chemical Education, v. 45, p. 581–587.

Pearson, R.G., 1968b, Hard and soft acids and bases, HSAB, Part II: Underlying theories: Journal of Chemical Education, v. 45, p. 643–648.

Perel'man, A.I., 1977, Geochemistry of elements in the supergene zone (translated from Russian): Keter Publishing House, Ltd., Jerusalem, 266 p.

Perel'man, A.I., 1986, Geochemical barriers: Theory and practical applications: Applied Geochemistry, v. 1, p. 669–680, doi: 10.1016/0883-2927(86)90088-0.

Plumlee, G.S., 1999, The environmental geology of mineral deposits, in Plumlee, G.S., and Logsdon, M.J., eds., The environmental geochemistry of mineral deposits, Reviews in Economic Geology, vol. 6A: Littleton, Colorado, Society of Economic Geologists, Inc., p. 71–116.

Plumlee, G.S., Morman, S.A., and Ziegler, T.L., 2006, The toxicological geochemistry of earth materials: An overview of processes and the interdisciplinary methods used to understand them, in Sahai, N., and Schoonen, M.A.A., eds., Medical Mineralogy and Geochemistry, Reviews in Mineralogy and Geochemistry, vol. 64: Mineralogical Society of America, Chantilly, Virginia, p. 5–57.

Plumlee, G.S., and Ziegler, T.L., 2003, The medical geochemistry of dusts, soils, and other earth materials, in Holland, H.D. and Turekian, K.K., eds., Treatise on Geochemistry, Volume 9: Pergamon, Oxford, p. 263–310, doi: 10.1016/B0-08-043751-6/09050-2.

Plumlee, G.S., Smith, K.S., Montour, M.R., Ficklin, W.H., and Mosier, E.L., 1999, Geologic controls on the composition of natural waters and mine waters draining diverse mineral-deposit types, in Filipek, L.H., and Plumlee, G.S., eds., The environmental geochemistry of mineral deposits, Reviews in Economic Geology, vol. 6B: Society of Economic Geologists, Inc., Littleton, Colorado, p. 373–432.

Price, W.A., 1997, Draft guidelines and recommended methods for the prediction of metal leaching and acid rock drainage at minesites in British Columbia: Reclamation Section, Energy and Minerals Division, Ministry of Employment and Investment, Smithers, B.C., Canada.

Price, W.A., 2005, List of potential information requirements in metal leaching and acid rock drainage assessment and mitigation work, MEND Report 5.10E: Natural Resources Canada, Ottawa, 23 p. (Available online at http://www.nrcan.gc.ca/mms/canmet-mtb/mmsl-lmsm/mend/reports/report510-e.pdf.)

Ranville, J.F., Blumenstein, E.P., Adams, M.K., Choate, L.M., Smith, K.S., and Wildeman, T.R., 2006, Integrating bioavailability approaches into waste rock evaluations, in Proceedings of the Seventh International Conference on Acid Rock Drainage (7th ICARD), St. Louis, Missouri, March 26–30, 2006.

Ritchie, A.I.M., 1994, The waste-rock environment, in Jambor, J.L., and Blowes, D.W., eds., Environmental geochemistry of sulfide mine-wastes, MAC Short Course Handbook, vol. 22: Mineralogical Association of Canada, Nepean, Ontario, p. 133–161.

Rose, A.W., Hawkes, H.E., and Webb, J.S., 1979, Geochemistry in mineral exploration (2nd edition): Academic Press, New York, 657 p.

Santore, R.C., Di Toro, D.M., Paquin, P.R., Allen, H.E., and Meyer, J.S., 2001, Biotic ligand model of the acute toxicity of metals, 2: Application to acute copper toxicity in freshwater fish and Daphnia: Environmental Toxicology and Chemistry, v. 20, p. 2397–2402, doi: 10.1897/1551-5028(2001)020<2397:BLMOTA>2.0.CO;2.

Sauvé, S., 2002, Speciation of metals in soils, in Allen, H.E., ed., Bioavailability of metals in terrestrial ecosystems: Importance of partitioning for bioavailability to invertebrates, microbes, and plants: SETAC Press, Pensacola, Florida.

Schindler, P.W. and Gamsjager, H., 1972, Acid-base reactions of the TiO_2 (anatase)-water interface and the point of zero charge of TiO_2 suspensions: Colloid & Polymer Science. v. 250, p. 759–763, doi: 10.1007/BF01498568.

Schindler, P.W., Furst, B., Dick, B., and Wolf, P.U., 1976, Ligand properties of surface silanol groups, I: Surface complex formation with Fe^{3+}, Cu^{2+}, Cd^{2+}, and Pb^{2+}: Journal of Colloid and Interface Science, v. 55, p. 469–475, doi: 10.1016/0021-9797(76)90057-6.

Schmiermund, R.L. and Drozd, M.A., eds., 1997, Acid mine drainage and other mining-influenced waters (MIW), in Marcus, J.J., ed., Mining environmental handbook: London, Imperial College Press, Chapter 13, p. 599–617.

Seal, R.R. II and Foley, N.K., eds., 2002, Progress on geoenvironmental models for selected mineral deposit types: U.S. Geological Survey Open-File Report 02-195. (Available online at http://pubs.usgs.gov/of/2002/of02-195/.)

Seal, R.R. II and Hammarstrom, J.M., 2003, Geoenvironmental models of mineral deposits: Examples from massive sulfide and gold deposits, in Jambor, J.L., Blowes, D.W., and Ritchie, A.I.M., eds., Environmental aspects of mine wastes, Short Course Handbook, vol. 31: Ottawa, Mineralogical Association of Canada.

Slaveykova, V.I. and Wilkinson, K.J., 2005, Predicting the bioavailability of metals and metal complexes: Critical review of the biotic ligand model: Environmental Chemistry, v. 2, p. 9–24, doi: 10.1071/EN04076.

Smith, K.S., 1999a, Geoavailability, in Marshall, C.P., and Fairbridge, R.W., eds., Encyclopedia of geochemistry: Kluwer Academic Publishers, Dordrecht, The Netherlands, p. 262–263.

Smith, K.S., 1999b, Metal sorption on mineral surfaces: An overview with examples relating to mineral deposits, in Plumlee, G.S., and Logsdon, M.J., eds., The environmental geochemistry of mineral deposits, Part A: Processes, techniques, and health issues, Reviews in Economic Geology, vol. 6A, Chapter 7: Littleton, Colorado, Society of Economic Geologists, Inc., p. 161–182.

Smith, K.S., 2005a, Acid rock drainage, in Price, L.G., Bland, D., McLemore, V.T., and Barker, J.M., eds., Mining in New Mexico: The environment, water, economics, and sustainable development, in Decision-makers field guide 2005, Chapter 2: New Mexico Bureau of Geology and Mineral Resources, p. 59–63. (Available online at http://geoinfo.nmt.edu/publications/decisionmakers/2005/DM_2005_Ch2.pdf.)

Smith, K.S., 2005b, Use of the biotic ligand model to predict metal toxicity to aquatic biota in areas of differing geology, in Proceedings of the 2005 National Meeting of the American Society of Mining and Reclamation, Breckenridge, Colorado, June 19–23, 2005, p. 1134–1154.

Smith, K.S., and Huyck, H.L.O., 1999, An overview of the abundance, relative mobility, bioavailability, and human toxicity of metals, in Plumlee, G.S., and Logsdon, M.J., eds., The environmental geochemistry of mineral deposits, Part A: Processes, techniques, and health issues, Reviews in Economic Geology, vol. 6A, Chapter 2: Littleton, Colorado, Society of Economic Geologists, Inc., p. 29–70.

Smith, K.S., and Macalady, D.L., 1991, Water/sediment partitioning of trace elements in a stream receiving acid-mine drainage, in Proceedings, Second International Conference on the Abatement of Acidic Drainage, Montreal, Canada, September 16–18, 1991, vol. 3, p. 435–450.

Smith, K.S., Ranville, J.F., Plumlee, G.S., and Macalady, D.L., 1998, Predictive double-layer modeling of metal sorption in mine-drainage systems, *in* Jenne, E.A., ed., Adsorption of metals by geomedia: Variables, mechanisms, and model applications, Chapter 24: San Diego, California, Academic Press, p. 521–547.

Smith, K.S., Walton-Day, K., and Ranville, J.F., 2000, Evaluating the effects of fluvial tailings deposits on water quality in the Upper Arkansas River Basin, Colorado: Observational scale considerations, *in* Proceedings of the Fifth International Conference on Acid Rock Drainage (ICARD 2000), Denver, Colorado, May 21–24, 2000, vol. 2: Littleton, Colorado, Society for Mining, Metallurgy, and Exploration, Inc., p. 1415–1424.

Smith, K.S., Campbell, D.L., Desborough, G.A., Hageman, P.L., Leinz, R.W., Stanton, M.R., Sutley, S.J., Swayze, G.A., and Yager, D.B., 2002, Toolkit for the rapid screening and characterization of waste piles on abandoned mine lands, *in* Seal, R.R. II, and Foley, N.K., eds., Progress on geoenvironmental models for selected mineral deposit types: U.S. Geological Survey Open-File Report 02-0195, p. 55–64. (Available online at http://pubs.usgs.gov/of/2002/of02-195/.)

Smith, K.S., Wildeman, T.R., Choate, L.M., Diehl, S.F., Fey, D.L., Hageman, P.L., Ranville, J.F., Rojas, R., and Smith, B.D., 2003, Determining the toxicity potential of mine-waste piles: U.S. Geological Survey Open-File Report 03-210. (Available online at http://pubs.usgs.gov/of/2003/ofr-03-210/.)

Smith, K.S., Ranville, J.F., Adams, M.K., Choate, L.M., Church, S.E., Fey, D.L., Wanty, R.B., and Crock, J.G., 2006, Predicting toxic effects of copper on aquatic biota in mineralized areas by using the biotic ligand model, *in* Proceedings of the Seventh International Conference on Acid Rock Drainage, St. Louis, Missouri, March 26–30, 2006, p. 2055–2077.

Spark, K.M., Johnson, B.B., and Wells, J.D., 1995, Characterizing heavy-metal adsorption on oxides and oxyhydroxides: European Journal of Soil Science, v. 46, p. 621–631, doi: 10.1111/j.1365-2389.1995.tb01358.x.

Sposito, G., 1986, Distinguishing adsorption from surface precipitation, *in* Davis, J.A., and Hayes, K.F., eds., Geochemical processes at mineral surfaces, ACS Symposium Series 323: Washington, D.C., American Chemical Society, p. 217–228.

Stumm, W., and Morgan, J.J., 1996, Aquatic chemistry: John Wiley & Sons, New York, 1022 p.

Stumm, W., Huang, C.P., and Jenkins, S.R., 1970, Specific chemical interaction affecting the stability of dispersed systems: Croatica Chemica Acta, v. 42, p. 223–245.

Stumm, W., Hohl, H., and Dalang, F., 1976, Interaction of metal ions with hydrous oxide surface: Croatica Chemica Acta, v. 48, p. 491–504.

Templeton, D.M., Ariese, F., Cornelis, R., Danielsson, L.-G., Muntau, H., Van Leeuwen, H.P., and Lobinski, R., 2000, Guidelines for terms related to chemical speciation and fractionation of elements: Definitions, structural aspects, and methodological approaches: Pure and Applied Chemistry, v. 72, p. 1453–1470.

Tessier, A., and Turner, D.R., eds., 1995, Metal speciation and bioavailability in aquatic systems: John Wiley & Sons, Chichester, 679 p.

Tessier, A., Campbell, P.C.G., and Bisson, M., 1979, Sequential extraction procedure for the speciation of particulate trace metals: Analytical Chemistry, v. 51, p. 844–851, doi: 10.1021/ac50043a017.

Ure, A.M., and Davidson, C.M., eds., 2002, Chemical speciation in the environment (2nd edition): Oxford, Blackwell Science, 452 p.

U.S. EPA, 1994, Water quality standards handbook, Appendix L: Interim guidance on determination and use of water-effect ratios for metals, EPA-823-B-94-005. (Available online at http://www.epa.gov/waterscience/standards/handbook/.)

U.S. EPA, 1998, Guidelines for ecological risk assessment, EPA 630-R-95-002F: U.S. Environmental Protection Agency. (Available online at http://oaspub.epa.gov/eims/eimscomm.getfile?p_download_id=36512.)

U.S. EPA, 2001, Abandoned mine site characterization and cleanup handbook, EPA 530-C-01-001, March 2001: U.S. Environmental Protection Agency. (Available online at http://www.ott.wrcc.osmre.gov/library/hbmanual/epa530c.htm#downloadhandbook.)

U.S. EPA, 2004, Test methods for evaluating solid waste, physical/chemical methods (SW-846), revision 6, November, 2004: U.S. Environmental Protection Agency, accessed on April 30, 2006 at http://www.epa.gov/epaoswer/hazwaste/test/main.htm.

U.S. EPA, 2007a, Aquatic life ambient freshwater quality criteria—copper, EPA-822-R-07-001, February 2007: U.S. Environmental Protection Agency. (Available online at http://ww.epa.gov/waterscience/criteria/copper/2007/index.htm.)

U.S. EPA, 2007b, Framework for metals risk assessment, EPA 120/R-07/001, March 2007: U.S. Environmental Protection Agency. (Available online at http://www.epa.gov/osa/metalsframework/.)

U.S. Geological Survey, variously dated, National field manual for the collection of water-quality data: U.S. Geological Survey Techniques of Water-Resources Investigations, book 9, chaps. A1–A9, (Available online at http://pubs.water.usgs.gov/twri9A.)

Villavicencio, G., Urrestarazu, P., Carvajal, C., De Schamphelaere, K.A.C., Janssen, C.R., Torres, J.C., and Rodriguez, P.H., 2005, Biotic ligand model prediction of copper toxicity to daphnids in a range of natural waters in Chile: Environmental Toxicology and Chemistry, v. 24, p. 1287–1299, doi: 10.1897/04-095R.1.

Vlasov, K.A., ed., 1966, Geochemistry and mineralogy of rare elements and genetic types of their deposits, vol. 1, Geochemistry of rare elements (translated from Russian): Israel Program for Scientific Translations, Ltd., Jerusalem, 688 p.

Waeterschoot, H., Van Assche, F., Regoli, L., Schoeters, I., and Delbeke, K., 2003, Metals—environmental risk assessment of metals for international chemicals safety programs: a recent and fast growing area of regulatory development, requiring new science concepts: Journal of Environmental Monitoring, v. 5, p. 96N–102N.

Walker, J.D., Enache, M., and Dearden, J.C., 2003, Quantitative cationic-activity relationships for predicting toxicity of metals: Environmental Toxicology and Chemistry, v. 22, p. 1916–1935, doi: 10.1897/02-568.

Wanty, R.B., Berger, B.R., and Tuttle, M.L., 2001, Scale versus detail in water-rock investigations, 1: A process-oriented framework for studies of natural systems, *in* Water-Rock Interaction 2001: Lisse, Swets & Zeitlinger, p. 221–225.

Webb, J.S., 1964, Geochemistry and life: New Scientist, v. 23, p. 504–507.

Whitfield, M. and Turner, D.R., 1983, Chemical periodicity and the speciation and cycling of the elements, *in* Wong, C.S., Boyle, E., Bruland, K.W., Burton, J.D., and Goldberg, E.D., eds., Trace metals in sea water: Plenum Press, New York, p. 719–750.

Wildeman, T.R., Smith, K.S., and Ranville, J.F., 2007, A simple scheme to determine potential aquatic metal toxicity from mining wastes: Environmental Forensics Journal, v. 8:1, p. 119–128, doi: 10.1080/15275920601180651.

Wood, J.M., 1988, Transport, bioaccumulation, and toxicity of metals and metalloids in microorganisms under environmental stress, *in* Kramer, J.R., and Allen, H.E., eds., Metal speciation: Theory, analysis, and application: Lewis Publishers, Chelsea, Michigan, p. 295–314.

MANUSCRIPT ACCEPTED BY THE SOCIETY 28 NOVEMBER 2006

The effects of acidic mine drainage from historical mines in the Animas River watershed, San Juan County, Colorado—What is being done and what can be done to improve water quality?

Stanley E. Church
U.S. Geological Survey, P.O. Box 25046, MS973, Denver, Colorado 80225, USA

J. Robert Owen
Animas River Stakeholders Group, 8185 C.R. 203, Durango, Colorado 81301, USA

Paul von Guerard
U.S. Geological Survey, 764 Horizon Dr., Grand Junction, Colorado 81506, USA

Philip L. Verplanck
U.S. Geological Survey, P.O. Box 25046, MS973, Denver, Colorado 80225, USA

Briant A. Kimball
U.S. Geological Survey, 2329 W. Orton Circle, Salt Lake City, Utah 84119, USA

Douglas B. Yager
U.S. Geological Survey, P.O. Box 25046, MS973, Denver, Colorado 80225, USA

ABSTRACT

Historical production of metals in the western United States has left a legacy of acidic drainage and toxic metals in many mountain watersheds that are a potential threat to human and ecosystem health. Studies of the effects of historical mining on surface water chemistry and riparian habitat in the Animas River watershed have shown that cost-effective remediation of mine sites must be carefully planned. Of the more than 5400 mine, mill, and prospect sites in the watershed, ~80 sites account for more than 90% of the metal loads to the surface drainages. Much of the low pH water and some of the metal loads are the result of weathering of hydrothermally altered rock that has not been disturbed by historical mining. Some stream reaches in areas underlain by hydrothermally altered rock contained no aquatic life prior to mining.

Scientific studies of the processes and metal-release pathways are necessary to develop effective remediation strategies, particularly in watersheds where there is little land available to build mine-waste repositories. Characterization of mine waste, development of runoff profiles, and evaluation of ground-water pathways all require rigorous study and are expensive upfront costs that land managers find difficult to justify. Tracer studies of water quality provide a detailed spatial analysis of processes affecting surface- and ground-water chemistry. Reactive transport models were used in conjunction with the best state-of-the-art engineering solutions to make informed and cost-effective remediation decisions.

Remediation of 23% of the high-priority sites identified in the watershed has resulted in steady improvement in water quality. More than $12 million, most contributed by private entities, has been spent on remediation in the Animas River watershed. The recovery curve for aquatic life in the Animas River system will require further documentation and long-term monitoring to evaluate the effectiveness of remediation projects implemented.

Keywords: acid mine drainage, watershed impacts, historical mining, environmental effects, remediation

INTRODUCTION

Thousands of inactive hardrock mines have left a legacy of acid drainage and toxic metals across mountain watersheds in the western United States. Many watersheds in or west of the Rocky Mountains have headwater streams in which the effects of historical hardrock mining are thought to represent a potential threat to human and ecosystem health (e.g., Fields, 2003). In many areas, weathering of unmined mineral deposits, waste rock, and mill tailings in areas of historical mining may increase metal concentrations and lower pH, thereby contaminating the surrounding watershed and ecosystem. Streams near abandoned inactive mines can be so acidic or metal laden that fish and aquatic insects cannot survive (e.g., Besser and Brumbaugh, 2007; Besser and Leib, 2007; Anderson, 2007), and birds are negatively affected by the uptake of metals through the food chain (e.g., Larison et al., 2000). Although estimates of the number of inactive mine sites in the West vary, observers agree that the scope of this problem is huge, particularly in the western United States, where public lands contain thousands of inactive mines and prospects.

Numerous inactive mines are located either on or adjacent to public lands or affect aquatic or wildlife habitat on federal land. In 1995, personnel from a U.S. Department of the Interior (DOI) and U.S. Department of Agriculture (USDA) interagency task force developed a coordinated strategy for the cleanup of environmental contamination from inactive mines associated with federal lands. Estimates of the number of inactive mines that affect surface water quality on National Forest (USDA-FS) and Bureau of Land Management (BLM) administered lands were low (6000 mine sites; Greeley, 1999) relative to those provided for the entire United States by the Minerals Policy Center (131,000 mine sites; Da Rosa and Lyon, 1997). As part of an interagency effort, the U.S. Geological Survey implemented an Abandoned Mine Lands (AML) Initiative to develop a strategy for gathering and communicating the scientific information needed to formulate effective and cost-efficient remediation of inactive, abandoned mines on federal land. Objectives of the AML Initiative included: (1) watershed-scale and site characterization, understanding of the effect and extent of sources of metals and acidity, and (2) communication of these results to stakeholders, land managers, and the general public. Additional objectives addressed included transfer of technologies developed within the AML Initiative into practical methods at the field scale and demonstration of their applicability to solve this national environmental problem in a timely manner within the framework of the watershed approach. Finally, developing working relationships with the private sector, local citizens, and state and federal land management and regulatory agencies will establish a scientific basis for consensus, providing an example for future investigations of watersheds affected by inactive historical mines (Buxton et al., 1997).

The combined AML interagency effort has been conducted in two pilot watersheds (Fig. 1), the Animas River watershed study area in Colorado (Church et al., 2007c) and the Boulder River watershed study area in Montana (Nimick et al., 2004). Land and resource-management agencies are faced with evaluating the risks associated with thousands of potentially harmful mine sites on federal lands. Comprehensive scientific investigations have been conducted in both AML watersheds. The level of scientific study conducted in the AML watersheds will not be feasible in every watershed affected by historical mining. Development of criteria for evaluating ecological and environmental effects of historical mining was a paramount objective of these studies. Clearly, remediation of federal lands affected by inactive historical mines will require substantial investments of resources.

Land management and regulatory agencies face two fundamental questions when they approach a region or watershed affected by inactive historical mines. First, with potentially hundreds of dispersed and potentially contaminated mine sites, how should limited federal resources for prioritizing, characterizing, and remediating the watershed be invested to achieve cost-effective and efficient cleanup? Second, how can realistic remediation targets be identified, considering:

- The potential for adverse effects from unmined mineralized deposits (including any effects that may have been present under premining conditions or still may persist from unmined deposits adjacent to existing abandoned mines)
- The possible impact of incomplete cleanup of specific inactive historical mine sites
- Other physical or environmental factors that may limit sustainability of desired ecosystems

similar inactive historical mine sites throughout the nation. The watershed approach:

- Gives high priority to actions most likely to most significantly improve water quality and ecosystem health
- Enables assessment of the cumulative effect of multiple and (or) nonpoint sources of contamination
- Encourages collaboration among federal, state, and local levels of government and stakeholders
- Provides information that will assist disposal-siting decisions
- Accelerates remediation and reduces total cost compared to remediation on a site-by-site basis
- Enables consideration of revenue generation from selected sites to supplement overall watershed remediation costs.

The report by Church et al. (2007c) provides a geologic description and summary of the field and laboratory work conducted by the U.S. Geological Survey in the Animas River watershed during 1996–2000. The objectives of this study were to:

- Estimate premining geochemical baseline (background) conditions
- Define current geochemical baseline conditions
- Characterize processes affecting contaminant dispersal and effects on ecosystem health
- Develop remediation goals on the basis of scientific study of watershed conditions
- Transfer data to users in a timely and effective manner

Investigations were coordinated with personnel from the Animas River Stakeholders Group (ARSG), Colorado Division of Wildlife, Colorado Department of Public Health and Environment, Colorado Division of Mines and Geology, Colorado Geological Survey, U.S. Environmental Protection Agency, U.S. Forest Service in the Department of Agriculture, and U.S. Bureau of Land Management in the Department of Interior, all of whom are coordinating the design and implementation of remediation activities within the watershed.

Figure 1. Map of the western United States showing the Boulder and Animas River watershed study areas. The Animas River watershed study area is shown in the inset map.

To answer these questions, the AML Initiative adopted a watershed approach rather than a site-by-site approach to characterize and remediate abandoned mines (Buxton et al., 1997). This approach is based on the premise that watersheds affected by acid mine drainage in a state or region should be prioritized on the basis of its effect on the biologic resources of the watershed so that the funds spent on remediation will have the greatest benefit on affected streams. Within these watersheds, contaminated sites that have the greatest impact on water quality and ecosystem health within the watershed would then be identified, characterized, and ranked for remediation. The watershed approach establishes a framework of interdisciplinary scientific knowledge and methods that can be employed at

DESCRIPTION OF STUDY AREA

The Animas River watershed study area is located in southwestern Colorado near Silverton, ~40 miles (65 km) north of Durango (Fig. 1). Four candidate watersheds were nominated in Colorado on the basis of geologic factors, metal loading, the status of ongoing remediation activities, general knowledge of the candidate watersheds, and the extent of federal lands within the watershed. The Animas River watershed study area was chosen as one of two pilot watersheds for the AML Initiative in May 1996.

The Animas River watershed study area, as defined in this study, is the drainage area of three large streams and their tributaries (Mineral Creek, Cement Creek, and the Animas River upstream from Silverton). Although the compliance point for water quality established by the Colorado Water Quality Control

Commission is the gauge downstream of the confluence of the Animas River with Mineral Creek (Fig. 2), the Animas River watershed study extends downstream from the confluence of Mineral Creek to an area known as Elk Park just upstream from the confluence of the Animas River with Elk Creek. Most of the watershed is in four mining districts: (1) the Silverton district, which covers the southeastern part of the watershed from the South Fork Mineral Creek to north of Howardsville; (2) the Eureka District, which covers the northern part of the Mineral and Cement Creek basins as well as the Animas River basin from Eureka north (Davis and Stewart, 1990); (3) the Red Mountain district, which extends up the Mineral Creek basin from Ohio Peak to the north and is largely outside the study area; and (4) the Ice Lake district, which is in the headwaters of South Fork Min-

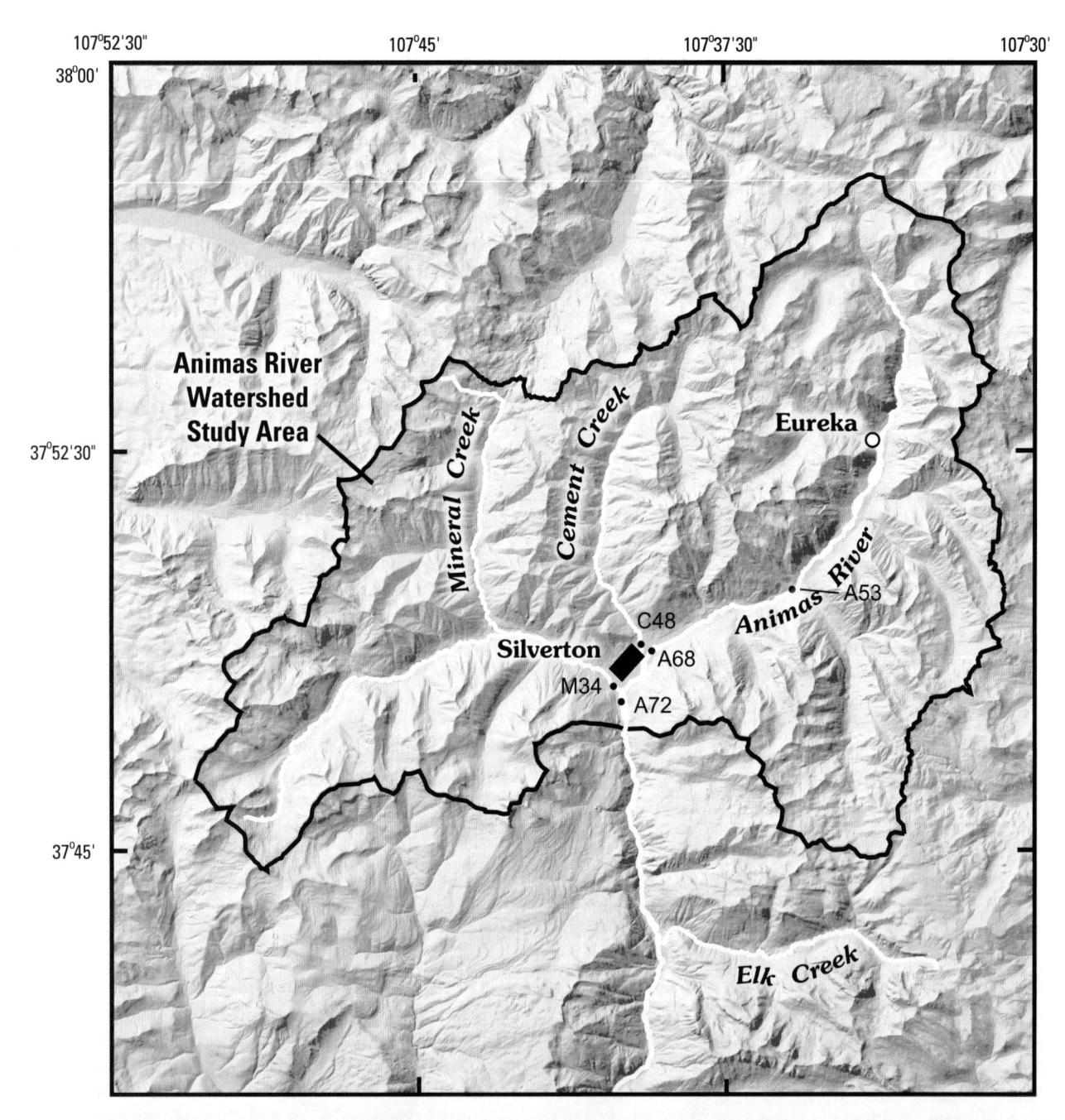

Figure 2. Shaded relief map of the Animas River watershed showing the Animas River and its main tributaries, Mineral and Cement Creeks, which are impacted by historical mining. The watershed study area boundary is outlined in black; gauging stations are shown as black dots on each of the major streams: A72, gauge 09359020; M34, gauge 09359010; A68, gauge 09358000; C48, 09358550; and A53, gauge 09357500.

eral Creek (Church et al., 2007b). During watershed studies, additional sampling and investigations were conducted downstream of the study area to document the extent of enriched trace-element concentrations downstream and to provide reference localities unaffected by historical mining (Church et al., 1997; Anderson, 2007).

The watershed is mountainous with elevations ranging from ~9300 feet (2830 m) at Silverton to more than 13,300 feet (4050 m) above sea level. The terrain is a rugged mountainous area with U-shaped and hanging valleys carved out during the last glaciation (e.g., Blair et al., 2002). Mean annual precipitation ranges from ~24–40 in/yr (600–1000 mm/yr) (NRCS, 2007). Although the population varies seasonally as temporary residents move into the area during the summer, ~400 people live in the Animas River study area throughout the year. Residents are engaged primarily in tourism, which is largely based on the historical nature of the quaint mining town of Silverton served by the historical narrow-gauge railroad from Durango (Sloan and Skowronski, 1975).

Hydrologic Setting

Stream flow in the Animas River study area is typical of high-gradient mountain streams throughout the southern Rocky Mountains. Stream flow is dominated by snowmelt runoff, which typically occurs between April and June. Snowmelt runoff often is augmented by rain during summer from July through September. Stream flow typically peaks in May or June and decreases as the shallow ground-water system drains. Spring runoff conditions extend into July. Low stream-flow conditions are typical from August to March (Fig. 3). The nearest U.S. Geological Survey stream-flow-gauging station 09359020, Animas River downstream from Silverton (period of record Oct. 1991–present, drainage area of 146 mi^2 [378 km^2]) is downstream from the confluence of Mineral Creek (Fig. 2). Stream gauges are located at the mouth of the major tributaries in the study area: the gauge on the Animas River at Silverton (09358000, period of record 1991–1993 and 1994–2004, drainage area 70.6 mi^2 [183 km^2]); the gauge on Cement Creek at Silverton (09358550, period of record 1991–1993 and 1994–2004, drainage area 20.1 mi^2 [52 km^2]); and the gauge on Mineral Creek at Silverton (09359010, period of record 1991–1993 and 1994–2004, drainage area 52.5 mi^2 [136 km^2]). These gauges measure 98% of the stream-flow drainage area upstream from Silverton. The State of Colorado operates a stream-flow-gauging station on the Animas River at Howardsville (09357500, period of record 1935–2002, drainage area 55.9 mi^2 [145 km^2]; Fig. 2). Real-time and historical stream-flow data are

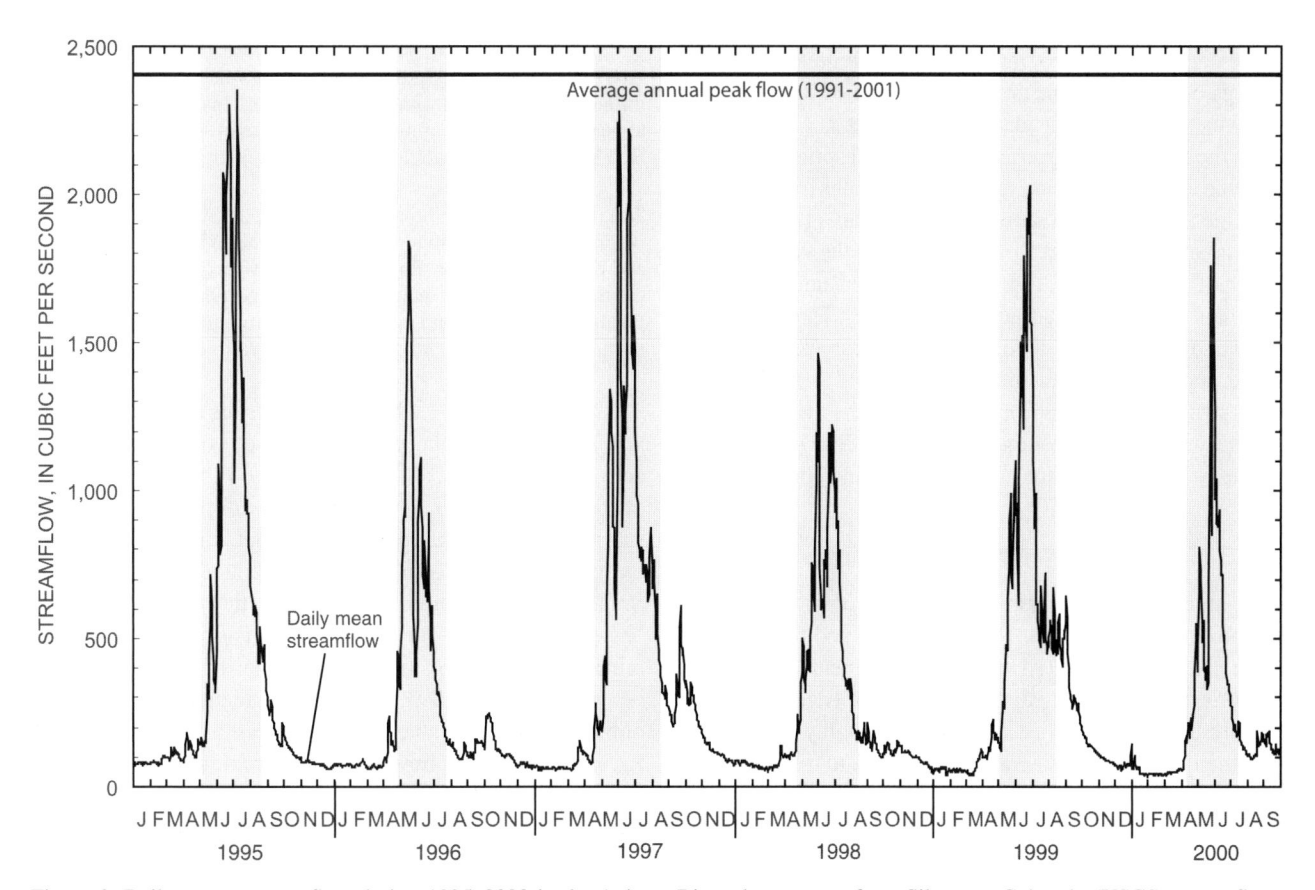

Figure 3. Daily mean stream flow during 1995–2000 in the Animas River downstream from Silverton, Colorado (USGS stream-flow-gauging station 09359020, A72, fig. 2). Average annual peak flow is 2420 ± 500 ft^3 per second for water years 1992–2001. Shaded areas indicate periods of spring runoff, which was arbitrarily defined as stream flow greater than 150 ft^3 per second.

available online (USGS, 2007b). Data are also published in the U.S. Geological Survey annual data report (Crowfoot et al., 2005).

The Animas River watershed study area is subdivided into three large basins, the Mineral and Cement Creek basins and the upper Animas River basin. Drainage basin areas of tributary streams to the main stem drainages are also referred to here as subbasins.

Geology of the watershed is the primary factor affecting the distribution of trace-element concentrations and pH in streams in the upper Animas River watershed. Although there are more than 300 mines and an estimated 5400 mining-related features in the study area (Church et al., 2007b), not all of these features contribute to water-quality degradation. Because of the combined effects of hydrothermal alteration that is directly associated with the mineral deposits and the widespread distribution of historical mine sites throughout the watershed, it is difficult to attribute low pH values and high trace-metal concentrations exclusively to either source. The pH of water samples collected at background sites ranged from 2.58 to 8.49 compared to pH of water from mine sites that ranged from 2.35 to 7.77 (Mast et al., 2000a).

Ground-water flow is largely controlled by topography, by the distribution of unconsolidated Quaternary deposits that overlie bedrock units, and by the decrease in hydraulic conductivity of geologic units with depth. Topography strongly controls the direction of ground-water flow and the location of discharge areas. Recharge occurs on all topographic highs, with greater amounts of recharge on areas with the greatest precipitation and hydraulic conductivity (McDougal et al., 2007; Mast et al., 2007). Discharge in the form of numerous seeps and small springs occurs in topographic lows and at breaks in slope. Flow paths from recharge to discharge areas are short, commonly less than a few thousand feet. Regional ground-water flow is limited by the very low permeability of the bedrock. The overlying, thin, unconsolidated deposits have the highest hydraulic conductivity. The uppermost, fractured and weathered zone in the igneous bedrock has a lower hydraulic conductivity than the unconsolidated deposits (R.H. Johnson, written comm., 2007). Fractures are the major conduits for ground-water flow in bedrock, with more flow in the uppermost zone, where the fractures are weathered and open.

Biologic Setting

The Animas River watershed study area consists entirely of alpine and subalpine habitats. The headwaters and tributaries of the three principal basins (Animas River, Cement and Mineral Creeks) originate in treeless alpine regions with vegetation cover ranging from essentially none, especially in highly mineralized areas of the Cement and Mineral Creek basins, to relatively lush alpine meadows. Streams follow steep, narrow valleys, with the exception of a few low-gradient areas, such as the Animas River between Eureka and Howardsville, Mineral Creek near Chattanooga, and the open valley near Silverton (Bakers Park), which have relatively wide valley floors (Blair et al., 2002). Vegetation on valley walls is restricted in many areas by extensive areas of exposed rock and talus, but some areas of sparse coniferous Engelmann spruce forest occur on north-facing slopes and in valley bottoms, where deciduous trees also occur. Riparian vegetation is often limited along high-gradient streams, but low-gradient reaches typically contain extensive areas of beaver ponds and associated willow thickets except where limited by mining activity and mill tailings disposal (e.g., Vincent and Elliott, 2007).

The native fish community of the watershed before European settlement was restricted by the severe climate and hydrology and by barriers to upstream movement of fish in the Animas River canyon downstream of Silverton. The only native fish species known to occur in the watershed would be the Colorado River cutthroat trout (*Onchorhynchus clarkii pleuriticus*), although one account suggests that the mottled sculpin (*Cottus bairdi*), a species that occurs commonly in downstream reaches, may have occurred in portions of the upper watershed.

There are few accounts of the aquatic biota of the upper Animas River watershed before the most active period of mining (ca. 1890 based on district production records summarized by Jones, 2007). Ichtyologist David Starr Jordan (1891) visited the Animas River watershed in 1889 and recorded the following references, based on second-hand accounts:

In the deep and narrow "Cañon de las Animas Perdidas" [Animas Canyon] are many deep pools, said to be full of trout.

Above its cañon of "Lost Souls," it is clear, shallow, and swift, flowing through an open cañon with a bottom of rocks. In its upper course it is said to be without fish, one of its principal tributaries, Mineral Creek, rising in Red Mountain and Uncompahgre Pass, being highly charged with iron.

The distribution of cutthroat trout in the upper Animas River watershed before settlement is unknown. However, the existence of water quality and habitat suitable for trout is indicated by reports of good populations of trout (undoubtedly including stocked, non-native trout) in artificial ponds near Silverton and in the Animas River upstream of Silverton near the turn of the century. (Silverton Standard, various accounts, 1903–1905)

Surveys conducted by federal and state agencies in the 1960s and 1970s indicate that the many decades of mining and milling activity had a significant adverse effect on stream biota. The reach of the Animas River upstream of Silverton (Fig. 2), which had supported trout in previous years, yielded only a single trout in an electro fishing survey in 1968 (U.S. Dept. of Interior, 1968). N.F. Smith (1976) declared this reach of the Animas River to be "essentially dead." The Colorado Division of Wildlife stocked rainbow trout, brook trout, and brown trout in the watershed between 1973 and 1993. There is no evidence that either rainbow or brown trout were able to reproduce upstream from the Animas River canyon reach; however, brook trout, which are more tolerant of pollution by acid and toxic metals than the other species, were more successful. Brook trout remains the predominant fish species in the Animas River watershed study area despite no documented stocking of this species since 1985.

Recent surveys of fish and benthic invertebrate communities (Butler et al., 2001; Anderson, 2007; Besser and Brumbaugh,

2007) indicate that the effects of poor water quality on stream communities vary widely among the three basins and suggest that stream biota have responded to some improvements in water quality. The headwaters of the Animas River upstream from Eureka (Fig. 2), the entire length of Cement and Mineral Creeks, and several smaller tributaries support little or no aquatic life due to the effects of mining and naturally acidic water draining from hydrothermally altered areas (Bove et al., 2007). The South Fork Mineral Creek and several tributaries of the upper Animas River, which drain subbasins that provide substantial acid-neutralizing capacity, support viable populations of brook trout and a few cutthroat trout. The Animas River between Maggie Gulch and the mouth of Cement Creek in Silverton supports brook trout and a substantial invertebrate community, suggesting that substantial improvements in water quality have occurred in this reach since the 1970s when the early electro fishing surveys were done (e.g., N.F. Smith, unpublished aquatic inventory: Animas-La Plata project, Colorado Division of Wildlife, Durango Colorado, 1976). Impacts of degraded water quality on stream biota persist in the Animas River for a substantial distance downstream of Silverton, although there is some evidence of recovery of fish and invertebrate communities since the Sunnyside mine closed in 1991 and remediation efforts in the watershed began immediately.

Geologic Setting

The geology of the rugged western San Juan Mountains is exceptional in that many diverse rock types representing every geologic era from the Proterozoic to the Cenozoic are preserved. It is also an area that has high topographic relief providing excellent bedrock exposures. The general stratigraphy of the San Juan Mountains near Silverton consists of a Precambrian crystalline basement overlain by Paleozoic to Tertiary sedimentary rocks and by a voluminous Tertiary volcanic cover (Fig. 4).

Precambrian rocks are exposed south of Silverton along the Animas River and in upper Cunningham Creek and are part of a broad uplifted and eroded surface (Fig. 4). The Precambrian section near the study area consists primarily of amphibolite, schist, and gneiss. South and west of Silverton, gently dipping Paleozoic to Tertiary age sedimentary strata of varying lithologies overlie Precambrian basement rocks. The sedimentary section, which crops out in subbasins in the headwaters of South Fork Mineral Creek and in other subbasins south of Silverton, is comprised mainly of marine and terrestrial limestone and mudstone in addition to terrestrial deposits of sandstone, siltstone, and conglomerate (Fig. 4). Many of these units contain calcite and are, therefore, important for their acid neutralization potential. A thick section of

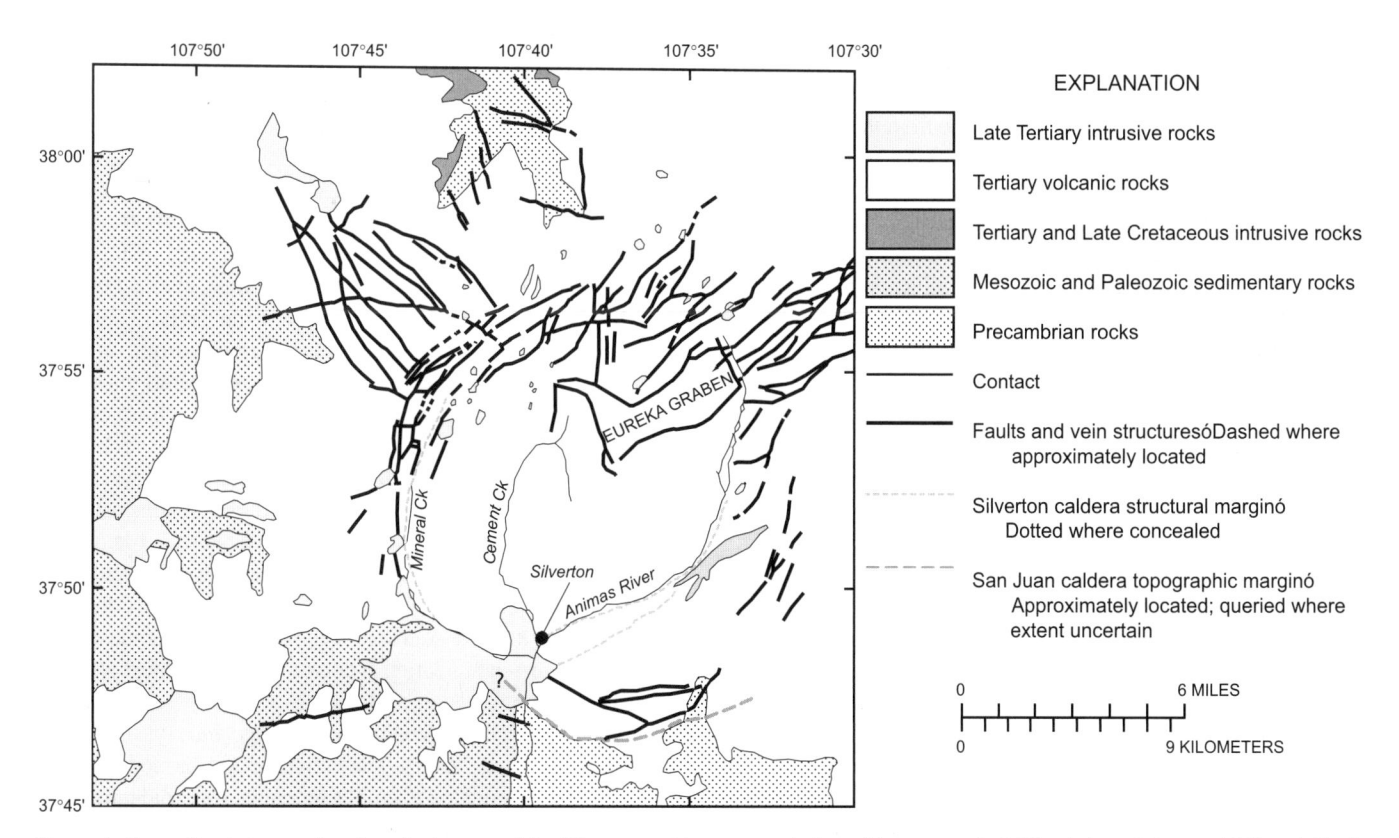

Figure 4. Generalized structural and geologic map of the Silverton caldera, upper Animas River watershed. The Animas River and Mineral Creek follow the structural margin of the Silverton caldera. In addition to the ring-fractures that were created when the Silverton and the earlier San Juan calderas formed, radial and graben faults, which host much of the subsequent vein mineralization, are shown schematically (modified from Casadevall and Ohmoto, 1977).

Tertiary volcanic rock caps the Paleozoic sedimentary rocks west and southwest of Silverton and covers most of the central part of the study area. The majority of Animas River headwater streams and tributaries originate in Tertiary volcanic and silicic (high silica content) intrusive rocks that have been deposited or emplaced in the area that is defined by Mineral Creek on the west and by the Animas River on the east, north of Silverton (Fig. 4). Subsequent hydrothermal alteration and mineralization resulted in the formation of economic mineral deposits that were exploited between 1871 and 1991. Thus, much of this study has focused on the Tertiary volcano-tectonic history and mineralization events that have contributed to present water-quality issues.

Onset of volcanism commenced between 35 and 30 Ma, with the eruption of intermediate-composition (52–63% SiO_2) lava flows and deposition of related volcaniclastic sedimentary rock forming a plateau that covered much of the San Juan Mountains area (Lipman et al., 1976). Following the early phase of intermediate-composition volcanism, silicic calderas began to form throughout the entire San Juan Mountains region. Two calderas formed in the Animas River watershed study area between ca. 28 and 27 Ma (Fig. 4). Eruption of the Oligocene (27.6 Ma) Silverton caldera created a large semicircular depression ~13 km (8 miles) in diameter, which is nested within the older (28.2 Ma) San Juan caldera (Lipman et al., 1976; Yager and Bove, 2007). The central part of the San Juan caldera is partially filled by ash-flow tuff and by later, intermediate-composition lava flows, volcaniclastic sedimentary rocks, and igneous intrusive rocks. Ash-flow tuff is a volcanic rock containing pumice, broken crystals, and wall rock fragments in a matrix of ash-size material that was ejected from the ring fracture zone of an actively forming caldera. Eruption of intermediate-composition lava flows and related volcaniclastic rocks filled the San Juan caldera volcanic depression; these rocks host the majority of the mineralization in the study area. Granitic igneous magmas intruded the southern margins of the Silverton and San Juan calderas shortly after the Silverton caldera formed. The intrusions south of Silverton formed along the caldera structural margins and are centered near the area between Sultan Mountain and peak 3792 m on the South Fork Mineral Creek (Fig. 5), in lower Cunningham Creek from Howardsville to lower Maggie Gulch, and near the mouth of Cement Creek (see Table 1).

An extensive bedrock fracture and fault network has developed in response to caldera development in the region. Structures related to caldera formation not only influence the hydrologic system today but also are largely responsible for controlling where postcaldera hydrothermal fluids altered the country rock and focused the emplacement of mineral deposits. Important faults related to caldera formation include the arcuate faults that form the caldera structural margin. In addition, the northeast-southwest trending faults and veins that comprise the Eureka graben and that cross the central core of the caldera are prominent structural features that have been extensively mineralized and mined for base and precious metals (Casadevall and Ohmoto, 1977; Yager and Bove, 2007). Northwest to southeast trending faults and veins that are radial to the caldera ring fault zone were extensively mineralized. Caldera-related faults, which in places were only partially closed by later mineralization, can extend laterally and vertically from tens of meters to a few kilometers. The structures related to the San Juan-Uncompahgre and Silverton calderas are pervasive features that were not sealed by mineralizing fluids and may be important ground-water flow paths at the basin-wide scale.

Pre- and postcaldera crustal stresses have also resulted in an extensive near surface fracture network. Fractures at the outcrop scale commonly have spacings of one centimeter to several meters. These fractures developed either as volcanic rocks cooled, forming cooling joints, or in response to regional and local tectonic stresses. Fractures that are densely spaced, unfilled by later mineralization, and interconnected focus near-surface ground-water flow at the local or subbasin scale.

Mineralization and Alteration

Multiple hydrothermal alteration and mineralization events that span a 17 m.y. history, from ca. 27 Ma to 10 Ma, were the culmination of a complex cycle of volcano-tectonic events that affected the region (Lipman et al., 1976; Bove et al., 2001). The first episode of hydrothermal alteration formed during the cooling of the San Juan caldera volcanic fill, when lava flows cooled, degassed, and released large quantities of CO_2, along with other volatile constituents such as SO_2 and H_2O. Geologic mapping and airborne geophysical surveys suggest that regional alteration extended from the surface to depths as great as 1 km (Smith et al., 2007; McDougal et al., 2007). This widespread hydrothermal alteration changed the primary minerals of the lava flows, forming an alteration assemblage that includes calcite, epidote, and chlorite (Burbank, 1960). This mineral suite is part of the preore propylitic hydrothermal assemblage, which has a high acid-neutralizing potential (Desborough and Yager, 2000). Near-surface spring and surface water that has interacted mainly with propylitic rock has a pH range of 6.0–7.5 (Mast, et al., 2000a).

Mineralization events that postdated the preore propylitic hydrothermal assemblage contained sulfur-rich hydrothermal fluids and metals that produced various vein and alteration mineral assemblages, all of which include abundant pyrite (Burbank and Luedke, 1968; Casadevall and Ohmoto, 1977). Host rock alteration in many places throughout the Animas River watershed study area effectively removed the acid-neutralizing mineral assemblage of calcite-epidote-chlorite from these subsequently altered areas. This later phase of mineralization was coincident in the timing and emplacement of multiple silicic intrusions that likely provided the heat sources for the mineralizing fluids.

Most of the mineralization events that overprint rocks affected by regional propylitization in the study area may be subdivided into three broad categories on the basis of age and style of mineralization (Bove et al., 2007). The earliest event formed between 26 and 25 Ma, was related to emplacement of granitoid intrusions near the southern margin of the San Juan and Silverton calderas in the area between Middle Fork and South Fork Mineral Creek, and consists of low-grade molybdenum-copper-porphyry mineralization (Ringrose, 1982; Bove et al., 2001). The central part of this

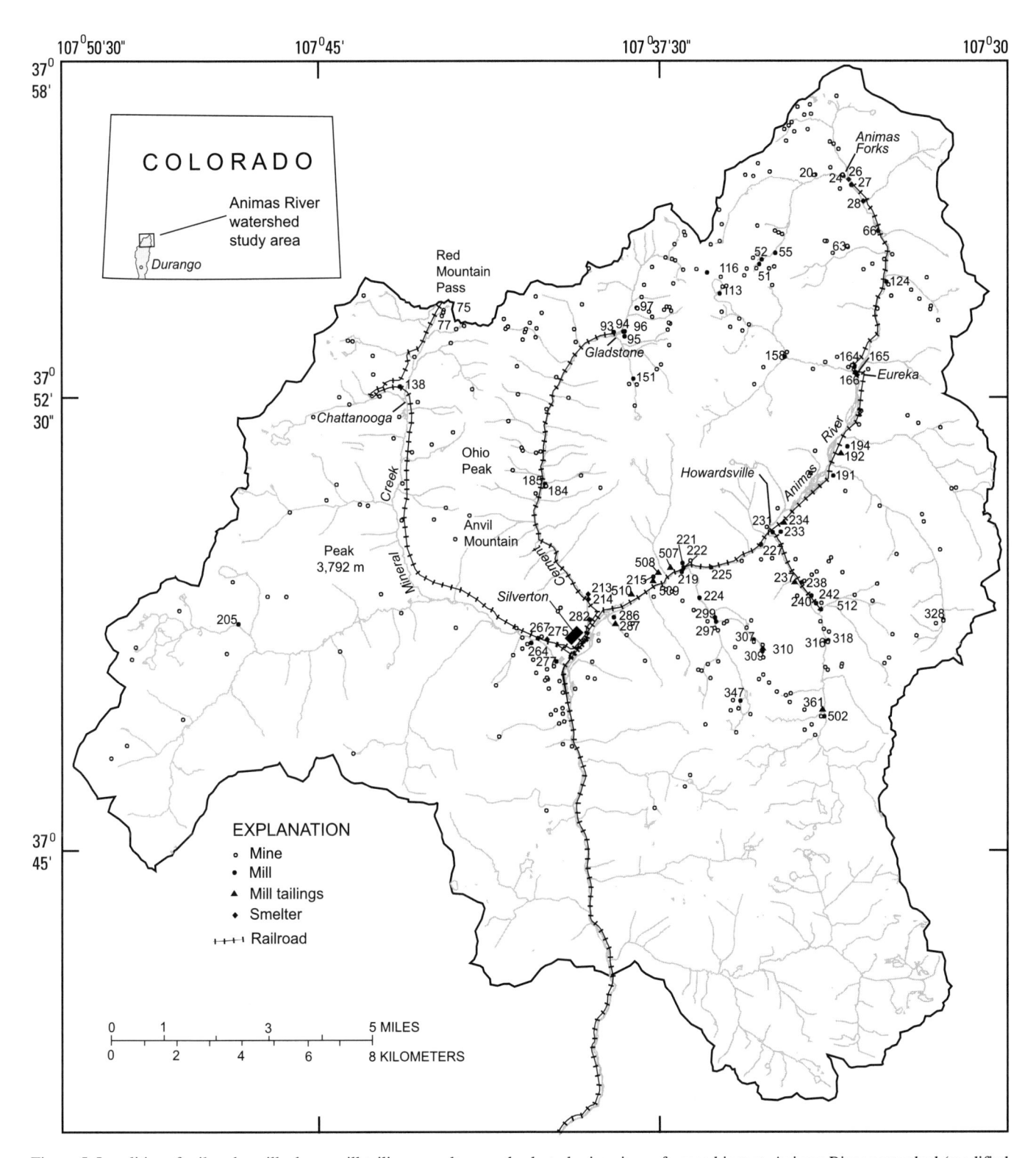

Figure 5. Localities of railroads, mills, large mill tailings, smelters, and selected mine sites referenced in text, Animas River watershed (modified from Jones, 2007). Selected mine sites referenced in text are in Table 1.

TABLE 1. MILL, LARGE MILL TAILINGS, AND SELECTED MINE SITES, ANIMAS RIVER WATERSHED

Site Name	Site No.	Site Name	Site No.
Mines		**Mills** (*continued*)	
Koehler tunnel	75	Ward and Shepard Mill	224
Junction mine	76	Contention Mill	225
Longfellow mine	77	Pride of the West Mill #2	227
American tunnel, Sunnyside mine	96	Little Nation Mill	231
Sunnyside mine	116	Pride of the West Mill #4	233
Mayday mine	181	Old Hundred Mill	238
Mills		Green Mountain Mill	240
Bagley Mill (Frisco)	20	Vertex Mill	242
Columbus Mill	24	North Star (Sultan) Mill	264
Gold Prince Mill	27	Victoria Mill	267
Hanson Mill (Sunnyside Extension Mill)	51	Hercules Mill (Empire)	277
Mastodon Mill	52	Lackawanna Mill	287
Sound Democrat Mill	55	Iowa Mill	297
Treasure Mountain Mill	63	Little Giant Mill	299
Mogul Mill	93	Big Giant Mill	307
Gold King Mill	94	North Star Mill #1	309
Lead Carbonate Mill	95	Pride of the West Mill #3	316
Red and Bonita Mill	97	Pride of the West Mill #1	318
Sunnyside-Thompson Mill	113	Intersection Mill	328
Silver Wing Mill	124	Silver Lake Mill #1	347
Silver Ledge Mill	138	Highland Mary Mill	502
Natalie/Occidental Mill	151	**Mill Tailings**	
Sunnyside Mill #1	158	Kittimack tailings	192
Sunnyside Eureka Mill	164	Pride of the West Mill tailings	234
Sunnyside Mill #2	165	Old Hundred Mill tailings	237
Yukon Mill	184	Lackawanna tailings (removed)	286
Hamlet Mill	191	North Star Mill tailings	310
Kittimack Mill	194	Highland Mary Mill tailings	361
Ice Lake Mill	205	Mayflower Mill tailings repository #1	507
William Crooke Mill	215	Mayflower Mill tailings repository #2	508
Silver Lake Mill #2	219	Mayflower Mill tailings repository #3	509
Mayflower Mill (S-D Mill)	221	Mayflower Mill tailings repository #4	510
Mears-Wilfley Mill	222		

Note: Data refer to sites in Figure 5; data from Jones (2007), Church et al. (2007a).

zoned mineralized system, centered near peak 3792 m (Fig. 5; Silverton 1:24,000-scale, USGS topographic map), is composed of bleached, quartz-stockwork-veined, quartz-sericite-pyrite altered intrusive rocks and volcanic rocks. Rock in the central part of this system is host to exposed molybdenum-copper mineralized rock (McCusker, 1982). Disseminated sulfides in this zone consist mainly of pyrite, lesser chalcopyrite, and traces of molybdenite and bornite, comprising as much as 5 volume percent of the host rock (McCusker, 1982). Progressively outward from the locus of mineralization, zones of weak-sericite-pyrite and propylitic altered igneous and volcaniclastic rocks, respectively, form the periphery of the hydrothermally altered and mineralized porphyry system.

A younger, acid-sulfate system formed at 23 Ma and developed in response to emplacement of coarsely porphyritic dacite intrusions. Dacite porphyry intrusive activity and formation of associated acid-sulfate alteration was mainly focused in two areas. One area of acid-sulfate alteration is centered in the vicinity of the Red Mountains. The Red Mountains form the headwaters of the Uncompahgre River, which flows north, and Mineral Creek, which flows south into the Animas River. The second area is located south of the Red Mountains near Ohio Peak and along Anvil Mountain, which form the drainage divide between Mineral and Cement Creeks. Acid-sulfate mineralization in the Red Mountain area is often characterized by breccia-pipe and fault-hosted vein ore with abundant copper-arsenic-antimony-rich minerals such as enargite-tetrahedrite-tennanite, in addition to copper ores of chalcocite, bornite, and covellite. This combined suite of minerals distinguishes the acid sulfate-related ores from more typical polymetallic vein deposits outside the Red Mountains area (Bove et al., 2007). Gangue minerals include barite, calcite, and fluorite. Breccia bodies and brecciated faults were commonly silicified and replaced with microcrystalline masses of quartz, alunite, pyrophyllite, natroalunite, dickite, diospore, pyrite, and traces of leucoxene. Hydrothermal sericitic assemblages that formed include quartz-sericite-pyrite and weak sericite-pyrite assemblages. The quartz-sericite-pyrite assemblage is typified by total replacement of primary host-rock minerals by quartz, abundant illite (sericite), and pyrite, whereas host-rocks affected by the weak sericite-pyrite assemblage still contain weakly altered primary plagioclase in addition to secondary chlorite derived from the earlier, regional propylitization event (Bove et al., 2007).

The third and most economically important episode of mineralization formed post 18 Ma and is closely associated with the emplacement of high-silica alkali rhyolite intrusions (Lipman, et al., 1973; Bartos, 1993). Mineral deposits formed during this episode consist of polymetallic, Cu-Pb-Zn base- and precious-metal veins deposited along caldera-related northwest-southeast trending fractures tangential to the Silverton and San Juan calderas and along primarily northeast-southwest trending graben faults and some northwesterly trending faults that originally developed during resurgence of the San Juan caldera (Varnes, 1963; Casadevall and Ohmoto, 1977). Six ore-forming stages are recognized in the Sunnyside mine, the largest producing mine developed in this youngest style of mineralization (Casadevall and Ohmoto, 1977). Ores of a massive sulfide stage formed early in the paragenetic sequence and consist of intergrown masses of spahlerite, galena, and lesser amounts of pyrite, chalcopyrite, and tetrahedrite (Casadevall and Ohmoto, 1977). Ores of gold-telluride-quartz, manganese and quartz-fluorite-carbonate-sulfate formed later. Deposition of manganese-rich ores postdate the principal gold-bearing mineralization phase and are composed of light pink bands of pyroxmangite ($MnSiO_3$) intergrown with quartz and rhodochrosite, among other manganese-bearing phases. Late-stage gangue minerals include anhydrite, fluorite, calcite, and gypsum (Casadevall and Ohmoto, 1977). Unlike the pervasive areas of alteration that are associated with both the porphyry molybdenum-copper mineralization and acid-sulfate mineralization systems that often affect entire mountain blocks, the style of post–18-Ma alteration tends to be focused adjacent to veins and vein structures. An assemblage of quartz-sericite-pyrite-zunite occurs proximal to veins. This assemblage grades laterally to assemblages of sericite-kaolinite along with increasing volume percentages of wall-rock chlorite derived from rocks affected by regional propylitization distal from the veins. Intermediate composition volcanic rocks that filled the San Juan caldera are host to 90% of this latest episode of mineralization (Bejnar, 1957).

Surface water quality that results from weathering of the highly altered areas is notable. One of many such examples is centered near peak 3792 m between South Fork Mineral Creek and Middle Fork Mineral Creek northwest of Silverton (Fig. 5). Headwater tributaries that originate in propylitic altered volcanic and volcaniclastic rocks west of peak 3792 m have near neutral pH between 6.5 and 6.8. However, as surface water and ground water interacts with hydrothermal alteration assemblages that contain abundant pyrite downstream, pH drops below 3.5 (Mast et al., 2000b; Yager et al., 2000; Mast et al., 2007).

Late Tertiary erosion exposed large areas of hydrothermally altered rock in the study area to weathering processes. Subsequent Pleistocene glaciation further sculpted the landscape, creating the classic U-shaped valleys, carving the cirque headwater subbasins, and depositing glacial moraine that is partly responsible for the spectacular scenery near Silverton. Multiple surficial deposits formed during and subsequent to glaciation and now cover over 25% of the bedrock with a veneer of porous and permeable material (Blair et al., 2002; Vincent et al., 2007; Vincent and Elliott, 2007). A several-thousand-year history of acidic drainage is recorded in many of the surficial deposits, where iron-rich ground water derived from pyrite weathering has infiltrated these deposits and cemented them with oxides of iron, forming what is referred to as ferricrete (Yager et al., 2003; Yager and Bove, 2007; Wirt et al., 2007). These recent geologic events have exposed mineral deposits to surface weathering prior to mining. Weathering reactions in areas underlain by more intensely altered rock produce greater acidity and release more metals to surface and ground water than areas underlain by propylitic altered rock (Bove et al., 2001; Mast et al., 2000b; Yager et al., 2000).

History of Mining

More than one hundred years of historical mining activity has created many miles of underground workings and produced large volumes of mine waste rock that were pulverized to remove ore metals. These mine workings provide flow paths for ground water that reacted with mineralized rock to produce acidic mine water that flows from mine adits (Mast et al., 2007; Church, et al., 2007b). The waste rock dumps have resulted in increased surface area and exposure of large amounts of pyrite to oxidation resulting in large anthropogenic sources of acidic drainage that affect water quality and aquatic and riparian habitats in the watershed. These anthropogenic conditions exacerbate the cumulative water-quality effect due to weathering the intensely hydrothermally altered rock in the watershed. Changes in the different drainage basins resulting directly from historical mining activities are apparent by comparison of data from streambed-sediment geochemical baselines prior to mining and today (Church et al., 2000; Church et al., 2007a).

Jones (2007) discusses the affects of government price supports and the demand for metals on historical mining practices in the area. These policies, as well as changes in mining practices, greatly affected mineral production in the Animas River watershed and the distribution of mill tailings in the streams. Jones (2007) identifies four major periods of production in the watershed: the smelting era (1871–1889), the gravity milling era (1890–1913), the early flotation era (1914–1935), and the modern flotation era (1936–1991). The following is summarized from his work:

- During the smelting era (1871–1889), ore was hand sorted and most was shipped directly to smelters. A few small gravity or stamp mills were in operation. Total ore production was small and the amount of mill tailings released to streams was small but unknown. There was no market for zinc, so sphalerite was left in the mines or on the mine waste dumps because the smelters charged a penalty to process ores containing more than 10% percent sphalerite (Ransome, 1901).

- During the gravity milling era (1890–1913), many small mines had a stamp mill to prepare concentrate for shipment to the smelter. Milling of ore occurred on site or trams transported ore to the mills or the railroad for shipment to smelters outside the watershed. Mill tailings were not impounded but rather were dumped in the riparian zone or directly into the streams. Sulfide recovery was ~60%. Copper was not recovered, and the demand for zinc was small, so it was generally not recovered. An estimated 4.3 million short tons of mill tailings were discharged directly into streams. The Animas River contained so much mill tailings that contaminated water downstream necessitated building a new reservoir and public water supply for the city of Durango (*Durango Democrat*, 1902).
- During the early flotation era (1914–1935), a few mills dominated ore processing in the basin. The Sunnyside Mill #2 at Eureka (site #165, Fig. 5) was a leader in applying flotation technology in the district. Ores were ground to finer grain size, and sphalerite and copper sulfide concentrates were recovered. The volume of mill tailings produced increased dramatically over the previous period. All the gravity or stamp mills in the watershed were closed down by 1921, and ore was processed at these large modern mills. In 1917, U.S. Smelting and Refining Co. built a large flotation mill at Eureka (Sunnyside Eureka Mill, site #164) to process ore from the Sunnyside mine (site #116). This mill was by far the largest in the basin; it processed ~2.5 million tons of ore from 1917 to 1930 (Bird, 1999). Sulfide recovery exceeded 80%. Mill tailings were impounded in retaining ponds on the Animas River flood plain immediately downstream from the mill to allow the fines to settle. The clear water was then allowed to decant over the tailings dams, along with the dissolved metal loads, directly into the Animas River. As evidenced by the dispersed mill tailings deposits present in the braided reach downstream from Eureka (Vincent and Elliott, 2007), floods periodically breached the tailings impoundments and released mill tailings into the Animas River. The Sunnyside Eureka Mill closed in 1930 during the Depression and reopened briefly in 1937 (Bird, 1999). An estimated 4.2 million tons of ore were processed by mills during this period (Jones, 2007). Most of the mill tailings were discharged into settling ponds in the riparian zone or directly into the streams. Frequently, floods breached these mill tailings impoundments (dams were formed using small wooden shipping barrels filled with rock) and released mill tailings into the Animas River.
- During the modern flotation era (1936–1991), the Mayflower Mill, built in 1929 (site #221, Fig. 5), was the primary mill operating in the watershed. This mill was designed not to release mill tailings to the Animas River. Although the tailings impoundment effort was not completely successful in the beginning, the majority of the mill tailings were retained after 1935 (Jones, 2007). Sulfide recovery using improved flotation technology was greater than 95% after 1940. Jones (2007) estimates that only 200,000 tons of mill tailings were released to the streams during this period. Furthermore, in support of both World War II and the Korean War, many of the old stamp mill tailings were reprocessed, and the stamp mills were burned to recover scrap iron.

Nash and Fey (2007) note that many of the old stamp mill sites contain little or no mill tailings. Given the historical mining practices just summarized above, this is not surprising. A substantial volume of mill tailings was released into the surface streams (an estimated 8.6 million short tons, or ~48% of the total district production; Jones, 2007). As a result, there has been a loss of productive aquatic and riparian habitat and a reduction in recreational and aesthetic values, values important to tourism. Furthermore, the increased acidity and metal loading constitutes a potential threat to downstream drinking water supplies.

SUMMARY OF WATER-QUALITY STUDIES

To evaluate the effects of historical mining on water quality in the Animas River watershed study area, three separate but overlapping investigations were undertaken. One study focused on sampling springs, streams, and mine water in subbasins upgradient of the major stream segments, Mineral Creek, Cement Creek, and the upper Animas River (Mast et al., 2007). Additional water-quality work was conducted by the ARSG (Butler et al., 2001). The second investigation compiled water-quality data throughout the study area (Wright et al., 2007). The third investigation encompassed a series of 13 stream tracers along the 3 major stream segments in the study area (Kimball et al., 2007). The objective of the work in the subbasins was to characterize premining baseline water quality for comparison of the results with mine adit water chemistry. The objective of the basin-wide data compilation was to evaluate spatial and temporal trends in water quality. The objective of the stream tracer studies was to quantify metal loading along the major stream segments, investigate the effects of instream processes on stream chemistry, and simulate remediation scenarios. In addition to these studies, two other process-related studies were undertaken: (1) investigation of the age, composition, and formation of ferricrete deposits (Verplanck et al., 2007; Wirt et al., 2007) and iron bogs (Stanton et al., 2007), and (2) an investigation of the seasonal variation in water quality (Leib et al., 2007).

The data set for the subbasin investigation included water samples collected from 241 spring and stream sites and 75 mine sites during summer low-flow conditions in 1997–1999. For the spring and stream sites, a ranking system primarily based on field observations was devised to evaluate the potential for mining activity effects. The ranking system consisted of four categories ranging from category I (no evidence of mining activity) to category IV (direct discharges from mine sites). Ranges and median values for pH, sulfate, and zinc concentrations for sites unaffected by mining (category I and II) are given in Table 2. The primary factor controlling the premining baseline water quality is the degree of hydrothermal alteration of the bedrock. For each site unaffected by mining, the dominant type of up-gradient hydrothermal alteration was determined.

Streams and springs draining propylitic altered rock had higher pH (5.74–8.49) and lower dissolved metal concentrations than water draining other alteration assemblages or mine adit water (Fig. 6). In addition, these sites are characterized by measurable alkalinity. Propylitically altered rock contains calcite, as well as chlorite and epidote, which dissolve and produce circumneutral water. Sulfate concentrations are slightly elevated because of the weathering of minor amounts of pyrite and gypsum (Mast et al., 2007; Nordstrom et al., 2007).

In contrast to propylitic altered rock, quartz-sericite pyrite altered rock produces water that is acidic and metal-rich (Fig. 6), reflecting an abundance of pyrite and a lack of acid-neutralizing minerals (Mast et al., 2007). The acidic water readily reacts with alumino-silicate minerals in the bedrock, producing surface and ground water with high concentrations of aluminum and silica. Quartz-sericite-pyrite altered bedrock tends to be located within or adjacent to base-metal mineralized areas, thus water draining from quartz-sericite-pyrite altered areas has relatively high concentrations of copper, manganese, and zinc (Fig. 6). Weak sericitic altered bedrock tends to produce water with intermediate compositions between that derived from propylitic or quartz-sericite-pyrite altered bedrock (Fig. 6).

Water draining from inactive mine sites generally has the greatest range in compositions and tends to have higher dissolved sulfate and metal concentrations than most sites unaffected by mining (Fig. 6). The wide range in water quality of mine water likely results from three factors: (1) chemistry of the water entering mines varies because of weathering of different hydrothermal alteration assemblages; (2) weathering of different ore and gangue minerals depends upon the mineral deposit type (e.g., Cox and Singer, 1986); and (3) some of the sites classified as "mining-affected" may actually have been metal-poor mine prospects.

Surface Water Quality

Wright et al. (2007) compiled water-quality data collected from 1991 to 1999 from multiple sources including data collected by ARSG and the State of Colorado (Butler et al., 2001) to evaluate the spatial variation in water quality throughout the study area. The distribution of pH values in streams during low-flow conditions (Fig. 7) shows that stream segments draining more intensely altered rocks have lower pH values than streams draining propylitic altered areas. For example, streams draining Ohio Peak, which is underlain by acid-sulfate alteration, tend to have pH values < 4.5. Near Red Mountain Pass in the headwaters of Mineral Creek basin, which is underlain by acid-sulfate alteration (Fig. 7), extremely low pH values (2.4–2.5) were measured in water draining from the Longfellow mine (site #77, Fig. 5) and Koehler tunnel (site #75, Fig. 5). Much of Cement Creek has low pH (<4.5), and the Cement Creek basin is characterized by both intensely altered rock and numerous mine sites. In contrast, although a substantial percent of the mining activity occurred in the upper Animas River basin upstream of Silverton, most of the stream segments there have pH values > 6.5. This area is primarily composed of propylitic altered bedrock (Fig. 7). Similarly, relatively high pH values were measured along the South Fork Mineral Creek, which is characterized by sedimentary rock that has been partially overprinted by propylitic alteration.

The distribution of dissolved zinc concentrations in streams during low-flow conditions (Fig. 8) shows that zinc in surface water is primarily derived from sphalerite dissolution. Since zinc has been identified as a trace element in other sulfides, particularly pyrite, dissolution of pyrite also contributes to the zinc load (Bove et al., 2007). In the upper parts of Mineral Creek, Cement Creek, and the Animas River, mine adit water tends to have high zinc concentrations. The highest zinc concentration (228 mg/L) was measured in mine-adit water draining the Koehler tunnel (site #77, Fig. 5) near Red Mountain Pass. In general, zinc concentrations decrease downstream because of dilution by stream water with relatively low zinc concentrations.

Both mined and unmined areas contribute substantial loads of constituents to the streams that drain the Animas River watershed because of the widespread, extensively altered and mineralized rock in the area. Prioritizing mine-site remediation at the watershed scale requires an understanding of how multiple sources of acidic, metal-rich drainage affect the streams in the

TABLE 2. RANGES OF CHEMISTRY OF SPRINGS UNAFFECTED BY MINING AND OF MINE WATER FROM ADITS

	pH Range	pH Median	SO$_4$ (mg/L) Range	SO$_4$ (mg/L) Median	Zn (µg/L) Range	Zn (µg/L) Median
Spring from sites unaffected by mining	2.58–8.49	4.89	1–1300	90	<20–14,300	28
Adit mine drainage	2.35–7.77	5.72	45–2720	310	<20–228,000	620

Note: Data summarized from Mast et al., 2000a.

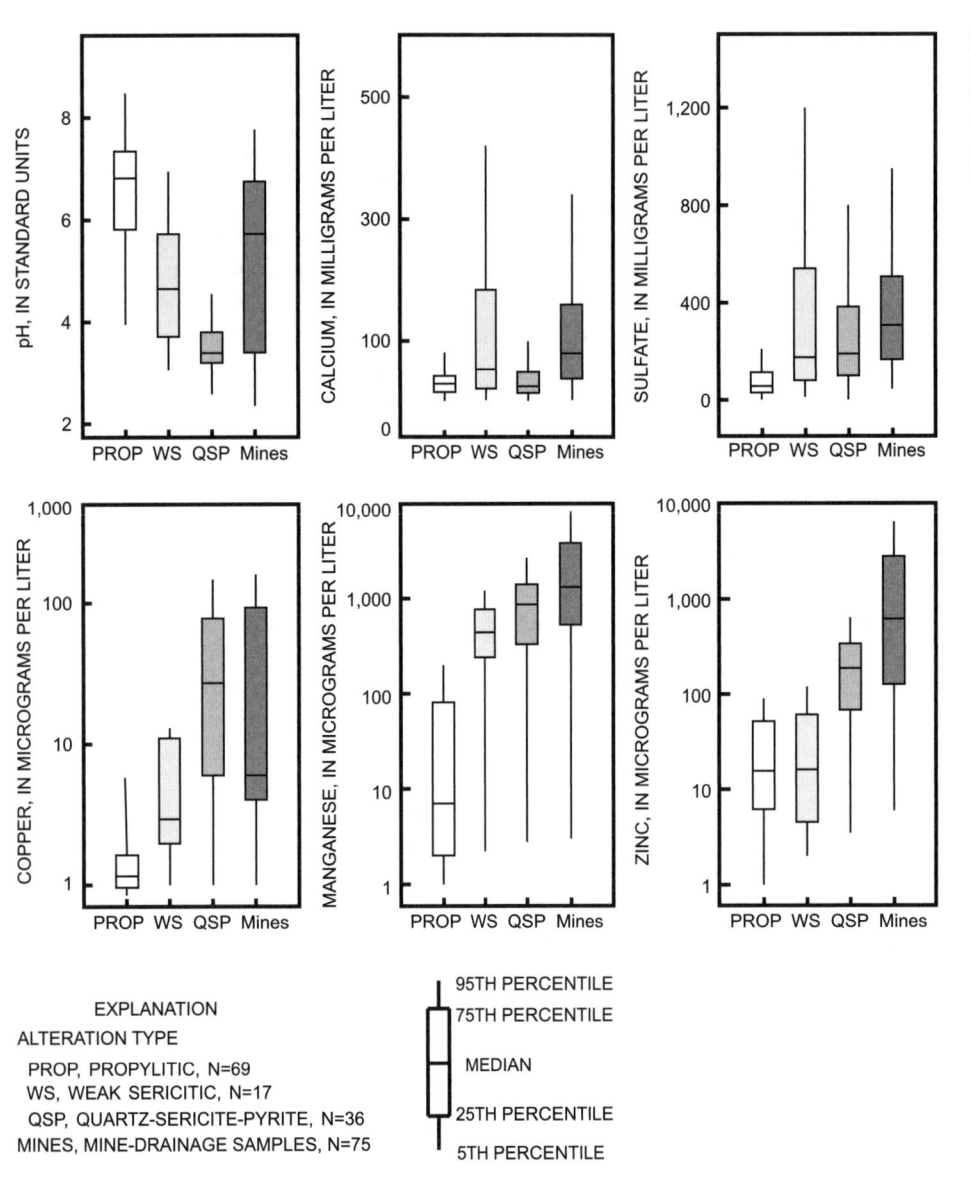

Figure 6. Comparison of dissolved-constituent concentrations in mine-adit discharge samples and in background water samples draining different alteration assemblages (from Mast et al., 2007).

EXPLANATION
ALTERATION TYPE
PROP, PROPYLITIC, N=69
WS, WEAK SERICITIC, N=17
QSP, QUARTZ-SERICITE-PYRITE, N=36
MINES, MINE-DRAINAGE SAMPLES, N=75

95TH PERCENTILE
75TH PERCENTILE
MEDIAN
25TH PERCENTILE
5TH PERCENTILE

watershed (Kimball et al., 2002). Contributions to the stream from all sources range from well-defined tributary contributions to dispersed, ground-water contributions. Mass-loading studies are useful to identify and compare the complex sources of loads in a watershed. A detailed mass-loading approach differs from a more traditional method of measuring load at a watershed outlet because it provides the necessary spatial detail to facilitate decisions about remediation. These studies are based on two well-established techniques: (1) the tracer-dilution method and (2) synoptic sampling. The tracer-dilution method provides estimates of stream discharge that are in turn used to quantify the amount of water entering the stream, both from tributaries and ground water, in a given stream segment (Kimball et al., 2002). Synoptic sampling of the instream flow and additional contributions from tributaries and ground-water chemistry provides a detailed longitudinal snapshot of stream water quality and chemistry (Bencala and McKnight, 1987). When used together, the tracer-dilution and synoptic sampling techniques provide discharge and concentration data that are used to determine the mass loading associated with various sources of water.

Tracer Studies

During the AML Initiative, a series of 13 tracer-injection studies established a hydrologic framework to quantify metal loading within the Animas River watershed (Kimball et al., 2007), providing a level of spatial and hydrologic detail never before collected anywhere. Within the three principal basins, 24 locations including both mined and unmined areas accounted for 73–87% of the total mass loading of aluminum, iron, copper, zinc, and manganese (Fig. 9). Weathering of extensive acid-sulfate and quartz-sericite-pyrite alteration zones in Mineral and Cement

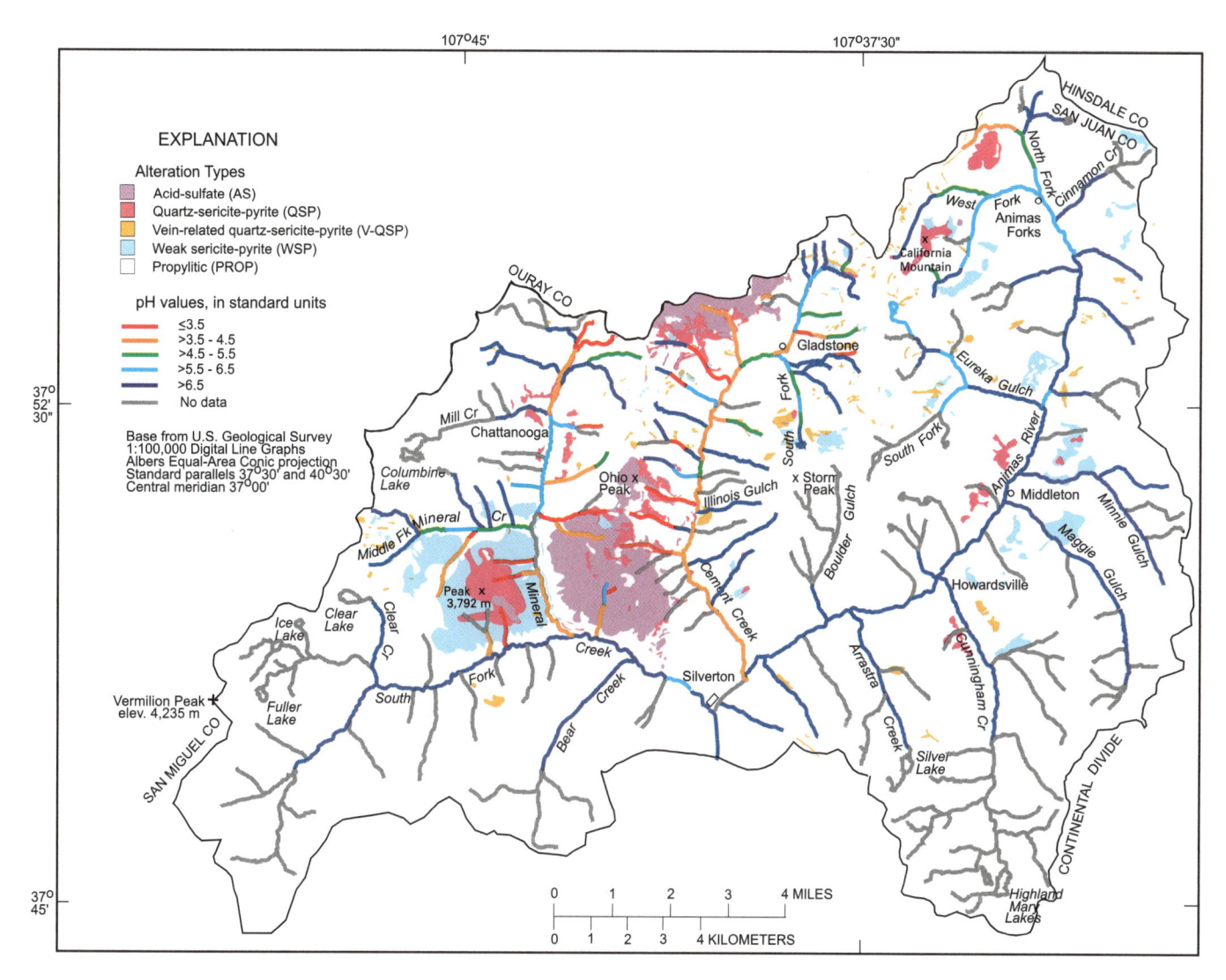

Figure 7. Distribution of pH measured in streams during low-flow conditions (1991–1999; from Wright et al., 2007). Hydrothermal alteration map from Bove et al. (2007) and Dalton et al. (2007).

Creek basins substantially contributes to loading of aluminum and iron (Figs. 9A and 9B). The location of greatest aluminum loading was the Middle Fork Mineral Creek, which drains extensive areas of quartz-sericite-pyrite alteration. Substantial aluminum and iron loads also entered Mineral Creek where it drains the acid-sulfate alteration of Ohio Peak and Anvil Mountain (Figs. 7 and 8). In Cement Creek basin, both Prospect and Minnesota Gulches drain acid-sulfate and quartz-sericite-pyrite alteration and contributed substantial aluminum and iron loads. Mineral Creek dominated the contribution of total copper load, whereas Cement Creek had the greatest contribution of total zinc load (Figs. 9C and 9D). Dispersed ground water added to the stream near Red Mountain Pass, as well as water draining the Koehler tunnel (site #75, Fig. 5), contributed substantial copper and zinc loads (Mineral Creek, site A, Fig. 9F). This is an area of acid-sulfate alteration. The Mogul mine in Cement Creek (site #97, Fig. 5), which likely drains the mine pool behind the bulkhead in the American tunnel (site #96, Fig. 5), and the North Fork Cement Creek also contributed large loads of copper and zinc (Fig. 8; Cement B and Cement C, Fig. 9F). In contrast to Cement and Mineral Creek basins, the Animas River basin drains mostly regional propylitic alteration. As a result, it does not have comparable loads of copper (Fig. 9C). Areas of vein related quartz-sericite-pyrite alteration in the headwaters of California Gulch, however, contributed substantial manganese and some zinc loading (Fig. 8; Figs. 9D and 9E). Substantial manganese and zinc loads were added to the Animas River downstream from Arrastra Creek (Fig. 8). This area drains historical mill tailings repositories (site #510, Fig. 5) that contain manganese gangue minerals from the milling of ore from the Sunnyside mine from 1961 to 1991 (Bird, 1999).

Zinc loading principally occurred in 24 areas (Fig. 9F). These areas are identified in the figure by basin, showing that

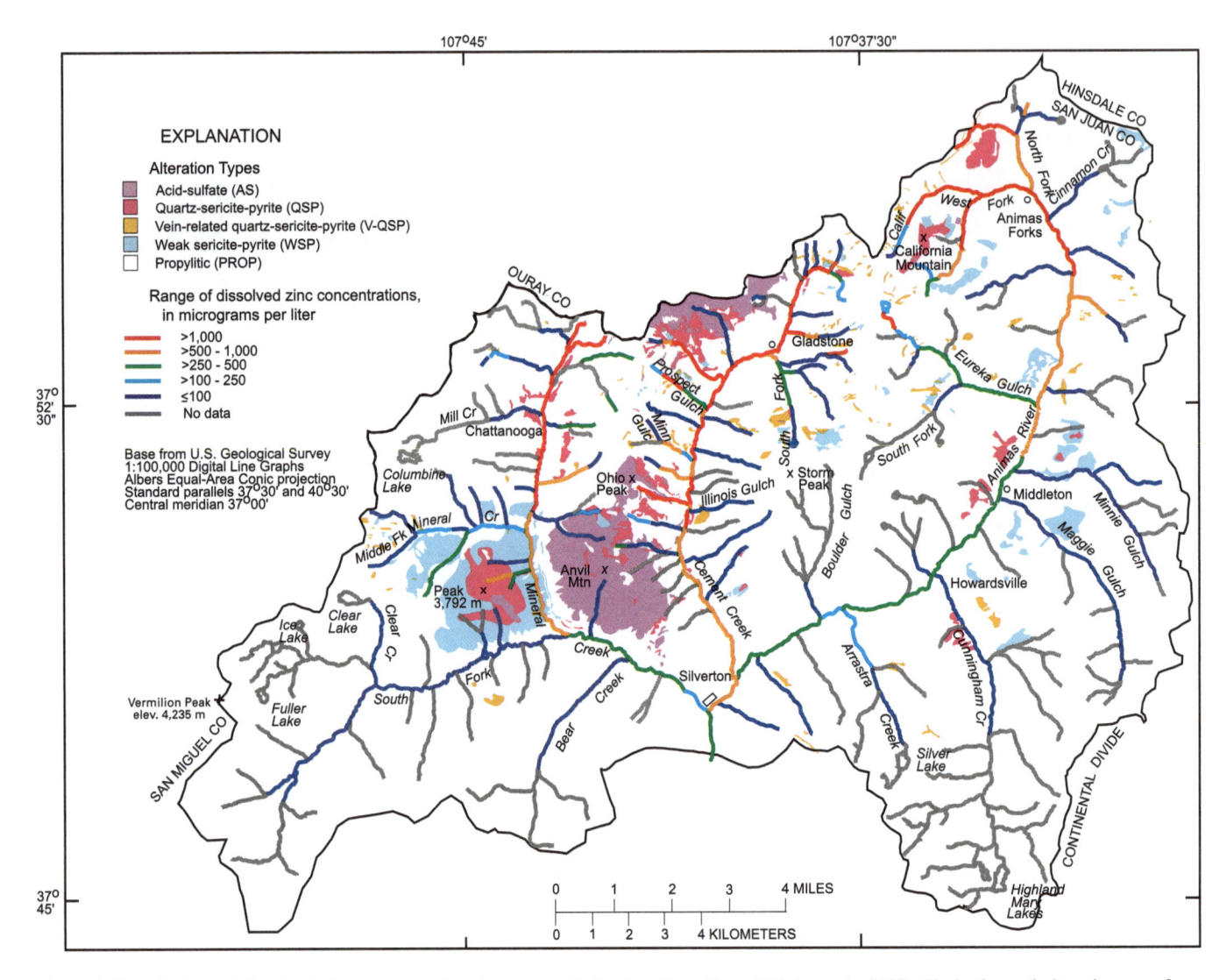

Figure 8. Distribution of dissolved zinc concentrations in streams during low flow (from Wright et al., 2007). Hydrothermal alteration map from Bove et al. (2007) and Dalton et al. (2007).

there were 4 principal areas in the Mineral Creek basin, 10 areas in the Cement Creek basin, 8 areas in the upper Animas River basin, and 1 area downstream from all three of these tributaries. These 24 areas accounted for 77% of the zinc loading in the watershed, but it is important to note that for 23% of the zinc load and for 13–29% of the load of the other metals (Fig. 9), the metal loading from other, dispersed sources could complicate remediation efforts because they would continue to release metals to the streams.

With the high cost of remediation, a predictive tool based upon sound science that could be used to anticipate results of various remediation options would be a desirable objective of any watershed characterization effort, particularly if it could be used in conjunction with the best state-of-the-art engineering solutions to make informed and cost-effective remediation decisions. Reactive transport models are an example of such a predictive tool developed during the AML Initiative. Solutions have been run on the study reaches in the Animas River watershed for selected stream reaches in the study area. These models integrate the tracer-dilution discharge with synoptic water chemistry from tracer studies. Runkel and Kimball (2002) provide an example by using a calibrated solute transport model to the downstream results. Two different active treatment options were evaluated to assess metal loading at Mineral Creek, site A (Fig. 9F). Option 1 removes ferric, but not ferrous, iron, whereas option 2 removes all iron from the stream at site A. Both options increase instream pH and substantially reduce total and dissolved concentrations of aluminum, arsenic, copper, and ferrous and ferric iron near the gauge on Mineral Creek. Dissolved lead concentrations are reduced by 18% with the first remediation plan. Both lead and iron are removed in an active treatment system with the second option, but this remediation option results in an increase in dissolved lead

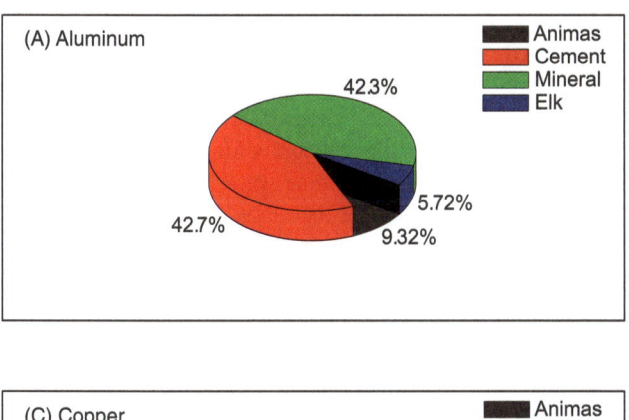

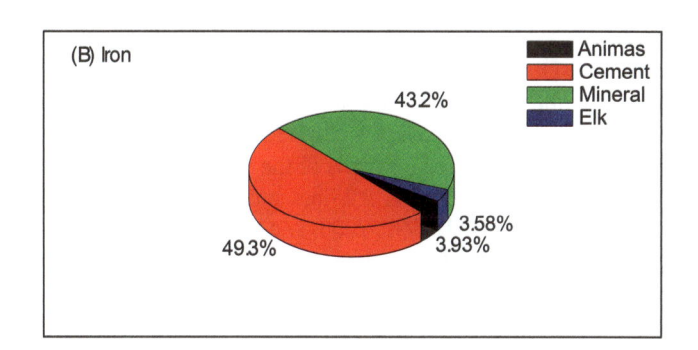

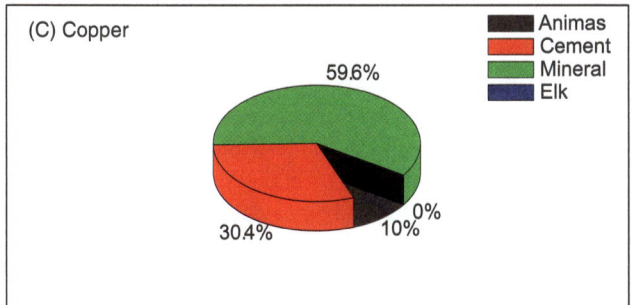

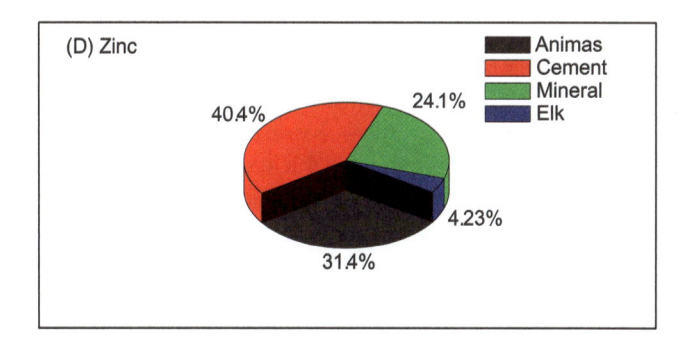

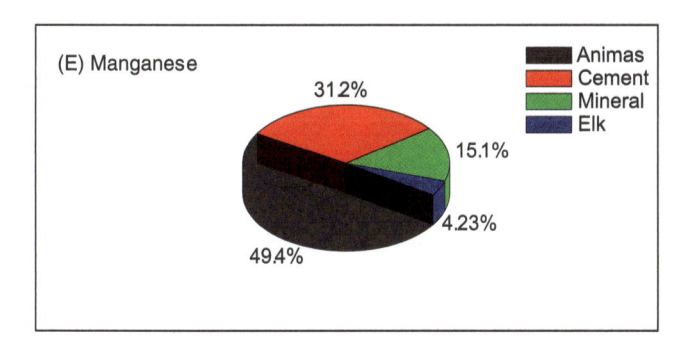

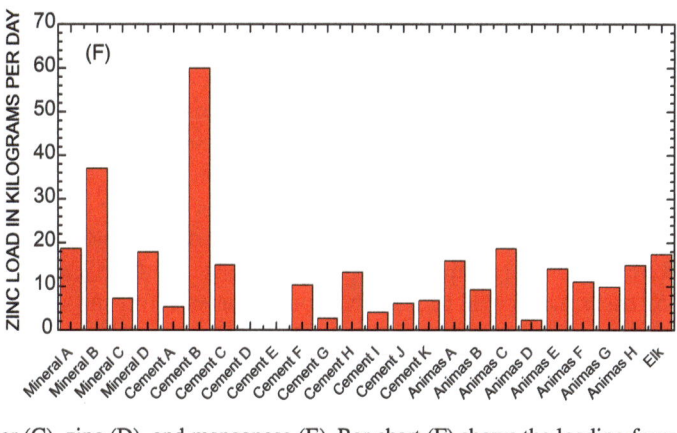

Figure 9. Pie diagrams showing loads for aluminum (A), iron (B), copper (C), zinc (D), and manganese (E). Bar chart (F) shows the loading from different mine sites for zinc.

concentrations over existing untreated conditions because additional downstream sources of lead are not attenuated by sorption to iron colloids. Neither of the proposed options, however, effectively reduces concentrations of zinc (Kimball et al., 2003).

MINES AND MILLS IDENTIFIED AS SOURCES

Large volumes of disturbed rock in the form of mine-waste dumps at historical mine sites have been implicated as the source of metals and acidity in many historical mining districts. A combination of high precipitation, extensive fracturing, and high topographic relief in many historical mining districts results in a substantial volume of ground water flowing through the rocks at the mine sites. Mining has forever changed the ground-water hydrology in historical mining districts. Historical mine adits and mine workings act as conduits for ground water, which is funneled to the surface once the mine pool is filled. The chemical reactions resulting from the interaction of ground water with these highly fractured and disturbed volumes of rock produce mine adit-water chemistries that vary only slightly throughout the year. Water chemistry from mine adits, however, is dependent upon the hydrothermal alteration type and the mineralogy of the ore (Bove et al., 2007). Flowing mine adits constitute an essentially constant source of contaminated ground water (Church et al., 2007b).

Interest in reducing the environmental effects of the many inactive mines and prospects in the Animas River watershed study area began in the early 1990s. In 1991–1992, a preliminary water-quality analysis of the Animas River watershed was coordinated by the Colorado Department of Health and Environment. The Colorado Division of Mines and Geology, U.S.D.A. Forest Service, and the U.S. Bureau of Land Management inventoried and ranked inactive mines in the study area. Unpublished reports of mine inventories in various parts of the watershed are available from the Colorado Division of Mines and Geology (CDMG Herron et al., 1997, 1998, 1999, 2000) and the Colorado Geological Survey (Lovekin et al., 1997).

Studies of the effects of mines on water quality by Nash and Fey (2007) and by the ARSG and the State of Colorado (summarized in Wright et al., 2007) were undertaken to determine which draining mine adits and mine-waste dumps most affect watershed water quality. These studies, coupled with the detailed mapping of altered areas by Bove et al. (2007) and by Dalton et al. (2007), were used to quantify the effects of historical mine wastes as sources of contaminants at a watershed scale (Walton-Day et al., 2007). Mast et al. (2007) examined acidic drainage from 75 mine sites in the watershed that contain flowing adits. Nash and Fey (2007) sampled and quantified the geochemistry of 97 mine-waste sites and 18 mill-tailings sites located on public land, each of which contained more than 100 tons of mine waste. Sampling protocols developed to obtain a representative sample are described in Smith et al. (2000), and analytical methods used are in Fey et al. (2000). The most acidic and metalliferous water (ranked 6, Fig. 10) drained mine waste from deposits containing acid-sulfate alteration, followed by quartz-sericite-pyrite alteration, and lastly by propylitic alteration (ranked 1, Fig. 10). The metal and acidity available from different mine-waste sites were quantified on the basis of the sum-of-metals concentration versus pH of the leachate water from the mine wastes. Leachate chemistry was determined using a 20:1 water/sample ratio in deionized water. The leachate equilibrated in a few minutes, and the leaching experiment could be repeated multiple times before the supply of water-soluble salts present in the mine-waste sample was sufficiently reduced such that the leachate water chemistry changed. The important conclusion here is that these mine wastes behave essentially as an infinite and constant source of potentially toxic metals and acidity. The chemistry of surface and ground water supplied by leaching of these mine wastes is constant, and the loads supplied are limited by the amount of precipitation and runoff rather than by the reaction rate of sulfides present with incident water on the mine-waste sites. Of the more than 500 mine sites on public land that were sampled, only 39 sites had a significant effect on surface water quality (Table 3, sites were classified from moderate to very high on the basis of the pH and sum of metals scores, Fig. 11; see Fey et al., 2000).

The State of Colorado, Division of Mines and Geology and the ARSG sampled private sites as well as those on public lands and using somewhat different criteria developed a different list of 31 draining mine adits and 30 mine-waste dumps that needed remediation (Wright et al., 2007). These sites are presented in Figure 12 and Table 4. Neither study included active mines sites that were undergoing remediation by Sunnyside Gold, Inc.. These 61 sites contribute ~90% of the metal loads to the three major drainages. The conclusions drawn from these studies, although they differ in scope, are very similar for the sites located on federal lands. Twenty of the 50 sites located on federal lands identified by Nash and Fey (2007) for remediation also occur on the list of 61 sites identified for remediation by the ARSG and the State of Colorado.

REMEDIATION ACTIVITIES

The Animas River Stakeholders Group and federal land-management agencies began planning for clean-up activities in the mid-1990s. Sunnyside Gold, Inc. reached an agreement with the

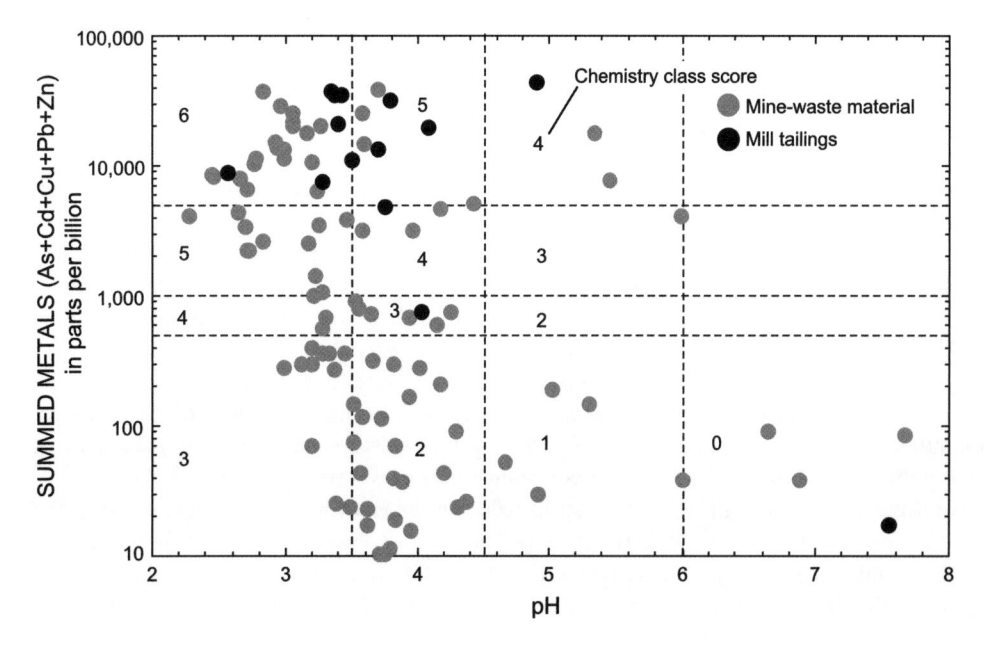

Figure 10. Composition of mine-waste leachate chemistry showing range of dissolved metal concentrations and pH from leachates from samples from 97 mine-waste sites and 18 mill-tailings sites. Dashed lines and numbers (0–6) delineate classification fields to categorize mine waters derived from mine and mill wastes (Nash and Fey, 2007).

State of Colorado to implement a number of remediation activities in the watershed as a condition for terminating its discharge permit at the American tunnel (site #96). Most of this remediation work on Sunnyside properties was completed by 1996. Remediation work completed throughout the Animas River watershed through 2004 is summarized in Table 5 and Figure 13. Funding for site remediation and watershed cleanup has come from both private and public sources. Remediation work done by Sunnyside Gold, Inc. has exceeded $10 million; most of the work has been done on permitted mine sites where they were involved directly in mining. Other remediation work has been done on mine sites identified as high priority (e.g., Koehler tunnel, site #75, Figs. 11 and 13). Funding for remediation of mine sites located on public lands has come from U.S. Bureau of Land Management and the U.S.D.A. Forest Service abandoned mine lands funds. Much of the funding for remediation work undertaken by the ARSG has come from U.S. EPA 319 grants. Funding has also been provided from various other sources: U.S. Office of Surface Mining, the San Juan Resource and Conservation District, Silver Wing Mining Company, Gold King mine, Inc., Salem Minerals, and Mining Remedial Recovery. Of the 39 priority sites identified by Nash and Fey (2007), 9 (23%) have been remediated. Of the 61 high-priority sites identified by the ARSG, 14 (23%) have been remediated. The ARSG is actively working with federal, state, and local funding sources to remediate these sites and reduce the metal loading in the Animas River watershed.

MONITORING

Data Collection and Methodology

Frequent water-quality sampling was done at the four gauges (Fig. 13). Water-quality data were collected at least monthly beginning in 1991 following the closure of the Sunnyside mine and more frequently between 1994 and 1999. Continuous stream-flow monitoring at the four stream gauges has been done since October 1, 1994. The U.S. Geological Survey, U.S. Bureau of Reclamation, Sunnyside Gold, Inc., and the Silverton Schools under the sponsorship of the Colorado River Watch Program have done most of the monthly chemical monitoring at the gauges. Monitoring was more frequent than monthly between 1994 and 1999 because all of these entities were involved in data collection. USGS and River Watch monitoring was discontinued after 1999. SGC continued monthly monitoring at A72 and monitored the other three gauges on alternate months with the U.S. Bureau of Reclamation through most of 2002. Since 2002, the River Watch program has resumed monitoring, alternating months with the U.S. Bureau of Reclamation at all four gauges.

Previous investigations in the basin have shown that stream flow and seasonality influence the concentration of most constituents (Butler et al., 2001; Leib et al., 2003). These exogenous factors, if not accounted for, may mask the effects of remediation. The effect of stream flow and seasonality on the concentration variation of solutes was removed through regression analysis. For a complete description of the methodology see Leib et al. (2003).

Moving average charts were prepared for the four gauge sites. The moving average chart smoothes irregularities owing to periodic factors not removed through the regression model. The "0," or centerline, is the expected concentration of any analyte, adjusted for stream flow and seasonal variations, observed during the baseline period. The charted variable is the difference between the observed concentration adjusted for stream flow and season and the expected concentration derived from the regression equation. If the moving average line remains between the first gridlines above and below the centerline, there is a 95% probability that there has been no shift in the 12-sample mean concentration.

The baseline period used data collected between October 12, 1994, and September 30, 1996, at A68, C48, and M34. The baseline period at Animas River downstream from Silverton, A72, uses the data collected between September 1991 and May 1996. A consent decree signed in May 1996 between the Colorado Department of Health and Environment and Sunnyside Gold, Inc. specified actions that Sunnyside Gold, Inc. was required to take before terminating their remediation permit. The valve on the American tunnel was closed in October 1996 and accelerated remediation projects were begun shortly thereafter (Table 5).

ANALYSIS OF THE MONITORING RESULTS

Mineral Creek Basin

Metals targeted for remediation in the Mineral Creek basin included cadmium, copper, and zinc. In the environmental risk assessment of surface water quality effects on aquatic life (Besser et al., 2007), dissolved copper posed the greatest risk to brook trout, the dominant species of fish in streams in the basin. Tracer loading studies (Fig. 9C) demonstrated that the Mineral Creek basin was a primary target for site remediation to reduce copper and zinc loads, particularly in the headwaters where acid-sulfate deposits containing enargite, tetrahedrite, and galena were produced from the 1880s through 1907 (Bove et al., 2007; Jones, 2007). Remediation began in Mineral Creek in November 1996 when the pond below the Koehler/Longfellow (sites #75 and #77) was drained and all sludge from the pond and mine waste from the Koehler was removed to Tailings Pond #4, site #510. In 1997 the Longfellow wastes were consolidated, neutralized with limestone, and runon/runoff controls were implemented. Mine waste was removed from the Congress mine, site #79, and Carbon Lake mine, site #80, both near the headwaters of Mineral Creek in 2000–2003. Hydrologic controls were implemented at the Bonner mine, site #172, on the Middle Fork of Mineral Creek in 2000. A bulkhead seal was placed in the Koehler tunnel in 2003. Mine-waste consolidation, neutralization, and hydrologic controls were implemented at the San Antonio mine, site SA (north of study area boundary, Fig. 13), in 2004. Data collected during and immediately after the remediation at sites 75–77 (Fig. 13) show sharp

TABLE 3. SELECTED MINE AND MILL-TAILINGS SITES ON FEDERAL LANDS RECOMMENDED FOR REMEDIATION (NASH AND FEY, 2007), ANIMAS RIVER WATERSHED

Name	Site No.	Ranking
Mines		
Early Bird Crosscut	8	Moderate
Ben Butler mine	9	High
Hermes Group	11	High
Eagle Chief mine	14	Low
Little Ida mine	15	High
Frisco tunnel	19	High
Grand Mogul	35	Very High
Koehler tunnel	75	Very High
Henrietta mine—#7 Level	85	Very High
Lark mine	86	Very High
Joe and Johns mine	87	High
Upper Joe and Johns mine	89	Low
Eveline mine	91	Low
Clipper mine	114	Very High
Silver Crown mine	133	High
Imogene mine	136	Moderate
Ferricrete mine	137	Low
Brooklyn mine	141	High
Kansas City mine—#1 Level	145	High
Elk tunnel	147	Low
Mammoth tunnel	148	High
Avalanche mine	149	Moderate
Paradise mine	168	Moderate
Ruby Trust	169	Very High
Independence mine	171	High
Bonner mine	172	High
Monarch	180	High
Mayday mine	181	Very High
Legal Tender mine	189	Moderate
Forest Queen mine	195	Moderate
Caledonia mine	198	Very High
Kittimack mine	201	High
Burbank mine	207	High
Columbine mine	260	Moderate
Sultan tunnel	266	High
Mighty Monarch mine	285	Low
Last Chance mine	289	Moderate
Bandora mine	332	Very High
Highland Mary—#7 Level	359	Low
Henrietta mine—#8 Level	505	Very High
Henrietta mine—#9 Level	506	Very High
Unnamed prospect (Mineral Creek basin)	X	Moderate
Unnamed prospect (upper Animas River basin)	X	Low
Unnamed prospect (upper Animas River basin)	X	Low
Unnamed prospect (upper Animas River basin)	X	Low
Unnamed prospect (upper Animas River basin)	X	Moderate
Mill Tailings		
Kittimack tailings	192	Very High
Lackawanna tailings (removed)	286	Very High
North Star Mill tailings	310	Moderate
Highland Mary Mill tailings	361	Low

Note: Data refer to sites in Figure 11; data from Jones (2007); Church et al. (2007b).

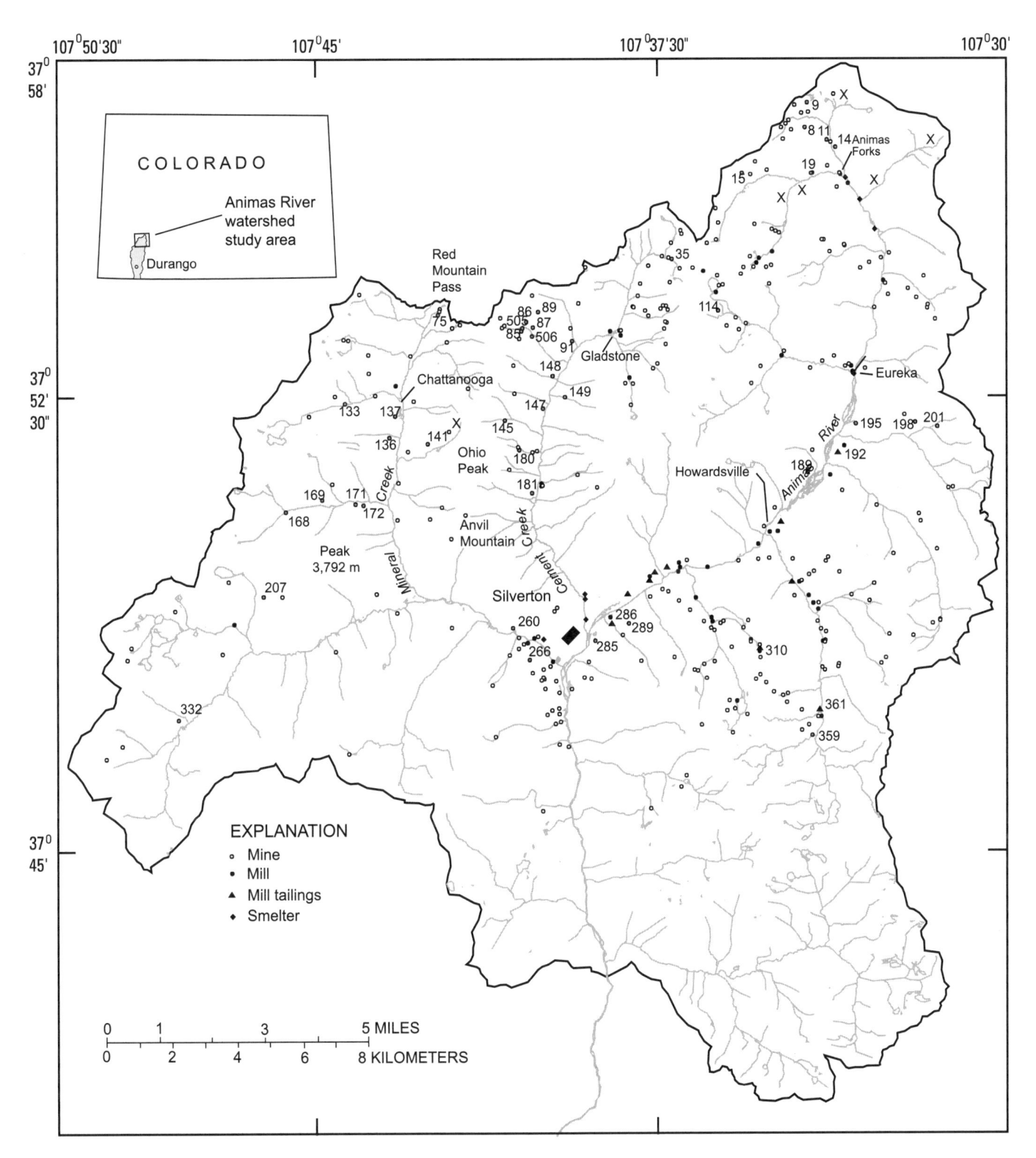

Figure 11. Location and ranking (Table 3) of mine waste, Animas River watershed. Rank is based on size, metal release, proximity to streams, and acid-generating potential (Fig. 10; modified from Nash and Fey, 2007); some sites contain multiple draining adits or mine-waste dumps. Prominent peaks discussed in text are located on map; unnamed peak between Middle and South Forks Mineral Creek is designated as peak 3792 m (USGS, 1955).

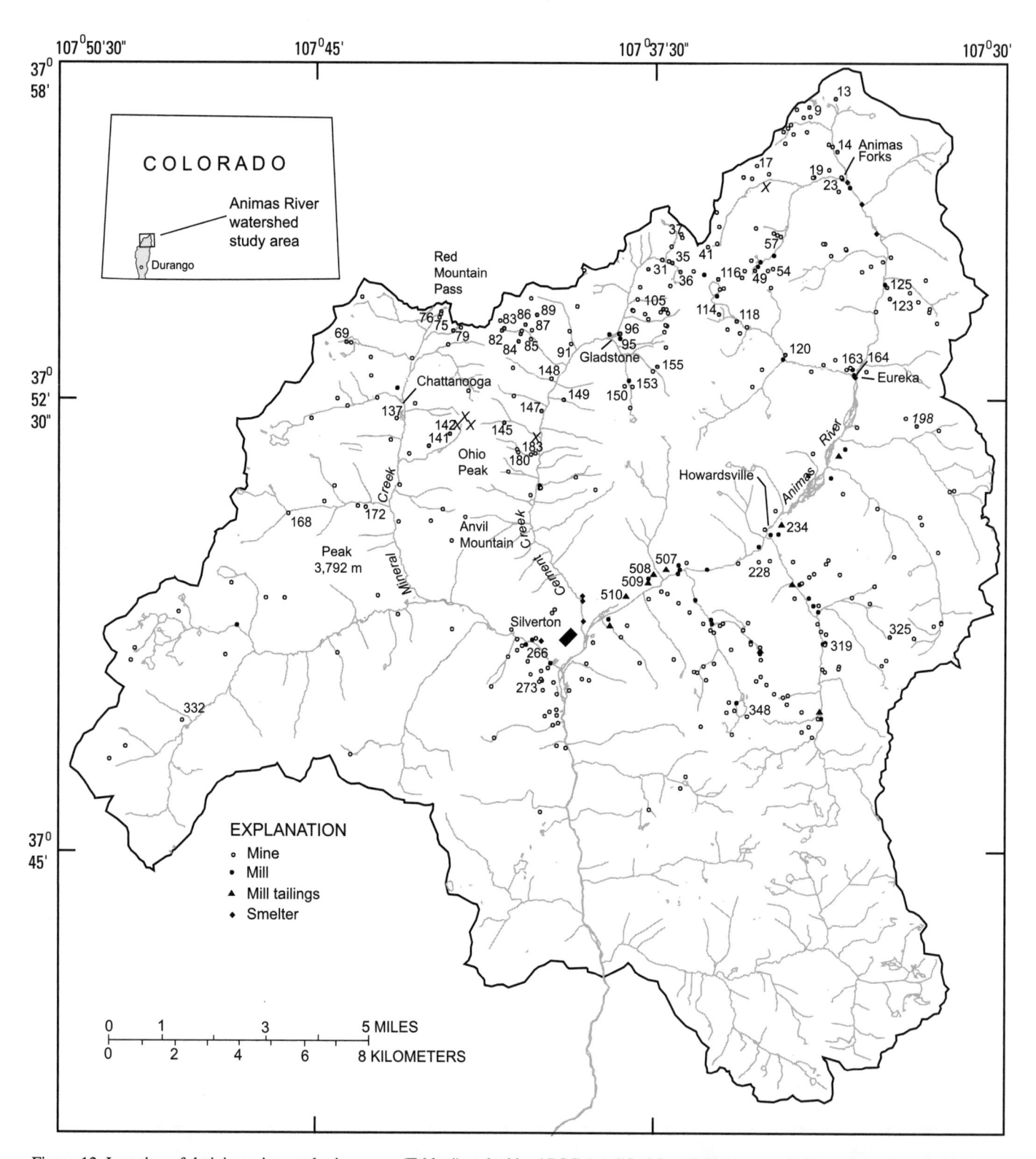

Figure 12. Location of draining mines and mine waste (Table 4) ranked by ARSG (modified from Wright et al., 2007), Animas River watershed. Rank is based on metal inventory and acid-generating potential of mine-waste dumps and metals released from draining adits; some sites contain multiple draining adits or mine-waste dumps. Prominent peaks discussed in text are located on map; unnamed peak between Middle and South Forks Mineral Creek is designated as peak 3792 m (USGS, 1955).

TABLE 4. SELECTED DRAINING MINES, WASTE ROCK PILES, AND PERMITTED MINE AND MILL SITES
IN THE UPPER ANIMAS RIVER WATERSHED

Site No.	Mine or Site Name	Site No.	Mine or Site Name
	Draining Mines		**Waste Rock Pile**
31	Mogul mine	82	Galena Queen mine
2	Silver Ledge mine	145	Kansas City #2 mine
35	Grand Mogul mine	83	Hercules mine, shaft
148	Mammoth tunnel	89	Upper Joe & Johns mine
183	Anglo-Saxon mine	35	Grand Mogul mine, East
87	Joe and Johns mine	145	Kansas City #1 mine
150	Big Colorado mine	155	Black Hawk mine
180	Monarch mine	95	Lead Carbonate Mill
91	Eveline mine	84	Henrietta mine (level 3)
37	Columbia mine	36	Ross Basin mine
75	Koehler tunnel	86	Lark mine
266	North Star mine (Sultan)	??	Pride of the Rockies mine
76	Junction mine	85	Henrietta mine (Level 7)
332	Bandora mine	31	Mogul mine
172	Upper Bonner mine	141	Brooklyn mine
172	Bonner mine	69	Upper Bullion King mine
172	Lower Bonner mine	X	Unnamed shaft mine, upper Browns Gulch
137	Ferricrete mine	79	Congress mine, shaft
168	Paradise mine	142	Brooklyn mine, upper waste rock pile
141	Brooklyn mine	X	Unnamed mine, upper Browns Gulch
273	Little Dora mine	273	Little Dora mine
17	Vermillion mine	141	Brooklyn mine, lower waste rock pile
23	Columbus mine	9	Ben Butler mine
X	Lower Comet mine	125	Silver Wing mine
X	Unnamed mine	123	Tom Moore mine
54	Sound Democrat mine	8	Eagle Chief mine
42	Mountain Queen mine	13	Lucky Jack mine
125	Silver Wing mine	114	Clipper mine
19	Frisco tunnel	325	Buffalo Boy mine
163	Senator mine	118	Ben Franklin mine
348	Royal Tiger mine	198	Caledonia mine
319	Pride of the West mine	116	Sunnyside mine
228	Little Nation mine		**Permitted Mine Sites**
		105	Upper Gold King mine
		96	American tunnel mine
		49	Gold Prince mine
		116	Sunnyside mine
		120	Terry tunnel mine
			Permitted Mill Sites
		234	Pride of the West Mill tailings
		507–510	Mayflower Mill tailings

Note: Data refer to sites in Figure 12; Wright et al. (2007).

TABLE 5. SUMMARY OF RECLAMATION PROJECTS COMPLETED AS OF 2004

Project Sponsor	Project Site	Location (Fig. 12)	Type of Remediation	Date Project Completed	Improvement (Actual or Anticipated)
Private Funds					
Sunnyside Gold, Inc.	Lead Carbonate Mill	Gladstone, South Fork Cement Creek, site #95	Removal of 27,000 yd^3 of mill tailings from stream bank	1991	Reduce loading of metals to Cement Creek and erosion transport of mill tailings
Sunnyside Gold, Inc.	Mayflower Mill	Mayflower Mill tailings, sites #507–509	Recontour inactive mill tailings ponds and cap, 625,000 yd^3 of mill tailings and overburden moved	1992	Mined land reclamation—reduce loading of metals to Animas River and erosion transport of mill tailings
Sunnyside Gold, Inc.	Lake Emma Sunnyside Basin	Sunnyside mine collapse, site #116	Fill mine subsidence at Lake Emma, remove 240,000 yd^3 mine waste and recontour	1993	Mined land reclamation and reduce loading of metals to Animas River
Sunnyside Gold, Inc.	American tunnel, waste dump	Gladstone, on bank of South Fork Cement Creek, site #96	Remove 90,000 yd^3 waste dump and underlying historical mill tailings	1995	Mined land reclamation and reduce loading of metals to Cement Creek and erosion transport of mill tailings
Sunnyside Gold, Inc.	Sunnyside Eureka Mill, tailings at town site	On banks and in flood plain of Animas River, downstream of site #164	Remove 112,000 yd^3 of mill tailings	1996	Reduce loading of metals to Animas River and erosion transport of mill tailings.
Sunnyside Gold, Inc.	Sunnyside mine hydraulic seal project	Sunnyside mine, sites #116, #120, #96	Bulkheads placed in Sunnyside mine to restore hydrologic regime to approximate premining hydrology and eliminate drainage from adits	1997	Place mine workings under water to reduce oxidation, restore groundwater movement around mine workings and eliminate need for perpetual water treatment
Sunnyside Gold, Inc.	American tunnel	American tunnel at Gladstone, site #96	Divert and treat Cement Creek, fill Sunnyside mine pool to mitigate any short-term impacts of reclamation projects	8/1996 to 12/1999	Reduce loading to Animas River to offset any short term impacts of reclamation of other sites
Sunnyside Gold, Inc.	Mayflower Mill tailings, Boulder Creek	Mill tailings at site #509, flood plain of Boulder Creek	Remove 5700 yd^3 of mill tailings	1997	Reduce loading of metals to Animas River and erosion transport of mill tailings
Sunnyside Gold, Inc.	Ransom mine adit	Eureka town site, site #161	Bulkhead seal to stop deep mine drainage and reclaim portal	1997	Restore hydrologic regime and reduce rate of ore oxidation by placing mine workings under water to reduce metal loading to Animas River
Sunnyside Gold, Inc.	Gold Prince mine waste and mill tailings	Placer Gulch, site #49	Bulkhead seals to stop deep mine drainage, consolidate mine waste and mill tailings, remove 6000 yd^3 mine waste and construct upland diversions	1997	Reduce exposure to water to reduce metal loading to Animas River
Sunnyside Gold, Inc.	Longfellow-Koehler	Headwaters of Mineral Creek, Longfellow Mine, site #77; Junction Mine, site #76; Koehler tunnel, site #75	Remove Koehler Mine-waste dump (32,100 yd^3), consolidate Junction mine dump and Longfellow mine dump and cap, capture adit drainage, construct diversions, conduct feasibility study of wetland treatment of Koehler tunnel acidic drainage	1997	Reduce metal loading to Mineral Creek and erosion transport of mine waste
Sunnyside Gold, Inc.	Pride of the West Mill tailings	Howardsville, site #234	Remove 84,000 yd^3 of mill tailings	1997	Reduce metal loading to Animas River and transport of mill tailings by erosion

(*continued*)

TABLE 5 (continued)

Sunnyside Gold, Inc.	Sunnyside mine	Sunnyside mine, American tunnel treatment, site #96	Inject 652 tons of hydrated lime into the Sunnyside mine pool to provide increased alkalinity and reduce oxygen available in mine for pyrite oxidation	1997	Improve initial conditions as water table is restored by installing bulkhead to stop acidic mine drainage
Sunnyside Gold, Inc.	Mayflower Hydrological Control	Mayflower Mill tailings pond #1, site #507	Capture and divert three upland drainages that provide supply to ground water up-gradient of mill tailings	1999	Minimize potential for contact of runoff with mill tailings and reduce potential for metal loading to Animas River
Sunnyside Gold, Inc.	Mayflower Mill tailings pond #4	Mayflower Mill tailings pond #4 drainage modification, site #510	Install lined diversion ditch to capture surface runoff and prevent infiltration through mill tailings	1999	Minimize potential for contact of runoff with mill tailings and reduce potential for metal loading to Animas River
Sunnyside Gold, Inc.	Mayflower Mill tailings pond #4	Divert ground water up-gradient from mill tailings pond #4, site #510	Capture ground water and divert around mill tailings impoundment, tailings pond #4	1995, 1999	Minimize potential for contact of groundwater with mill tailings and reduce potential for metal loading to Animas River
Sunnyside Gold, Inc.	Power plant flats	Animas River near old power plant, east of site #509	Remove mill tailings to tailings pond #4, site #510	2003	Reduce metal loading to Animas River
Sunnyside Gold, Inc.	Mogul mine bulkhead	Cement Creek, site #31	Install bulkhead in Mogul mine to stop acidic drainage	2003	Reduce metal loading to Cement Creek
Sunnyside Gold, Inc.	Koehler mine bulkhead	Koehler mine, (site #75)	Install bulkhead in Koehler mine to stop acidic drainage	2003	Reduce metal loading to Mineral Creek
Sunnyside Gold, Inc.	Reactive barrier	Animas River floodplain below Mayflower Mill, mill tailings pond #4 (site #510)	Passive treatment of contaminated ground water before entering Animas River	2003	Reduce metal loading to Animas River
Gold King mines, Inc.	Gold King mine	North Fork Cement Creek, site #111	Hydrologic controls for workings and mine waste	1998	Reduce metal loading to North Fork Cement Creek
Gold King mines, Inc.	Gold King mine	North Fork Cement Creek, site #111	Pipe Gold King adit discharge to Gladstone for active treatment of acidic drainage	2002	Reduce metal loading to Cement Creek
Mining Remedial Recovery	Sunbank Group	Placer Gulch, site #57	Anoxic drain, settling pond, waste consolidation, bulkhead	1995	Raise pH from draining adit, reduce metal loading to Animas River from adits and mine waste
Salem Minerals	Mammoth tunnel	Cement Creek, site #148	Settling ponds for mine drainage	1999	Reduce iron load to Cement Creek
Silver Wing Mining Co.	Silver Wing mine	Animas River, site #125	Collect acidic mine water, install hydrological controls	1995	Divert acidic drainage around mine-waste dump, reduce metals loading to Animas River
Private and Public Funds					
Silver Wing Mining Co.	Silver Wing mine	Animas River, site #125	Install anoxic drain, settling pond, and bioreactor	2000	Reduce metal loading to Animas River
San Juan R. C. & D., ARSG	Carbon Lake—Phase I	Carbon Lake mine, Mineral Creek, site #80	Removal of 1900 yd³ of waste rock from stream channel	Phase I—1999	Reduce loading of metals, especially cadmium, copper, iron, lead, manganese, and zinc, to Mineral Creek
San Juan R. C. & D., ARSG	Carbon Lake—Phase II	Koehler tunnel, Mineral Creek, site #75	Reduce flows from Koehler tunnel by reducing infiltration into surface mine workings	Phase II—2001	Reduce metals loading to Mineral Creek by reducing infiltration of water into old mine workings

(continued)

TABLE 5 (*continued*)

Project Sponsor	Project Site	Location	Type of Remediation	Date Project Completed	Improvement (Actual or Anticipated)
San Juan R. C. & D., ARSG	Carbon Lake—Phase III	Congress mine, Mineral Creek, site #79	Complete removal of Congress mine-waste dump	Phase III—2003	Reduce metals loading to Mineral Creek by removal of mine wastes and beneficiation
San Juan R. C. & D., ARSG	Carbon Lake—Phase III	Carbon Lake, site #80, ditch restoration	Diversion ditch, wetlands, and stream restoration, water rights purchased and water diverted to Uncompahgre River watershed restored to Mineral Creek drainage	Phase III—2003	Reduce metals loading to Mineral Creek by removal of mine wastes and site beneficiation
San Juan R. C. & D., ARSG	Galena Queen and Hercules mines	Prospect Gulch, sites #82 and #83	Remove mine waste, install hydrological controls, add soil amendments, and revegetation	2001	Elimination of surface water leaching of toxic metals. Reduce metal loading to Cement Creek
San Juan R. C. & D., ARSG	San Antonio Project	San Antonio mine	Install hydrological controls, remove wastes from stream, consolidate wastes and neutralize, revegetation	2004	Reduce metal loading and acidity to Mineral Creek; stabilize site, remove mine wastes, restore streambed and riparian habitat
San Juan R. C. & D., ARSG	Handies Peak Project	Lucky Jack mine (site #13)	Hydrological controls, remove wastes from fen, consolidate, neutralize, and revegetation; adit and shaft closures	2004	Reduce metal loading and acidity to upper Animas River; uncover fen and restore
U.S. BLM and Duke Energy	Henrietta mine	Prospect Gulch, site #84	Hydrologic controls and mine-waste removal	2004	Reduce metal loading to Cement Creek
Public Funds					
U.S. OSM	Galena Queen mine	Prospect Gulch, sites #82	Waste consolidation and hydrological controls	1998	Reduce surface water leaching of toxic metal loading to Cement Creek
U.S. BLM	Joe & Johns mine	Prospect Gulch, site #87	Mine drainage collection and diversion	1999	Collect acidic drainage for later treatment project development, reduce metal loading to Cement Creek
U.S. BLM	Lark mine	Prospect Gulch, site #86	Install collection system for acidic water, hydrological controls	1999	Collect acidic drainage for possible treatment, remove surface water from site, reduce metal loading to Cement Creek
U.S. BLM	Forest Queen mine	Animas River near Eureka, site #195	Install collection system for acidic water and passive wetland treatment	1999	Reduce metal loading to Animas River
U.S. BLM	Mayday mine	Cement Creek, site #181	Hydrological controls, cap top of mine-waste pile	1999	Reduce surface water leaching of toxic metals
U.S. BLM	Lackawanna Mill tailings	Animas River near Silverton, site #286, site #181	Removal of mill tailings from flood plain to Mayday dump for consolidation and capping	2000	Reduce metal loading to Animas River
U.S. BLM	Elk tunnel	Cement Creek, site #147	Install limestone drain	2003	Reduce metal loading to Cement Creek
U.S. F.S.	Bonner mine	Middle Fork Mineral Creek, site #172	Install collection system for acidic water and diversion, move waste rock from avalanche path	2000	Reduce metal loading to Mineral Creek
U.S. F.S.	Brooklyn mine	Mineral Creek, site #141	Hydrologic controls and mine-waste removal	2004	Reduce metal loading to Mineral Creek

Note: San Juan R. C. & D., San Juan County Resource and Conservation District; ARSG, Animas River Stakeholders Group; U.S. OSM, U.S. Dept. of Interior, Office of Surface Mining; U.S. BLM, U.S. Dept. of Interior, Bureau of Land Management; U.S. F.S., U.S. Dept. of Agriculture, Forest Service; mine site locations from Church (2000b); mill site locations from Jones (2006); yd^3, cubic yards of material; data provided by W. Simon, written comm., Animas River Stake Holders Group, Oct. 2004, and by R. Robinson, pers. comm., U.S. BLM, Apr. 2006.

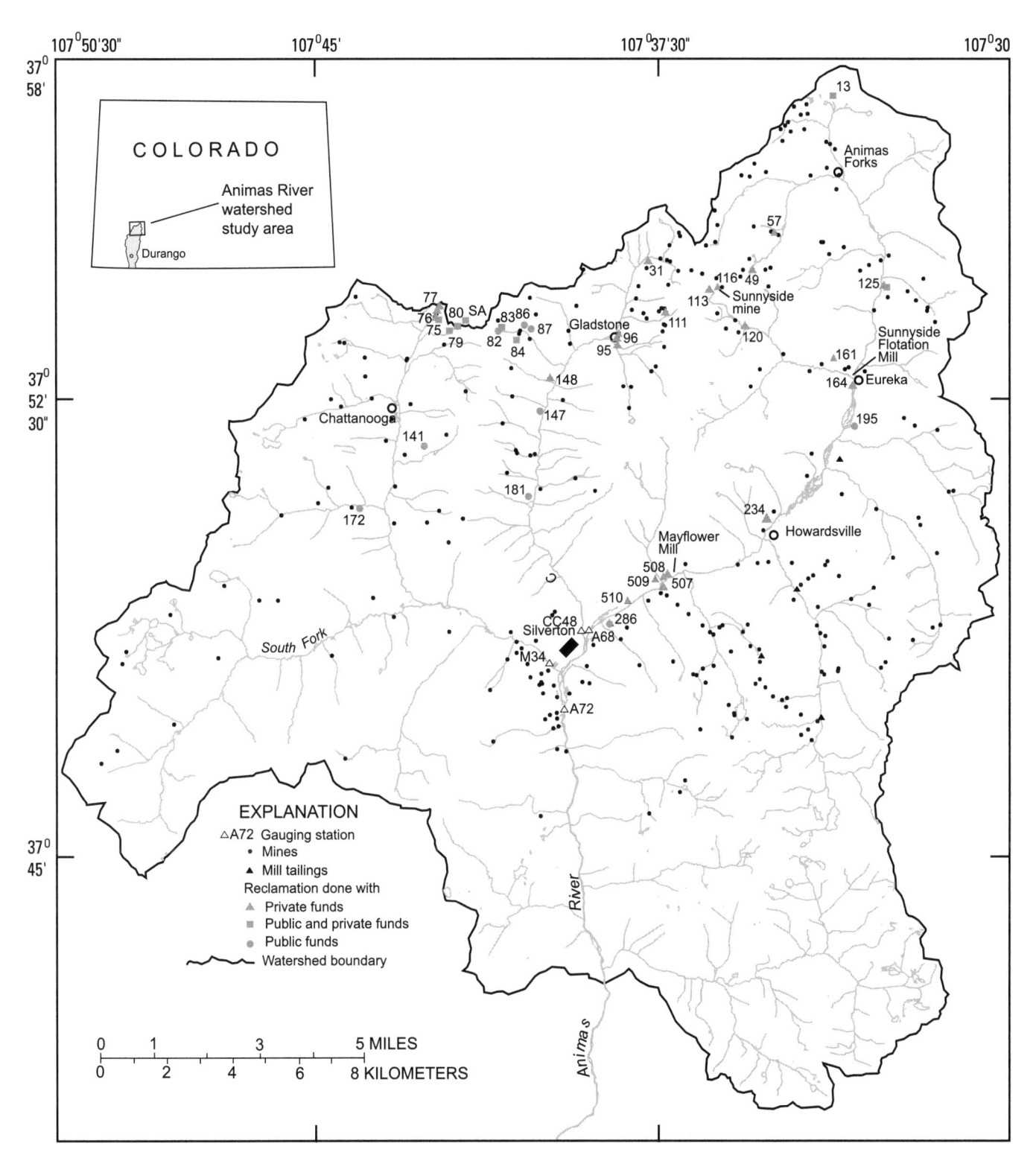

Figure 13. Map showing locations of sites remediated through October 2004 (Table 5). Sites designated by sources of funding for remediation work (private, public, or both).

increases in metal concentrations resulting from the removal of contaminated mine waste followed by continued reduction in the copper concentration in Mineral Creek. The second major remediation project at sites 79–80 (Fig. 11) resulted in long-term reduction of the copper concentration and significant improvement in copper load in Mineral Creek (Fig. 14A). A similar reduction in zinc (and cadmium) concentrations is also evident (Fig. 14B).

The total recoverable aluminum concentration (not shown) fluctuated around the baseline concentration until 2002, when it started into a steep decline. This change may be driven, in part, by the drought conditions experienced in 2002. The 12-period moving average through November 10, 2004, indicates that total recoverable aluminum concentration was 667 µg/L lower than during the 1994–1996 baseline period.

Total recoverable iron (not shown) and dissolved manganese concentrations were generally higher than the baseline condition from 1997 through most of 2002. Since 2002, concentrations of both constituents are approaching the concentrations of the baseline period. Remediation initiated at site #79 appears to have resulted in a substantial increase in the concentration of manganese (Fig. 14C).

Sulfate concentration (Fig. 14D) increased following the initial remediation work, but has been dropping since 2002. Evidence of the improved water-quality conditions downstream are indicated in the reach of Mineral Creek downstream from the confluence of South Fork Mineral Creek, which has since begun to show recovery of some invertebrates (W. Simon, pers. comm., 2004). Since 2002, the concentration of sulfate has dropped below the concentration of the baseline period.

Dissolved cadmium (not shown), copper, and zinc concentrations have all declined since late 1997 (Figs. 14A and B). The 12-period moving average indicates that through November 10, 2004, dissolved cadmium, copper, and zinc were 0.38 µg/L, 14 µg/L, and 94 µg/L, respectively, less than the concentrations during the 1994–1996 baseline period.

Cement Creek Basin

Cement Creek, which carries high concentrations of cadmium, copper, zinc, aluminum, and iron (Fig. 9), is not capable of supporting aquatic life even with remediation. However, remediation in the Cement Creek basin is vital to downstream water quality. The relationship between cadmium, copper, and zinc concentration, stream flow, and seasonality is weak in the Cement Creek basin. Remediation activities, which have been under way in Cement Creek since 1991, accompanied by treatment of the discharge from the American tunnel (site #96, Fig. 13) may explain the weak relationship among concentration, stream flow, and seasonality for the target metals—cadmium, copper, and zinc. The valve on the first American tunnel bulkhead (site #96) was closed in September 1996. The second bulkhead in the American tunnel (site #116) was sealed in August 2002. Treatment of Cement Creek upstream from the American tunnel (site #96) began in the fall of 1996 and continued through the non-runoff periods through 1999. Treatment during this period resulted in very significant reductions in mean metal concentrations (Fig. 15). The permit for the American tunnel was transferred to the Gold King mine in December 2002. Gold King continued to treat the remaining discharge from the American tunnel through May 2003. The mine pool in the Sunnyside mine reached equilibrium by November 2000; however, this was preceded by a large increase in the volume of flow from the Mogul mine (site #31) in 1999 causing a bulkhead to be placed in that portal in 2003. The Sunnyside mine pool related mitigations were completed in 2001. Remediation has occurred at 15 sites in the basin (Fig. 13; Table 5). Projects in the Cement Creek basin since October 1996 include hydrologic runon/runoff controls at Gold King mine (site #111), Joe and Johns mine (site #87), Lark mine (site #86), and Mayday mine (site #181). Settling ponds and runon/runoff controls were constructed in 1998 at the Mammoth mine (site #148). Runon/runoff controls and complete removal of mine wastes at the Hercules and Galena Queen mines (sites #82 and #83) were completed in 2001. A passive treatment system consisted of aerobic limestone drains, and settling ponds were implemented at the Elk tunnel (site #147) in 2003.

Dissolved cadmium (not shown) decreased significantly following initial treatment of Cement Creek. A series of high cadmium concentrations from July 1999 through November 1999 caused the 12-period moving average to rise above the baseline condition. This was repeated in 2000 but to a lesser degree. The near baseline condition was reached in 2002, but a steady trend upward beginning in 2003 found the average cadmium concentration to be 1.7 µg/L higher through November 10, 2004, than during the baseline period.

The dissolved copper concentration (Fig. 15A) has fluctuated around the baseline condition except for short periods in 1997 and 1999. Exceptionally high copper concentrations were measured from August 1999 through November 1999. Dissolved copper exceeded three standard deviations on four out of seven sampled dates in that time period.

The dissolved zinc concentration (Fig. 15B) decreased over 250 µg/L following closure of the American tunnel and treatment of upper Cement Creek through late 1998. The zinc concentration increased through 1999 reaching a maximum of nearly 300 µg/L higher than the baseline concentration by early 2000. The zinc concentration then declined to baseline concentration through early 2003. The zinc concentration was 341 µg/L higher than the baseline concentration in November 2004.

The concentration of dissolved manganese (Fig. 15C) was reduced over 500 µg/L, on the average, at C48 from the time treatment began until the summer of 2002. By November 10, 2004, manganese concentrations had returned to the baseline concentration.

The 12-period moving average total recoverable aluminum (not shown) concentration dropped more than 900 µg/L below the baseline condition for the first eight months following the closing of the American tunnel and initiation of treatment of Cement Creek. The average concentration remained less than

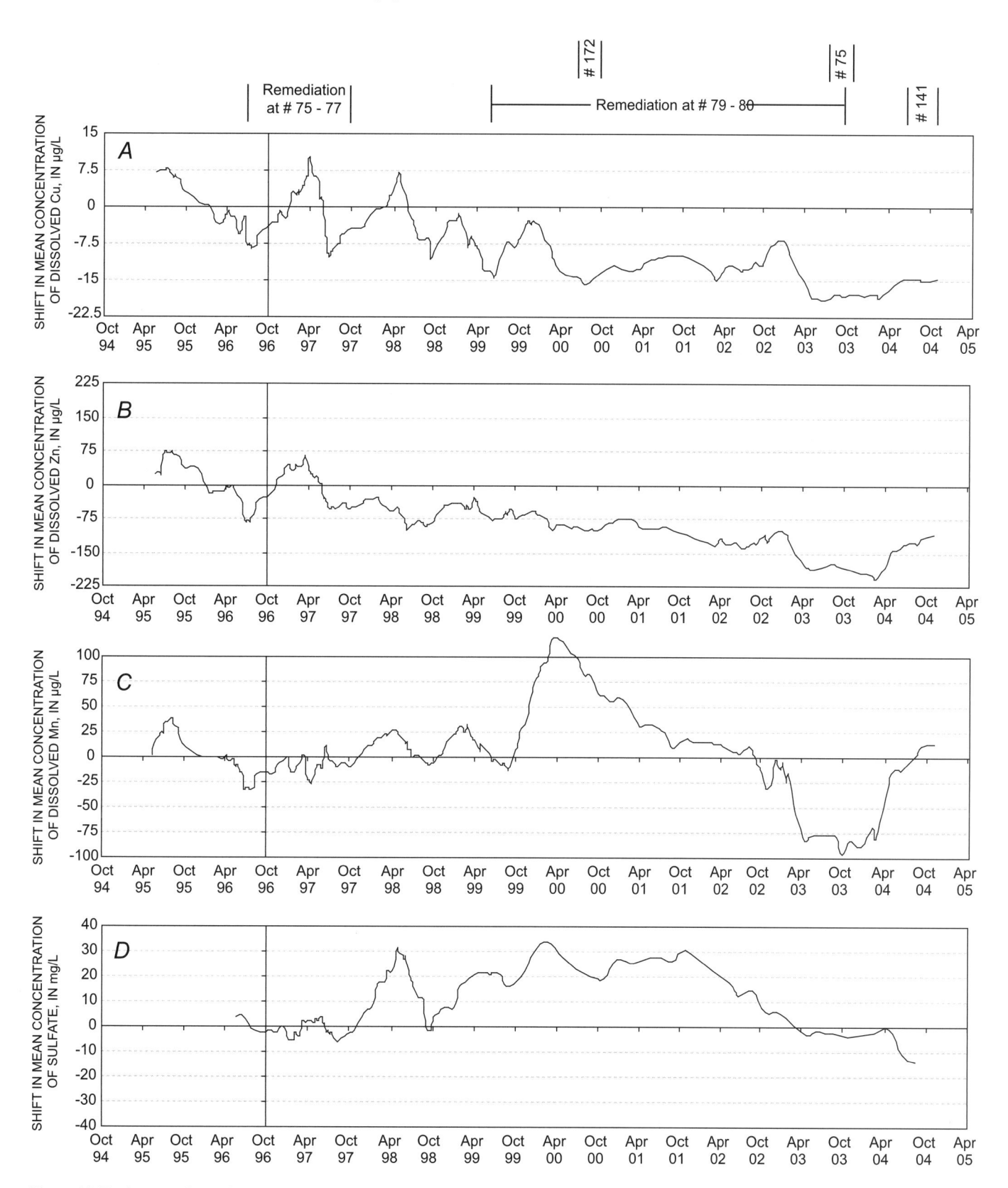

Figure 14. Twelve-sample moving average concentration determined from monitoring data: dissolved copper concentrations (A); dissolved zinc concentrations (B); dissolved manganese concentrations (C); and sulfate concentrations (D) measured at the gauge on Mineral Creek (M34, Fig. 13), 1994–2004. Baseline concentrations determined from monitoring data collected from 1994 to 1996.

76   Church et al.

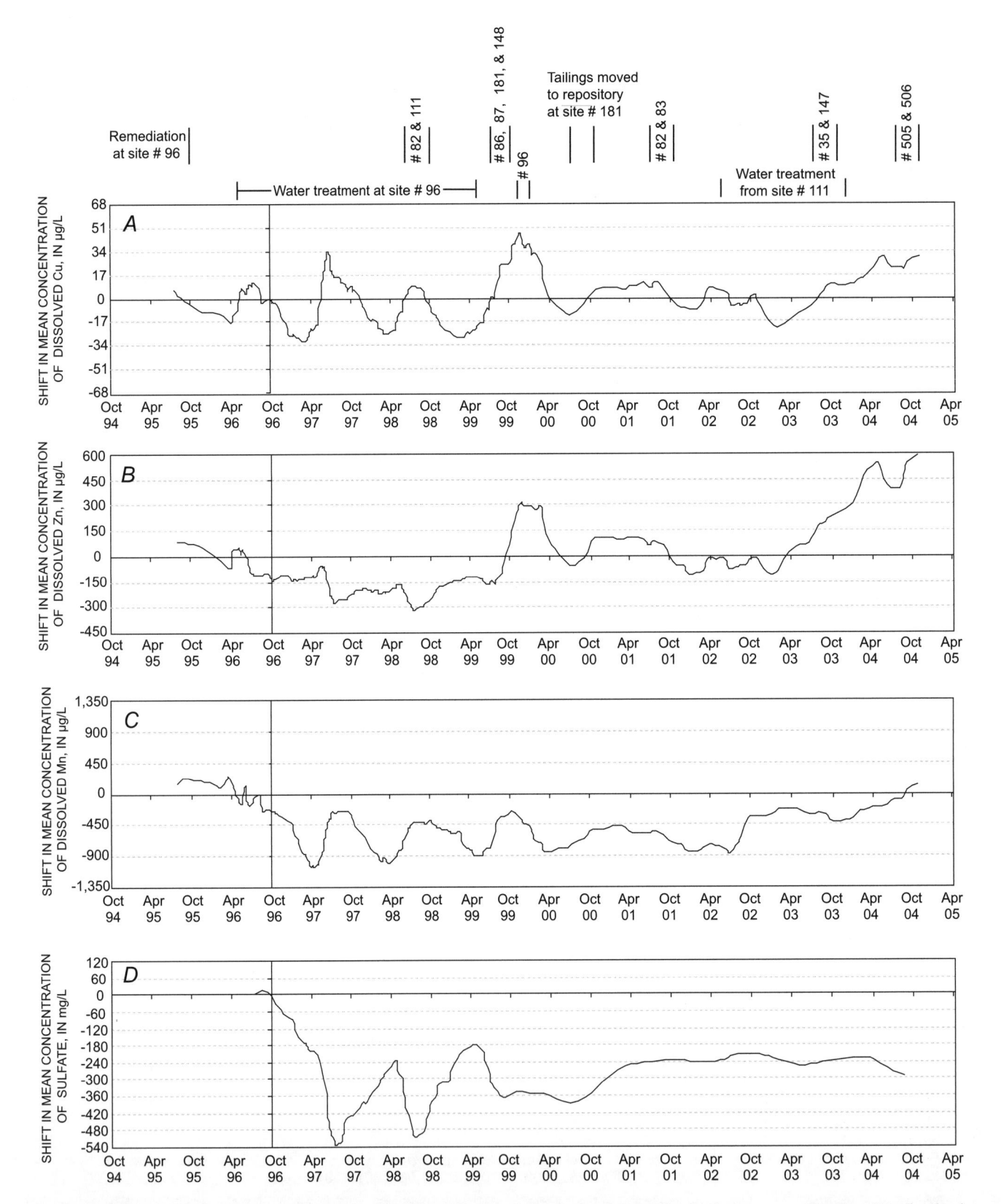

Figure 15. Twelve-sample moving average concentration determined from monitoring data for dissolved copper concentrations (A); dissolved zinc concentrations (B); dissolved manganese concentrations (C); and sulfate concentrations (D) measured at the gauge on Cement Creek (C48, Fig. 13), 1994–2004. Baseline concentrations determined from monitoring data collected from 1994 to 1996.

the baseline period through the spring of 1999. By the fall of 1999 the average concentration peaked at over 1000 µg/L higher than the baseline period. Total recoverable aluminum has fluctuated around the baseline concentration since it reached a peak in 1999 following cessation of treatment at Gladstone (site #96).

Total recoverable iron (not shown) has exceeded the baseline condition since the spring of 1999. Peak iron concentrations were reached in the spring of 2001. Although the 12-period moving average iron concentration through November 10, 2004, was nearly 1900 µg/L higher than the baseline period, total recoverable iron concentration has been decreasing toward the baseline condition since the peak was reached in 1999.

The concentration of sulfate (Fig. 15D) dropped substantially with the closing of the valve on the bulkhead in the American tunnel and remained at −240 mg/L as of November 10, 2004.

Upper Animas River Basin

The upper Animas River supports several species of trout; however, tracer studies have shown that the upper Animas River basin is the major loader for manganese and zinc (Figs. 8 and 9), both of which are soluble at the pH of water (Fig. 7). Remedial activities initiated by Sunnyside Gold Corp. began in 1991 (Table 5). These include closure of mill-tailings sites at their Mayflower Mill (sites 507–509; Fig. 13), removal of mill wastes from the Gladstone area (site #95, Fig. 13), and repair of the collapse of Lake Emma into the Sunnyside mine, which occurred in 1978 (site #116, Fig. 13; Jones, 2007). Major long-term remediation work has continued at the Mayflower Mill tailings sites 507–510 (Table 5; Fig. 13) to reduce the amount of metal added to the ground-water table by leaching from the mill tailings repositories. In addition, remediation work has been conducted at 11 additional sites (Fig. 13; Table 5) since October 1996. They include the removal of 112,000 cubic yards of tailings from the floodplain around Eureka (site #164), which was completed in 1997, and removal of 84,000 cubic yards of tailings from the floodplain at Howardsville (site #234), also completed in 1997. Mine waste or mill tailings were removed from Boulder Creek (released from site #509), Gold Prince mine (site #49), and Lakawanna Mill (sites #181 and #286). Passive treatment of adit discharge from the Forest Queen (site #195) was completed in 1999 (Table 5). Bulkhead seals were placed in the Ransom mine (site #161), the Terry tunnel (site #120), and the Gold Prince mine (site #49). Hydrologic runon/runoff controls were implemented at the Silver Wing mine (site #125).

Dissolved copper (Fig. 16A) has generally fluctuated around the baseline condition except for summer periods in 1997 and 1998 when the concentration was lower than expected.

The peaks and valleys of dissolved zinc at A68 (Fig. 16B) follow the same general pattern as those of cadmium and manganese; however, for the most part it has stayed relatively close to the baseline condition. The largest shift in the 12-period mean was nearly +300 µg/L following implementation of the consent decree (1997). The second year following the consent decree, a high of around +200 µg/L was reached. The concentration fluctuated around the mean through 2001 and then declined dramatically during the 2002 drought. This was followed by an increase to over +150 µg/L in late 2002 and early 2003. Since then the concentration has decreased to the baseline condition.

Dissolved manganese (Fig. 16C) has consistently been higher than the baseline condition since the start of remediation activities in 1996. Brief declines were noted in the summers of 1997, 1998, and 2002; however, for most of the post-consent-decree period concentrations have been more than 800 µg/L higher than baseline. The concentration appears to be in a steady decline since the spring of 2003, but through November 10, 2004, it remained over 600 µg/L above the baseline.

The declines in dissolved cadmium, copper, and zinc following the 2002 low runoff followed by steep increases in 2003 suggest that several years of weathered material accumulated and was washed off during the more normal runoff in 2003. It does not appear that either cadmium or manganese have reached an equilibrium condition following the remediation activities that have been accomplished upstream from the gauge at A68.

Total recoverable aluminum and iron are well within water-quality standards and were not analyzed at A68. Dissolved cadmium (not shown) has consistently been higher than the baseline condition except during the summers of 1998 and 2002. The highest concentrations were reached in 1997 following implementation of the consent decree, when significant remediation related disturbance occurred, and in the spring of 2003 following the 2002 drought. Since the spring of 2003, the cadmium concentration appears to be declining toward the baseline; however, cadmium was still ~0.7 µg/L higher than baseline through November 10, 2004.

Animas River Watershed

The gauge at A72 (Fig. 13) is the compliance point established by the State of Colorado for the Sunnyside Gold, Inc. consent decree. Four segments in the watershed were identified in Colorado's water-quality impaired 1998 303(d) list. The Colorado Water Quality Control Commission established water-quality goals for six parameters (Al, Cd, Cu, Fe, Mn, and Zn) and adopted total maximum daily loads in 2002 aimed at attaining aquatic life use classifications for the segments where they adopted those classifications. Water quality at A72 integrates the effects of remediation upstream from A68, C48, and M34. Total recoverable aluminum and iron were not monitored at A72 during 2000, 2001, and most of 2002; however, the data collected in 2003 and 2004 suggest that the concentration of these constituents has not changed from the baseline condition.

The dissolved cadmium concentration (not shown) generally fluctuated around the baseline period except immediately following the start of the remediation period and during the 2002 drought. By November 2004, the cadmium concentration was at the approximate baseline concentration (1.9 ± 0.64 µg/L).

The dissolved copper concentration (Fig. 17A) was generally lower than the baseline concentration in the late 1990s and

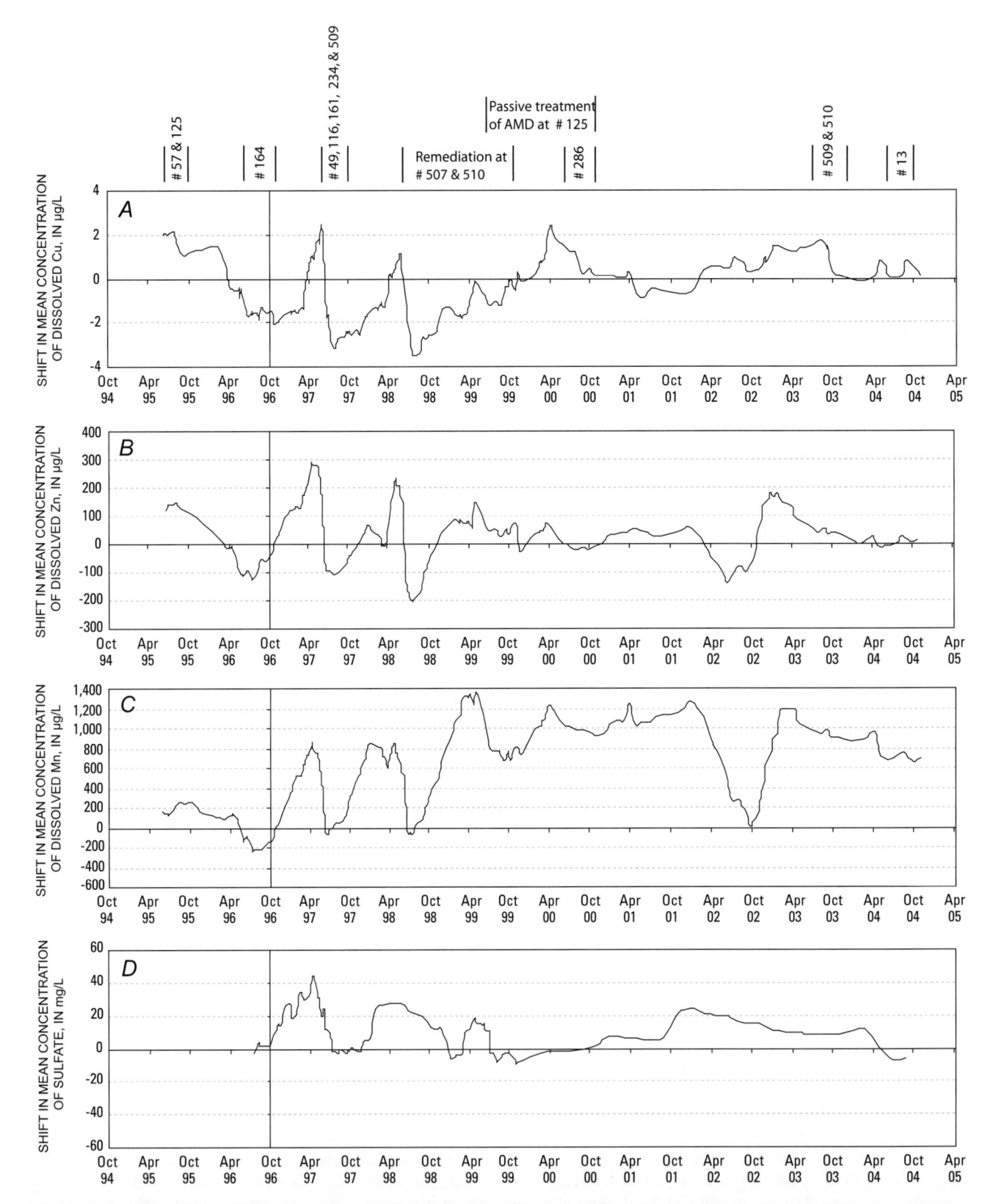

Figure 16. Twelve-sample moving average concentration determined from monitoring data for dissolved copper concentrations (A); dissolved zinc concentrations (B); dissolved manganese concentrations (C); and sulfate concentrations (D), measured at the gauge on the upper Animas River upstream from the confluence with Cement Creek (A68, Fig. 13), 1994–2004. Baseline concentrations determined from monitoring data collected from 1994 to 1996.

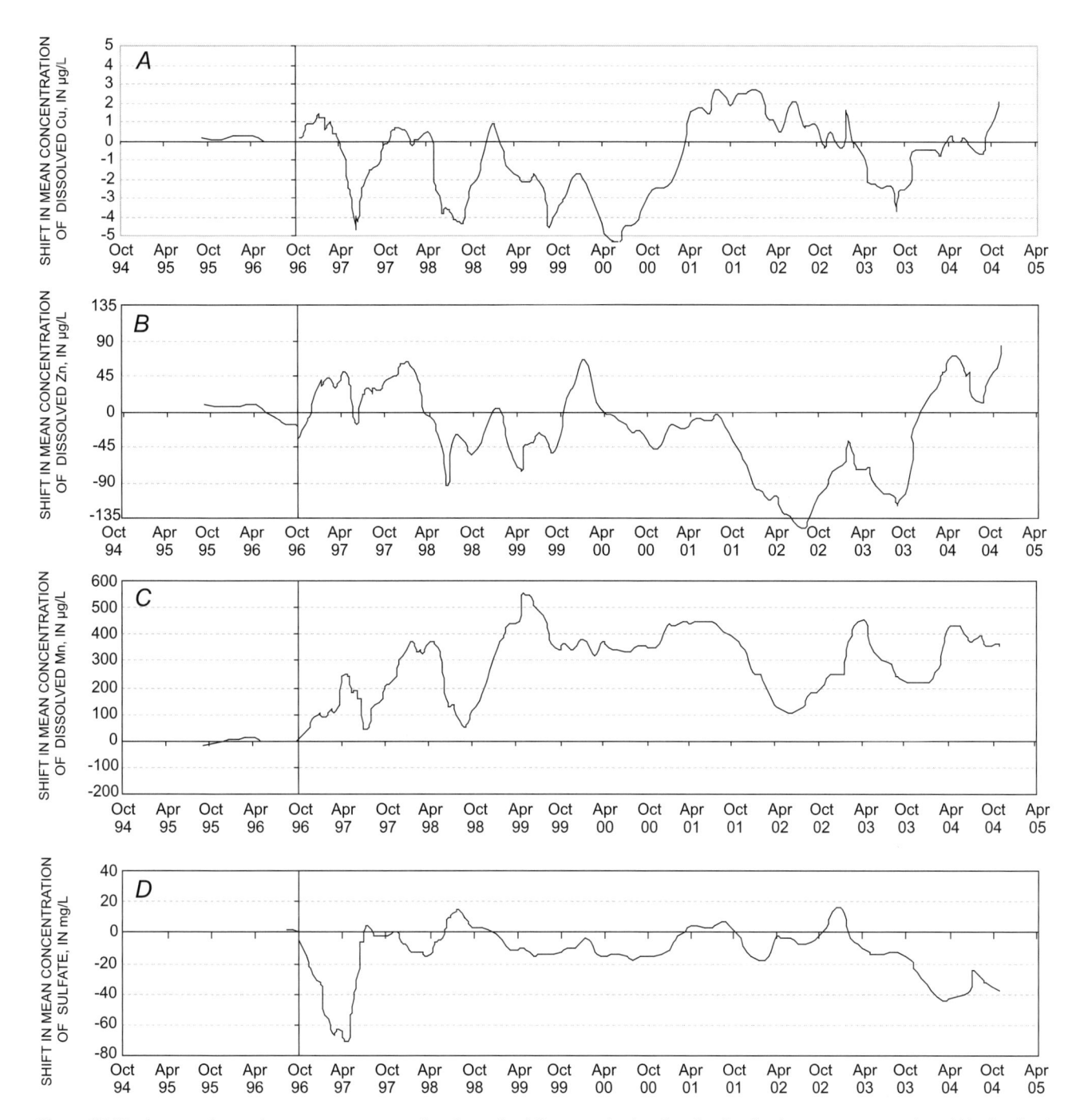

Figure 17. Twelve-sample moving average concentration determined from monitoring data for dissolved copper concentrations (A); dissolved zinc concentrations (B); dissolved manganese concentrations (C); and sulfate concentrations (D) measured at the gauge on the Animas River upstream from the confluence with Mineral Creek (A72, Fig. 13), 1994–2004. A72 is the compliance point for water-quality standards established by the Colorado Water Quality Control Commission. Baseline concentrations determined from monitoring data collected from 1991 to 1996.

during the 2002 drought, but it has increased slightly with time to +2 µg/L in November 2004.

There has been a large upward shift in the zinc concentration at A72 (Fig. 17B). The zinc concentration spiked in the late summer of 1999 and has gone from showing improvement of −90 µg/L in 2002/2003 to nearly +90 µg/L during 2004. The most likely source of the increase in zinc concentration is the increased zinc load from Cement Creek. Timing of the increase in zinc concentration corresponds well with the increase in zinc concentration in Cement Creek (Fig. 15B). Moreover, there has been no measurable change in zinc concentration at A68 (Fig. 16B), and the zinc concentration has been significantly reduced in Mineral Creek (M34; Fig. 14B).

Manganese concentration continues to be higher than those observed before remediation activities began (Fig. 17C), but recently the manganese concentration has declined in the direction of previous baseline concentration. The increase in manganese concentration at A72 is most likely the result of remediation and removal of the high-manganese tailings at site #164 and at the Mayflower Mill tailings impoundments (sites #507–510, Fig. 13) upstream from A68. Changes in the manganese concentration at C48 (Fig. 15C) and M34 (Fig. 14C) have been small.

Sulfate concentration (Fig. 17D) at A72 showed only minor changes relative to the 1991–1996 baseline concentration through early 2003. However, the sulfate concentration has decreased by about 30 mg/L in 2004.

CONCLUSIONS

The monitoring results indicate that remediation using the watershed approach, that is, focusing remediation on mine sites where it will have the largest effect on water quality, is the most cost-effective approach to improvement of water quality in a watershed affected by historical mining. Although half of the very-high priority sites on federal lands identified by Nash and Fey (2007) and by Kimball et al. (2007) have been addressed, and 23% of sites identified by both the ARSG and the USGS have been remediated through various engineering options, only small long-term gains in water quality have been demonstrated to date by the water-quality data analysis. Individual sites have responded to the remediation work done, as evidenced by the improved water chemistry at individual sites (W. Simon, pers. comm., 2003). Remediation work conducted by the federal agencies, the ARSG, and Sunnyside Gold, Inc. has addressed water-quality issues caused by some of the largest sites in the watershed. Steady improvement in water quality should continue as the effect of disturbances caused by the remediation activities' decline. Some recovery of the aquatic life in designated stream reaches has begun to occur. Recovery of the watershed to premining conditions is not an attainable goal because not all anthropogenic sources can be treated in a cost-effective manner. A substantial amount of the metal load in the surface drainages is derived simply from weathering of hydrothermally altered rock that has not been disturbed by mining (i.e., simply from weathering). As shown in the discussion of the tracer results (Fig. 9), 13–29% of the copper and zinc loads comes from these dispersed sources. Given that so many point sources have not been remediated, documenting the recovery curve for aquatic life in surface streams remains elusive while public funding for remediation activities has become more difficult to obtain due to increased demand. Continued remediation efforts should result in progressively smaller gains in water quality. Continued remediation work by the ARSG is necessary to achieve the water-quality standards set by the Colorado Water Quality Control Commission at site A72.

REFERENCES CITED

Anderson, C.R., 2007, Effects of mining on benthic macroinvertebrate communities and recommendations for monitoring, in Church, S.E., von Guerard, P., and Finger, S.E., eds., Integrated investigations of environmental effects of historical mining in the Animas River watershed, San Juan County, Colorado: U.S. Geological Survey Professional Paper 1651, p. 851–872.

Bartos, P.J., 1993, Comparison of gold-rich and gold-poor quartz-base metal veins, western San Juan Mountains, Colorado: The Mineral Point area as an example: Society of Economic Geologists Newsletter, no. 15, 11 p.

Bejnar, W., 1957, Lithologic control of ore deposits in the southwestern San Juan Mountains, in Kottlowski, F E., and Baldwin, B., eds., New Mexico Geological Society 8th Annual Field Conference Guidebook, p. 162–173.

Bencala, K.E., and McKnight, D.M., 1987, Identifying in-stream variability: Sampling iron in an acidic stream, in Averett, R.C., and McKnight, D.M., eds., Chemical quality of water and the hydrologic cycle: Chelsea, Michigan, Lewis Publishers, Inc., p. 255–269.

Besser, J.M., and Brumbaugh, W.G., 2007, Status of stream biotic communities in relation to metal exposure, in Church, S.E., von Guerard, P., and Finger, S.E., eds., Integrated investigations of environmental effects of historical mining in the Animas River watershed, San Juan County, Colorado: U.S. Geological Survey Professional Paper 1651, p. 823–835.

Besser, J.M., and Leib, K.J., 2007, Toxicity of metals in water and sediment to aquatic biota, in Church, S.E., von Guerard, P., and Finger, S.E., eds., Integrated investigations of environmental effects of historical mining in the Animas River watershed, San Juan County, Colorado: U.S. Geological Survey Professional Paper 1651, p. 837–849.

Besser, J.M., Finger, S.E., and Church, S.E., 2007, Impacts of historical mining on aquatic ecosystems: An ecological risk assessment, in Church, S.E., von Guerard, P., and Finger, S.E., eds., Integrated investigations of environmental effects of historical mining in the Animas River watershed, San Juan County, Colorado: U.S. Geological Survey Professional Paper 1651, p. 87–106.

Bird, A.G., 1999, Silverton gold (rev. edition), privately published, Lakewood, Colo., 223 p.

Blair, R.W. Jr., Yager, D.B., and Church, S.E., 2002, Surficial geologic maps along the riparian zone of the Animas River and its headwater tributaries, Silverton to Durango, Colorado, with upper Animas River watershed gradient profiles: U.S. Geological Survey Digital Data Series 071.

Bove, D.J., Hon, K., Budding, K.E., Slack, J.F., Snee, L.W., and Yeoman, R.A., 2001, Geochronology and geology of late Oligocene through Miocene volcanism and mineralization in the western San Juan Mountains, Colorado: U.S. Geological Survey Professional Paper 1642, 30 p.

Bove, D.J., Mast, M.A., Dalton, J.B., Wright, W.G., and Yager, D.B., 2007, Major styles of mineralization and hydrothermal alteration and related solid- and aqueous-phase geochemical signatures, in Church, S.E., von Guerard, P., and Finger, S.E., eds., Integrated investigations of environmental effects of historical mining in the Animas River watershed, San Juan County, Colorado: U.S. Geological Survey Professional Paper 1651, p. 161–230.

Burbank, W.S., 1960, Pre-ore propylization, Silverton caldera, Colorado, Journal of Research in the Geological Survey: U.S. Geological Survey Professional Paper 400-B, p. B12–B13.

Burbank, W.S., and Luedke, R.G., 1968, Geology and ore deposits of the San Juan Mountains, Colorado, in Ridge, J.D., ed., Ore deposits of the United States, 1933–1967: American Institute of Mining, Metallurgical, and Petroleum Engineers, New York, p. 714–733.

Butler, P., Owen, R., and Simon, W., 2001, Use Attainability Analysis for the Animas River Watershed, unpublished report prepared for the Colorado Water Quality Control Commission, Animas River Stakeholders Group.

Buxton, H.T., Nimick, D.A., von Guerard, P., Church, S.E., Frazier, A., Gray, J.R., Lipin, B.R., Marsh, S.P., Woodward, D., Kimball, B.T, Finger, S., Ischinger, L., Fordham, J.C., Power, M.S., Bunck, C., and Jones, J.W., 1997, A science-based, watershed strategy to support effective remediation of abandoned mine lands: Proceedings of the Fourth International Conference on Acid Rock Drainage, Vancouver, B.C., May 31–June 6, 1997, p. 1869–1880.

Casadevall, T., and Ohmoto, H., 1977, Sunnyside mine, Eureka mining district, San Juan County, Colorado: Geochemistry of gold and base metal ore deposition in a volcanic environment: Economic Geology and the Bulletin of the Society of Economic Geologists, v. 92, p. 1285–1320.

Church, S.E., Kimball, B.A., Fey, D.L., Ferderer, D.A., Yager, T.J., and Vaughn, R.B., 1997, Source, transport, and partitioning of metals between water, colloids, and bed sediments of the Animas River, Colorado: U.S. Geological Survey Open-File Report 97-151, 135 p.

Church, S.E., Fey, D.L., and Blair, R., 2000, Pre-mining bed sediment geochemical baseline in the Animas River watershed, southwestern Colorado: Proceedings of the Fifth International Conference on Acid Rock Drainage, Society for Mining, Metallurgy, and Exploration, Inc., Littleton, Colo., p. 499–512.

Church, S.E., Fey, D.L., and Unruh, D.M., 2007a, Trace elements and lead isotopes in modern streambed and terrace sediment: Determination of current and premining geochemical baselines, in Church, S.E., von Guerard, P., and Finger, S.E., eds., Integrated investigations of environmental effects of historical mining in the Animas River watershed, San Juan County, Colorado: U.S. Geological Survey Professional Paper 1651, p. 571–642.

Church, S.E., Mast, M.A., Martin, E.P., and Rich, C.L., 2007b, Mine inventory and compilation of mine-adit chemistry data, in Church, S.E., von Guerard, P., and Finger, S.E., eds., Integrated investigations of environmental effects of historical mining in the Animas River watershed, San Juan County, Colorado: U.S. Geological Survey Professional Paper 1651, p. 255–310.

Church, S.E., von Guerard, P., and Finger, S.E., 2007c, eds., Integrated investigations of environmental effects of historical mining in the Animas River watershed, San Juan County, Colorado: U.S. Geological Survey Professional Paper 1651, 1096 p.

Cox, D.P., and Singer, D.A., eds., 1986, Mineral Deposit Models: U.S. Geological Survey Bulletin 1693, 379 p.

Crowfoot, R.M., Payne, W.F., O'Neill, G.B., Boulger, R.W., and Sullivan, J.R., 2005, Water resources data for Colorado, water year 2005: U.S. Geological Survey, URL http://web10capp.er.gov/adr_lookup/adr-co-05/index.jsp.

Dalton, J.B., Bove, D.J., Mladinich, C.S., and Rockwell, B.W., 2007, Imaging spectroscopy applied to the Animas River watershed and Silverton caldera, in Church, S.E., von Guerard, P., and Finger, S.E., eds., Integrated investigations of environmental effects of historical mining in the Animas River watershed, San Juan County, Colorado: U.S. Geological Survey Professional Paper 1651, p. 141–159.

Da Rosa, J.D., and Lyon, J.S., 1997, Golden dreams, poisoned streams: Mineral Policy Center, Washington, D.C., 269 p.

Davis, M.W., and Stewart, R.K., 1990, Gold occurrences of Colorado: Colorado Geological Survey Resources Series 28, 101 p.

Desborough, G.A., and Yager, D.B., 2000, Acid neutralization potential of igneous bedrocks in the Animas River headwaters, San Juan County, Colorado: U.S. Geological Survey Open-File Report 00-165, 14 p.

Durango Democrat, 1902, Nov. 15, Article on approval by the City Council for a new public water supply reservoir on the Florida River.

Fey, D.L., Desborough, G.A., and Church, S.E., 2000, Comparison of two leach procedures applied to metal-mining related wastes in Colorado and Montana and a relative ranking method for mine wastes, in ICARD 2000: Proceedings of the Fifth International Conference on Acid Rock Drainage, Society for Mining, Metallurgy, and Exploration, Inc., Littleton, Colorado, p. 1477–1487.

Fields, S., 2003, The earth's open wounds: Environmental Health Perspectives, v. 111, p. A154–A161.

Greeley, M.N., 1999, National reclamation of abandoned mine lands, http://www.fs.fed.us/geology/amlpaper.htm.

Herron, J., Stover, B., Krabacher, P., and Bucknam, D., 1997, Mineral Creek reclamation feasibility report: Colorado Division of Mines and Geology unpublished report, 65 p.

Herron, J., Stover, B., and Krabacher, P., 1998, Cement Creek reclamation feasibility report, Upper Animas River Basin: Colorado Division of Mines and Geology unpublished report, 139 p., appendicies.

Herron, J., Stover, B., and Krabacher, P., 1999, Reclamation feasibility report, Upper Animas River above Eureka: Colorado Division of Mines and Geology unpublished report, 113 p., appendices.

Herron, J., Stover, B., and Krabacher, P., 2000, Reclamation feasibility report Animas River below Eureka: Colorado Division of Mines and Geology unpublished report, 148 p., appendicies.

Johnson, R.H., 2007, Ground water flow modeling with sensitivity analyses to guide field data collection in a mountain watershed: Ground Water Monitoring and Remediation, v. 27, n. 1, p. 1–9.

Jones, W.R., 2007, History of mining and milling practices and production in San Juan County, Colorado, 1871–1991, in Church, S.E., von Guerard, P., and Finger, S.E., eds., Integrated investigations of environmental effects of historical mining in the Animas River watershed, San Juan County, Colorado: U.S. Geological Survey Professional Paper 1651, p. 39–86.

Jordan, D.S., 1891, Report of explorations in Colorado and Utah during the summer of 1889, with an account of the fishes found in each of the river basins examined: Bulletin of the U.S. Fish Commission, v. 9, p. 1–40.

Kimball, B.A., Runkel, R.L., Walton-Day, K., and Bencala, K.E., 2002, Assessment of metal loads in watersheds affected by acid mine drainage by using tracer injection and synoptic sampling: Cement Creek, Colorado, USA: Applied Geochemistry, v. 17, p. 1183–1207, doi: 10.1016/S0883-2927(02)00017-3.

Kimball, B.A., Runkel, R.L., and Walton-Day, K., 2003, Use of field-scale experiments and reactive solute-transport modeling to evaluate remediation alternatives in streams affected by acid mine drainage, in Jambor, J.L., Blowes, D.W., and Ritchie, A.I.M., eds., Environmental aspects of mine wastes: Mineralogical Association of Canada, Vancouver, British Columbia, p. 261–282.

Kimball, B.A., Walton-Day, K., and Runkel, R.L., 2007, Quantification of metal loading by tracer injection and synoptic sampling, 1996–2000, in Church, S.E., von Guerard, P., and Finger, S.E., eds., Integrated investigations of environmental effects of historical mining in the Animas River watershed, San Juan County, Colorado: U.S. Geological Survey Professional Paper 1651, p. 417–495.

Larison, J.R., Likens, G.E., Fitzpatrick, J.W., and Crock, J.G., 2000, Cadmium toxicity among wildlife in the Colorado Rocky Mountains: Nature, v. 406, p. 181–183, doi: 10.1038/35018068.

Leib, K.J., Mast, M.A., and Wright, W.G., 2003, Using water-quality profiles to characterize seasonal water quality and loading in the upper Animas River basin, southwestern Colorado: U.S. Geological Survey Water-Resources Investigations Report 02-4230, 43 p.

Leib, K.J., Mast, M.A., and Wright, W.G., 2007, Characterization of mainstem streams using water-quality profiles, in Church, S.E., von Guerard, P., and Finger, S.E., eds., Integrated investigations of environmental effects of historical mining in the Animas River watershed, San Juan County, Colorado: U.S. Geological Survey Professional Paper 1651, p. 543–570.

Lipman, P.W., Steven, T.A., Luedke, R.G., and Burbank, W.S., 1973, Revised volcanic history of the San Juan, Uncompahgre, Silverton, and Lake City calderas in the western San Juan Mountains, Colorado: Journal of Research of the Geological Survey, v. 1, no. 6, p. 627–642.

Lipman, P.W., Fisher, W.S., Mehnert, H.H., Naeser, C.W., Luedke, R.G., and Steven, T.A., 1976, Multiple ages of mid-Tertiary mineralization and alteration in the western San Juan Mountains, Colorado: Economic Geology and the Bulletin of the Society of Economic Geologists, v. 71, p. 571–588.

Lovekin, J., Satre, M., Sheriff, W., and Sares, M., 1997, USFS-abandoned mine land inventory project final summary report for San Juan Forest, Columbine Ranger District: Colorado Geological Survey, unpublished report, 67 p.

Mast, M.A., Evans, J.B., Leib, K.J., and Wright, W.G., 2000a, Hydrologic and water-quality data at selected sites in the upper Animas River watershed, Southwestern Colorado, 1997–99: U.S. Geological Survey Open-File Report 00-53, 30 p.

Mast, M.A., Verplanck, P.L., Yager, D.B., Wright, W.G., and Bove, D.J., 2000b, Natural sources of metals to surface waters in the upper Animas River watershed: Proceedings of the Fifth International Conference on Acid Rock Drainage, Society for Mining, Metallurgy, and Exploration, Inc., Littleton, Colo., p. 513–522.

Mast, M.A., Verplanck, P.L., Wright, W.G., and Bove, D.J., 2007, Characterization of background water quality, in Church, S.E., von Guerard, P., and Finger, S.E., eds., Integrated investigations of environmental effects of historical mining in the Animas River watershed, San Juan County, Colorado: U.S. Geological Survey Professional Paper 1651, p. 347–386.

McCusker, R.T., 1982, Mount Moly progress report, 1979–1980, drill holes 1–6: Amax Exploration, Inc., unpublished report, 24 p.

McDougal, R.R., McCafferty, A.E., Smith, B.D., and Yager, D.B., 2007, Topographic, geophysical, and mineralogical characterization of geologic structures using a statistical modeling approach, in Church, S.E., von Guerard, P., and Finger, S.E., eds., Integrated investigations of environmental effects of historical mining in the Animas River watershed, San Juan County, Colorado: U.S. Geological Survey Professional Paper 1651, p. 643–687.

Nash, J.T., and Fey, D.L., 2007, Mine adits, mine-waste dumps, and mill tailings as sources of contamination, in Church, S.E., von Guerard, P., and Finger, S.E., eds., Integrated investigations of environmental effects of historical mining in the Animas River watershed, San Juan County, Colorado: U.S. Geological Survey Professional Paper 1651, p. 311–345.

Natural Resources Conservation Service (NRCS), 2007, Prism Mean Annual Map: U. S. Department of Agriculture (accessed July, 2005) http://www.wcc.nrcs.usda.gov/climate/prism.html.

Nimick, D.A., Church, S.E., and Finger, S.E., eds., 2004, Integrated investigation of environmental effects of historical mining in the Basin and Boulder mining districts, Boulder River watershed, Jefferson County, Montana: U.S. Geological Survey Professional Paper 1652, 523 p., one CD-ROM.

Nordstrom, D.K., Wright, W.G., Mast, M.A., Bove, D.J., and Rye, R.O., 2007, Aqueous-sulfate stable isotopes: A study of mining-affected and undisturbed acidic drainage, in Church, S.E., von Guerard, P., and Finger, S.E., eds., Integrated investigations of environmental effects of historical mining in the Animas River watershed, San Juan County, Colorado: U.S. Geological Survey Professional Paper 1651, p. 387–416.

Ransome, F.L., 1901, A report on the economic geology of the Silverton quadrangle, Colorado: U.S. Geological Survey Bulletin 182, 265 p.

Ringrose, C.R., 1982, Geology, geochemistry, and stable isotope studies of a porphyry-style hydrothermal system, west Silverton district, San Juan Mountains, Colorado: University of Aberdeen, Ph.D. dissertation, 257 p., 22 tables., 48 figs., 19 pl.

Runkel, R.L., and Kimball, B.A., 2002, Evaluating remedial alternatives for an acid mine drainage stream: Application of a reactive transport model: Environmental Science & Technology, v. 36, p. 1093–1101, doi: 10.1021/es0109794, doi: 10.1021/es0109794.

Silverton Standard, 1903, Various articles about opening a fish hatchery on the upper Animas River to provide fresh fish for local merchants.

Sloan, R.E., and Skowronski, C.A., 1975, The Rainbow Route: An illustrated history of the Silverton Railroad, the Northern Silverton Railroad, and the Silverton, Gladstone, & Northerly Railroad, Sundance Publications Limited, Denver, Colorado, 416 p.

Smith, N.F., 1976, Aquatic inventory: Animas-La Plata project: Colorado Division of Wildlife, Durango Colorado, unpublished report.

Smith, K.S., Ramsey, C.A., and Hageman, P.L., 2000, Sampling strategy for the rapid screening of mine-waste dumps on abandoned mine lands: Proceedings of the Fifth International Conference on Acid Rock Drainage, Society for Mining, Metallurgy, and Exploration, Inc., Littleton, Colo., p. 1453–1461.

Smith, B.D., McDougal, R.R., Deszcz-Pan, M., and Yager, D.B., 2007, Helicopter electromagnetic and magnetic surveys, in Church, S.E., von Guerard, P., and Finger, S.E., eds., Integrated investigations of environmental effects of historical mining in the Animas River watershed, San Juan County, Colorado: U.S. Geological Survey Professional Paper 1651, p. 231–254.

Stanton, M.R., Yager, D.B., Fey, D.L., and Wright, W.G., 2007, Formation and geochemical significance of iron bog deposits, in Church, S.E., von Guerard, P., and Finger, S.E., eds., Integrated investigations of environmental effects of historical mining in the Animas River watershed, San Juan County, Colorado: U.S. Geological Survey Professional Paper 1651, p. 689–720.

U.S. Dept. of Interior, 1968, Biological studies of selected reaches and tributaries of the Colorado River; Colorado River Basin Water Quality Control Project, U. S. Department of Interior, Water Pollution Control Administration, Technical Advisory and Investigations Branch, unpublished report.

U.S. Geological Survey, 1955, Topographic map of Silverton, Colorado: U.S. Geological Survey, scale: 1:24,000.

U.S. Geological Survey, 2007a, USGS Abandoned Mine Land Initiative: U.S. Geological Survey (accessed July 2005) URL: http://amli.usgs.gov/.

U.S. Geological Survey (USGS), 2007b, National water information system, real-time data fro Colorado: streamflow, U.S. Geological Survey (accessed May 2005) URL: http://waterdata.usgs.gov/co/nwis/current/?type=flow&group_key=huc_cd.

Varnes, D.J., 1963, Geology and ore deposits of the south Silverton mining area, San Juan County, Colorado: U.S. Geological Survey Professional Paper 378-A, 56 p.

Verplanck, P.L., Yager, D.B., Church, S.E., and Stanton, M.R., 2007, Ferricrete classification, morphology, distribution, and ^{14}C age constraints, in Church, S.E., von Guerard, P., and Finger, S.E., eds., Integrated investigations of environmental effects of historical mining in the Animas River watershed, San Juan County, Colorado: U.S. Geological Survey Professional Paper 1651, p. 721–744.

Vincent, K.R., and Elliott, J.G., 2007, Response of the upper Animas River downstream from Eureka to discharge of mill tailings, in Church, S.E., von Guerard, P., and Finger, S.E., eds., Integrated investigations of environmental effects of historical mining in the Animas River watershed, San Juan County, Colorado: U.S. Geological Survey Professional Paper 1651, p. 889–941.

Vincent, K.R., Church, S.E., and Wirt, L., 2007, Geomorphology of Cement Creek and its relation to ferricrete deposits, in Church, S.E., von Guerard, P., and Finger, S.E., eds., Integrated investigations of environmental effects of historical mining in the Animas River watershed, San Juan County, Colorado: U.S. Geological Survey Professional Paper 1651, p. 745–773.

Walton-Day, K., Paschke, S.S., Runkel, R.L., and Kimball, B.A., 2007, Using the OTIS solute-transport model to evaluate remediation scenarios in Cement Creek and the upper Animas River, in Church, S.E., von Guerard, P., and Finger, S.E., eds., Integrated investigations of environmental effects of historical mining in the Animas River watershed, San Juan County, Colorado: U.S. Geological Survey Professional Paper 1651, p. 973–1028.

Wirt, L., Vincent, K.R., Verplanck, P.L., Yager, D.B., Church, S.E., and Fey, D.L., 2007, Geochemical and hydrologic processes controlling formation of ferricrete, in Church, S.E., von Guerard, P., and Finger, S.E., eds., Integrated investigations of environmental effects of historical mining in the Animas River watershed, San Juan County, Colorado: U.S. Geological Survey Professional Paper 1651, p. 823–835.

Wright, W.G., Simon, W., Bove, D.J., Mast, M.A., and Leib, K.J., 2007, Distribution of pH values and dissolved trace-metal concentrations in streams, in Church, S.E., von Guerard, P., and Finger, S.E., eds., Integrated investigations of environmental effects of historical mining in the Animas River watershed, San Juan County, Colorado: U.S. Geological Survey Professional Paper 1651, p. 497–541.

Yager, D.B., and Bove, D.J., 2007, Geologic framework, *in* Church, S.E., von Guerard, P., and Finger, S.E., eds., Integrated investigations of environmental effects of historical mining in the Animas River watershed, San Juan County, Colorado: U.S. Geological Survey Professional Paper 1651, p. 107–140.

Yager, D.B., Mast, M.A., Verplanck, P.L., Bove, D.J., Wright, W.G., and Hageman, P.L., 2000, Natural versus mining-related water-quality degradation to tributaries draining Mount Moly, Silverton, Colo.: Proceedings of the Fifth International Conference on Acid Rock Drainage, Society for Mining, Metallurgy, and Exploration, Inc., Littleton, Colo., p. 535–547.

Yager, D.B., Church, S.E., Verplanck, P.L., and Wirt, L., 2003, Ferricrete, manganocrete, and bog iron occurrences with selected sedge bogs and active iron bogs and springs in the upper Animas River watershed, San Juan County, Colorado: U.S. Geological Survey, Miscellaneous Field Studies Map, MF-2406.

MANUSCRIPT ACCEPTED BY THE SOCIETY 28 NOVEMBER 2006

Mining-impacted sources of metal loading to an alpine stream based on a tracer-injection study, Clear Creek County, Colorado

David L. Fey
Laurie Wirt*
U.S. Geological Survey, Denver Federal Center, Lakewood, Colorado 80225-0046, USA

Dedicated to my wonderful colleague, Laurie Wirt.

ABSTRACT

Base flow water in Leavenworth Creek, a tributary to South Clear Creek in Clear Creek County, Colorado, contains copper and zinc at levels toxic to aquatic life. The metals are predominantly derived from the historical Waldorf mine, and sources include an adit, a mine-waste dump, and mill-tailings deposits. Tracer-injection and water-chemistry synoptic studies were conducted during low-flow conditions to quantify metal loads of mining-impacted inflows and their relative contributions to nearby Leavenworth Creek. During the 2-year investigation, the adit was rerouted in an attempt to reduce metal loading to the stream. During the first year, a lithium-bromide tracer was injected continuously into the stream to achieve steady-state conditions prior to synoptic sampling. Synoptic samples were collected from Leavenworth Creek and from discrete surface inflows. One year later, synoptic sampling was repeated at selected sites to evaluate whether rerouting of the adit flow had improved water quality.

The largest sources of copper and zinc to the creek were from surface inflows from the adit, diffuse inflows from wetland areas, and leaching of dispersed mill tailings. Major instream processes included mixing between mining- and non-mining-impacted waters and the attenuation of iron, aluminum, manganese, and other metals by precipitation or sorption. One year after the rerouting, the Zn and Cu loads in Leavenworth Creek from the adit discharge versus those from leaching of a large volume of dispersed mill tailings were approximately equal to, if not greater than, those before. The mine-waste dump does not appear to be a major source of metal loading. Any improvement that may have resulted from the elimination of adit flow across the dump was masked by higher adit discharge attributed to a larger snow pack. Although many mine remediation activities commonly proceed without prior scientific studies to identify the sources and pathways of metal transport, such strategies do not always translate to water-quality improvements in the stream. Assessment of sources and pathways to gain better understanding of the system is a necessary investment in the outcome of any successful remediation strategy.

Keywords: metal loading, tracer injection, remediation, toxicity, aquatic life

---

*Deceased

Fey, D.L., and Wirt, L., 2007, Mining-impacted sources of metal loading to an alpine stream based on a tracer-injection study, Clear Creek County, Colorado, *in* DeGraff, J.V., ed., Understanding and Responding to Hazardous Substances at Mine Sites in the Western United States: Geological Society of America Reviews in Engineering Geology, v. XVII, p. 85–103, doi: 10.1130/2007.4017(05). For permission to copy, contact editing@geosociety.org.

INTRODUCTION

This paper is the second of two companion papers that apply an integrated suite of techniques to investigate the movement of trace metals through the Waldorf mine waste dump and adjacent wetland area to a receiving stream. The first paper by McDougal and Wirt (this volume) describes the environmental setting and uses geophysical surveys, graphical information systems (GIS) analysis, a sodium chloride (NaCl) tracer study, and discharge measurements to measure infiltration, trace water flow paths, and map physical properties of the waste dump. Figure 1 in McDougal and Wirt (this volume) shows the regional setting of the study area. In this paper, the main focus is the identification of sources and quantification of loads to Leavenworth Creek and its inflows. A lithium bromide (LiBr) tracer-injection study and synoptic water-quality sampling were used to identify mine-related inflows and develop a detailed loading profile in a reach of Leavenworth Creek. In a synoptic study, many water-quality samples are collected from a tracer reach within a short period, providing a "snapshot" of water quality in time.

Remediation without the guidance of scientific studies often yields ineffective or inconclusive results. Before the study, discharge from the adit formed two braided channels over the top of the mine-waste dump. Following the tracer study on July 31, 2002, discharge from the collapsed adit was rerouted through a culvert and around the dump as part of a low-cost effort intended to improve water quality in Leavenworth Creek. One year later, on August 6, 2003, synoptic water-quality sampling of the adit and stream was repeated to determine whether metal loads to the creek had decreased as a result of this activity. The objective of these tracer and synoptic studies was to gain a better understanding of the sources of water and contaminants and metal loading, as well as to identify typical geochemical processes that affect concentrations of metals and loads in the creek.

The Waldorf adit and mine-waste dump were selected for study because the site is typical of many small abandoned mine sites located within the Colorado Mineral Belt and Rocky Mountain region. The tracer reach selected for study was 930 m in length (Fig. 1). Ground-water inflows to the creek typically emerge from wetland areas; consequently, inflows were not always evident, often diffuse, and thus difficult to measure directly.

APPROACH AND METHODS

Tracer-Injection Approach

The tracer-injection method (Kimball, 1997; Bencala et al., 1990; Broshears et al., 1993) was used to quantify the discharge of Leavenworth Creek and to locate and quantify inflows and metal loads from the historical Waldorf mine site. Dilution of a tracer, cou-

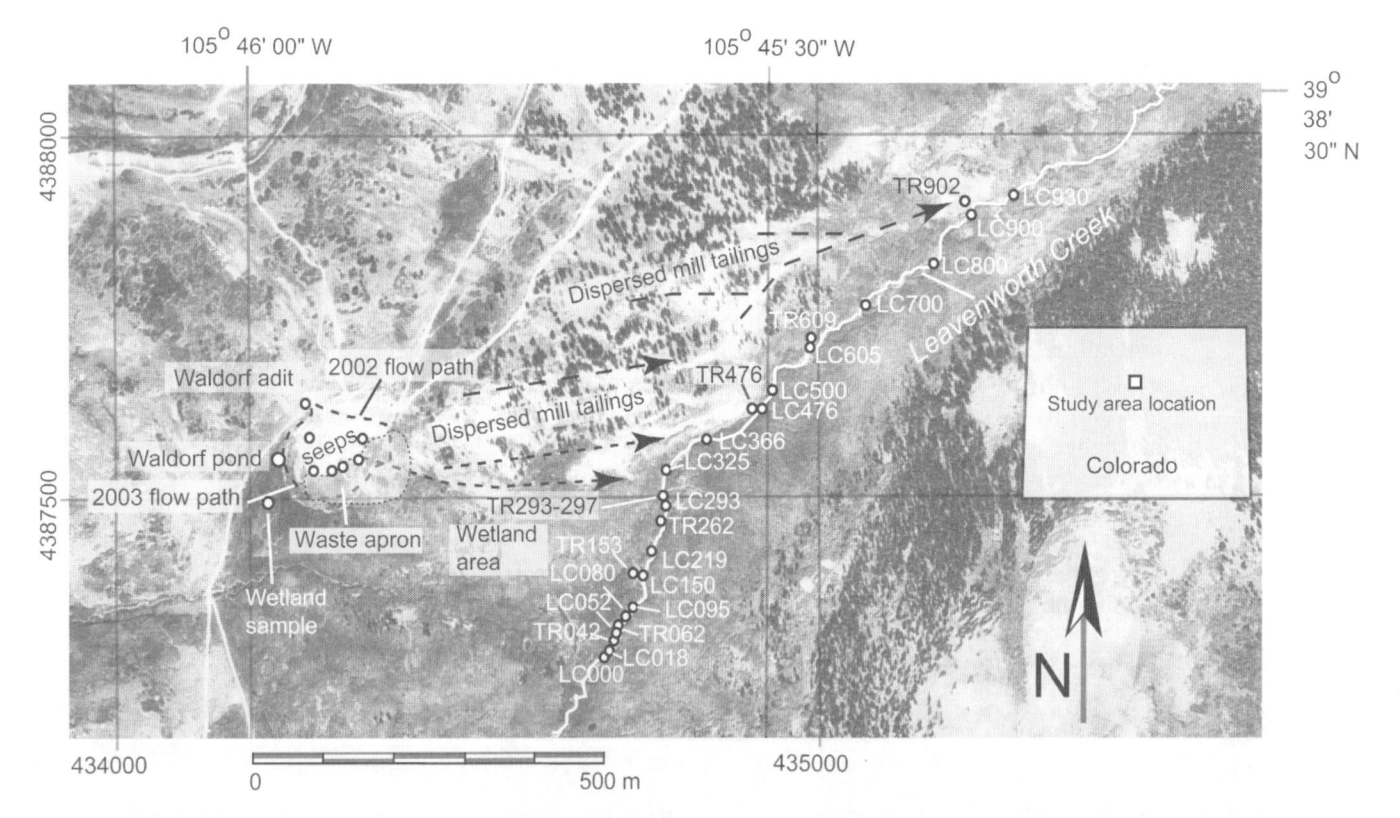

Figure 1. Digital orthophoto quad (DOQ) showing Waldorf mine study area and tracer-study water-quality sampling locations, Leavenworth Creek watershed, Colorado. Dump seep sites WM-101 through WM-106 are shown in more detail in Figure 6 of McDougal and Wirt (this volume).

pled with synoptic sampling, provides an accurate means to measure discharge in less-than-ideal stream cross sections. Current-meter measurements work well where the channel bottom and banks are smooth but tend to be less accurate when the channel is irregular owing to large boulders (Kimball, 1997) or dense aquatic vegetation (Wirt, 2004), or when a large fraction of flow moves through the hyporheic zone (Bencala et al., 1990). Traditional measurements of discharge can thus miss a substantial percentage of the flow (Kimball, 1997; Kimball et al., 2000). In Leavenworth Creek, the marshy banks, dense willow thickets, and steep cobble bottom created irregular cross sections, making it difficult to measure flow by current meter. Another advantage is that numerous synoptic samples can be collected much faster than it takes to complete the same number of current-meter measurements, allowing many discharge determinations to be calculated from samples collected in a short amount of time over a long stream reach.

In the tracer-injection approach, discharge is determined by adding a known quantity of salt tracer, such as LiBr, to a stream. The tracer method assumes that mixing of the tracer is rapid and uniform, that the behavior of the tracer is conservative, that the tracer-study segment has no losing reaches, and that background tracer constituent concentrations from tributaries and inflows are less than the injected tracer concentrations. Discharge is calculated by measuring the amount of dilution that occurs as the tracer moves downstream (Kimball, 1997). This technique is illustrated in Figure 2 and described by the following mass-balance equation:

$$Q_s = \left(\frac{C_{INJ} Q_{INJ}}{C_B - C_A}\right) \quad (1)$$

where:

Q_s = stream discharge, in cubic ft per second;

C_{INJ} = tracer concentration in the injection solution, in mg/L;

Q_{INJ} = rate of tracer injection to the stream, in cubic ft per second;

C_B = tracer concentration downstream from injection point, in mg/L; and

C_A = tracer concentration upstream from injection point, in mg/L.

Stream-flow discharge can be calculated at any site downstream from the injection site using the instream tracer concentration and the concentration and injection rate of the tracer. The background concentration of Li in Leavenworth Creek measured above the injection point was 4 µg/L and generally was ~5 µg/L in the tributaries. In this report, the term *background* is used to indicate sources of dissolved constituents that are unrelated to mining activities.

Lithium bromide was chosen for the tracer because background levels of Li and Br were low and varied little over the stream reach. Lithium bromide is nontoxic and has little effect on the stream environment at low concentrations. All samples

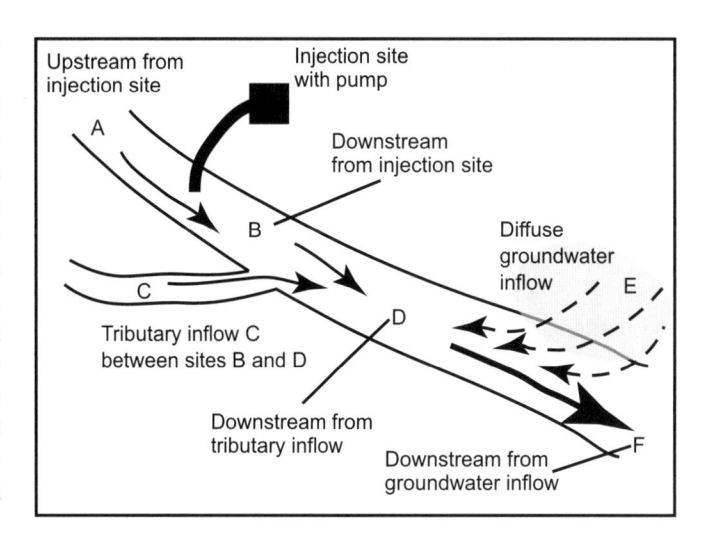

Figure 2. Schematic diagram of tracer-dilution setup on a gaining reach of stream with tributary and diffuse ground-water inflows. After Kimball et al. (2000).

were analyzed for lithium (a cation) and bromide (an anion). Although lithium can be sorbed at the pH values of Leavenworth Creek, changes in lithium concentrations were used to calculate the discharge because of greater confidence in the lithium analyses by ICP-AES (inductively coupled plasma-atomic emission spectroscopy) than the bromide analyses by IC (ion chromatography). With only one exception of 8%, discharge calculations using both elements were within 5% of each other. The discharge of each inflow was determined by the difference in calculated discharge between the mainstem sites immediately downstream and upstream from the inflow.

Calculation of Instream and Inflow Loads

Sampled instream loads are calculated by multiplying the respective discharges by concentrations of specific constituents. The sampled instream load at any point is the mass per unit time (typically g/day or kg/day) of a constituent being carried by the stream past that point. The increase in an instream load may be attributed to sampled inflows or to diffuse, unsampled or subsurface inflows. An instream load can also decrease downstream due to instream processes such as precipitation or sorption.

Sampled inflow loads are similarly calculated by multiplying sampled tributary inflow concentrations by their calculated discharges. When inflows are both visible and diffuse, a comparison of the instream load with the inflow load can help determine whether the sampled inflow load accounts for the increase in a stream segment or whether the increased instream load is in part due to unsampled diffuse inflow (Kimball et al., 2002).

Calculated cumulative instream and cumulative inflow loads help identify both losses due to instream processes and gains due to unsampled inflow reaches. The cumulative instream load is the sum of all the positive instream load increases, that is, all positive increases along the stream between synoptic sampling sites). When a load decreases due to instream losses, the cumulative load remains the same until the next increase. Load increases are, however, net increases because some losses can occur that are obscured by a larger increase (Kimball et al., 2002).

The cumulative inflow load at a given point is simply the sum of the sampled inflow loads at a given point. The cumulative inflow load compared with the cumulative instream load can reveal unsampled inflow loads or areas where diffuse inflow concentrations of particular constituents are higher than the sampled inflow concentrations. Loading profiles for selected constituents are presented in the results and discussion section below.

Field Reconnaissance

In the 2 days preceding the synoptic sampling in 2002, detailed field reconnaissance was conducted in order to select the synoptic water-chemistry sample sites (Fig. 1). The injection site (site LC-000) was up-gradient from all observed inflows draining from the Waldorf mine site. The lower end of the study reach (site LC-930) was downstream from a tributary impacted by mill tailings (site TR-902). Historically, a mill at the Waldorf mine site processed ore from the Wilcox (Waldorf) tunnel and the Santiago mine located 150 m higher in elevation. Ore from that mine complex was transported down to the mill by aerial tram (Lovering, 1935). Field observations indicate that mill tailings were carried away from the mill toward Leavenworth Creek by a ditch that runs 800 m to the northeast of the mine-waste dump (Fig. 1).

The study reach was measured and flagged by stretching a 100-m tape measure along the center of the stream. Latitude and longitude locations were determined using a handheld global positioning system (GPS). Field parameters (pH, specific conductance, temperature, dissolved oxygen) were measured above and below observed inflows and at 100-m intervals. Station numbers were assigned according to the taped distance and type of site. For example, site LC-930 is 930 m downstream from the injection point, and site TR-295 is a tributary inflow that is 295 m downstream from the injection point.

Synoptic sample sites were chosen to bracket known surface inflows in order to obtain discharge values above and below each inflow and at 100-m intervals to detect unobserved inflows along the study reach.

Field Activities and Methods

Field activities and equipment described in this section include: (a) portable flume readings and current-meter measurements; (b) the setup and operation of tracer-injection equipment; (c) the setup and timing of an automatic sampler; (d) synoptic and dump water sampling; and (e) CFC sampling.

Discharge Measurements

The value of discharge (Q_s) at the beginning of the study reach (measured with a portable cutthroat flume) was used in Equation (1) to estimate the concentration of injectate (C_{INJ}) and the flow rate (Q_{INJ}) needed. In 2003, discharge was measured by current-meter measurements instead of by tracer-dilution calculations.

Tracer-Injection Equipment

To prepare the lithium bromide tracer solution, powdered LiBr was mixed with stream water for several hours in a 35-gallon tank before the injection. The injection apparatus consisted of a piston-core pump driven by a battery-powered electric motor. Tracer solution was drawn from the reservoir through plastic tubing to a prepump filter capsule and then through the pump to the stream. The injection rate was measured with a volumetric flask and stopwatch periodically throughout the study to ensure proper pump operation and constant flux. An injection rate of 180 mL per minute was maintained, and the injectate concentration was 10,400 mg/L Li (130,000 mg/L LiBr).

Synoptic Sampling

Synoptic samples were collected by two teams, between 11:00 and 13:00 h on July 31, 2002. The injection pump was shut off at 13:30 h, after the synoptic sampling was completed. Water-quality samples were collected at all 14 mainstem sites and 10 inflow sites using standard USGS methods (Wilde and Radtke, 1998). The width of the stream ranged from ~2 to ~3 m over the reach. A representative sample was collected at each site by immersing a handheld open 1-L plastic bottle in the centroid of flow or at multiple verticals as described by Shelton (1994).

Ground-water samples were collected from small seeps at the base of the dump by shoveling out a small amount of waste material from where the water issued to make a small (10 cm) pool and withdrawing water with a plastic syringe. Discharge measurements of the seeps were made using a graduated cylinder and stopwatch. Samples were also collected from the dump surface flows, and these discharge measurements were made with the cutthroat flume.

Automatic Sampler

Stream samples at site LC-930 were collected by an automatic sampler before, during, and after the LiBr injection to monitor the concentration of the tracer solution in Leavenworth Creek. Samples were collected hourly for 24 hours before the tracer injection, and for 8 hours after. During the next 7 days, samples were collected once every 8 hours, and for the following 23 days, samples were collected daily. These samples were later analyzed at a USGS laboratory in Denver to assess whether there were any diel variations in constituent concentrations, to determine tracer arrival and recovery times, and to verify that the salt tracer was at steady-state conditions during the synoptic sampling. Lithium bromide concentrations at site LC-930 (Fig. 3) were around 1000 µg/L throughout the synoptic sampling, decreased by two orders of

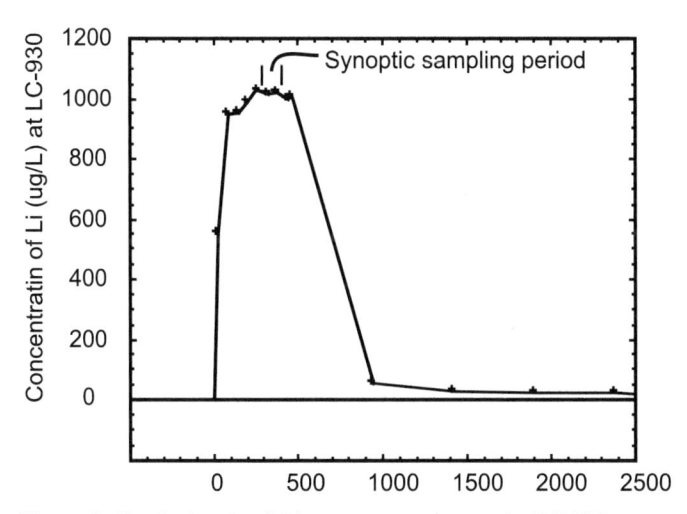

Figure 3. Graph showing lithium concentration at site LC-930 versus time (July 31, 2002, through August 2, 2002).

magnitude (to 50 μg/L) after ~15 hours and returned to background levels (5 μg/L) after 3 days.

Chlorofluorocarbon Sampling

Analyses for chlorofluorocarbons (CFCs) were used to determine the age of ground water discharging from the adit, an adjacent fen, and seeps at the base of the mine-waste dump. Herein, the term *fen* is used to refer to a wetland area that is less acidic than a bog and is predominantly supplied by ground water. CFCs were introduced in the 1930s as alternatives to ammonia and sulfur dioxide for refrigeration and now provide excellent tracers and dating tools for ground water that has recharged within a 50-year timescale (Plummer and Busenberg, 1996). Production of CFC-12 (dichlorodifluoromethane, CF_2Cl_2) began in 1930, followed by CFC-11 (trichlorofluoromethane, $CFCl_3$) in 1936. Samples for CFC dating were collected in triplicate in 125-mL glass bottles with aluminum-foil-lined metal caps. Analyses were conducted at the Department of Geology and Geophysics at the University of Utah in Salt Lake City.

Laboratory Processing and Analytical Methods

Water-chemistry samples collected in 2002 were processed in a Denver USGS laboratory within 24 hours of collection. Experimental centrifuge-tube filters were used to produce 0.45-μm dissolved-fraction samples for cation and sulfate (SO_4) analyses. Inconsistencies in the performance of these filters (as determined after analyses) resulted in our having greater confidence in the total-recoverable (unfiltered) results over the dissolved-fraction results, although the dissolved results, in general, could be used to provide qualitative comparisons with the total-recoverable results. Samples collected in 2003 were processed with conventional syringe-mounted filters and displayed no contamination. Concentrations of major and trace elements were determined by inductively coupled plasma-atomic emission spectroscopy (ICP-AES) (Briggs, 2002) and by inductively coupled plasma-mass spectrometry (ICP-MS) (Lamothe et al., 1999). Alkalinity was measured by incremental titration with 1.6 N H_2SO_4 (Wilde and Radtke, 1998). The quality of the laboratory analyses was assessed through analysis of laboratory blanks, sample duplicates, and USGS standard reference water samples (Long and Farrar, 1995).

RESULTS AND DISCUSSION

Gains and Losses as an Artifact of Discharge Method

Gains and losses within the study reach in different years appear related to the discharge measurement method. In 2002, discharge for synoptic instream samples was calculated using the tracer-injection method (Fig. 4). In 2003, discharge at synoptic sites was measured manually with a current meter. Spatial patterns in discharge along the tracer reach from sites LC-000 to LC-930 differed in 2002 versus 2003. In 2002, discharge more than doubled over the reach from 14 to 34.2 L/s, whereas in 2003 discharge appeared to increase only slightly from 102 to 118 L/s over the same reach. In 2003, surface water discharge from the middle to

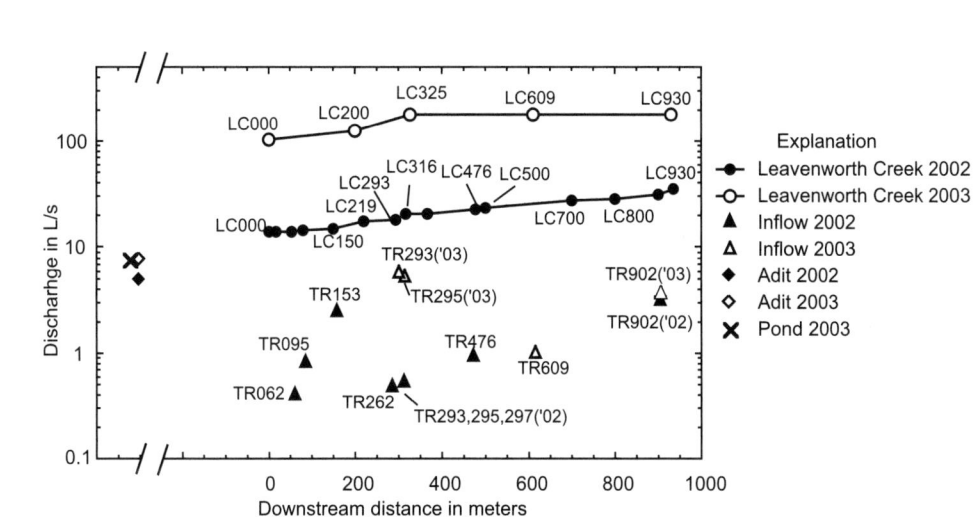

Figure 4. Graph showing calculated discharge from tracer-dilution study versus distance along tracer reach on July 31, 2002, and discharge from current meter measurements on August 6, 2003. The discharge values for the losing reach LC-325 to LC-930 were set equal to the discharge at LC-325 in 2003. The adit/pond discharge points are at an arbitrary distance upstream from LC-000, as indicated by the break in the *x*-axis.

the end of the reach (between sites LC-325 and LC-930) decreased from 179 L/s to 105 L/s.

The apparent result of a losing reach in the second year is attributed to having used different methods to measure discharge. The tracer method includes hyporheic flow, whereas the current-meter method does not (Kimball, 1997). Spatial changes in discharge over the reach probably were proportionate (followed a similar pattern) in both years. The increasing discharge indicated by the tracer method was real because the tracer solution continued to be diluted by both surface and hyporheic flow. The losing discharge trend (40% between LC-325 and LC-630 in 2003) indicated by the current-meter measurements was also real because hyporheic flow beneath the cobble-bottom stream was substantial and not measured in the lower part of the reach. However, there is an inherently greater error associated with manual discharge measurements, so the manual measurements in 2003 were repeated 1 week later. Those results still showed a 30% loss of discharge between the same two sites, so the losing reach was confirmed. Most of the hyporheic flow is thought to return as surface flow near the end of the tracer reach. For loading calculations, the discharge value for each site in the losing reach was set equal to the discharge at the beginning of the losing reach (site LC-325); consequently, the calculated loads for the losing reach in 2003 do not quantitatively reflect the total loads but rather a minimum. The most important finding was that discharge and, consequently, the calculated loads of copper and zinc (discussed below) were substantially higher in 2003.

Annual trends in discharge in the study reach are presumably similar to those for daily discharge at the U.S. Geological Survey stream-flow gauging station near the mouth of Leavenworth Creek near Georgetown (station number 06714800). Base flow at this gauge was 76 L/s on July 31, 2002, compared with 370 L/s on August 6, 2003 (Fig. 5), ~5 times higher in 2003. Discharge at site LC-000 in Leavenworth Creek was 14 L/s in 2002 versus 102 L/s in 2003, about seven times higher in 2003 (Fig. 5). The difference in base-flow discharge between the two years is attributed to differences in precipitation and snowpack of the previous winter. Correspondingly, the adit discharge also increased from 2002 (4.8 L/s) to 2003 (7.5 L/s).

Water Chemistry of the Mine-Waste Dump Seeps Before and After Remediation

Water-quality samples were collected from seeps at the base of the dump in 2002 and 2003. In addition in 2003, CFC samples were collected at selected sites to determine the age of groundwater recharge for the various stream inflows. Field parameters and analytical results are given in Tables 1 and 2. The recharge dates for the seep samples, adit discharge, the wetland, and Leavenworth Creek at LC-000 are in Table 1. In 2002, water issued from several discrete seeps at the base of the waste dump (sites WM102, WM103 (not sampled), WM104, and WM105). Sites WM102 and WM105 were dry in 2003, but samples were collected from sites WM103 and WM104.

A sodium chloride tracer was injected into the adit discharge in 2002 (McDougal and Wirt, this volume), which demonstrated that the source of seeps at sites WM104 and WM105 was adit water flowing across the dump. The surface flow had a chloride concentration of ~40 mg/L, whereas site WM104 had a concentration of ~10 mg/L and site WM105 had a concentration of ~30 mg/L. The redirection of the adit discharge around the dump in 2002 dried up the large seep at site WM105. However, a small amount of water was still issuing from sites WM103 and WM104 in 2003 (Table 1).

The discharge of site WM104 in 2003 was about half that of 2002 (0.14 L/s versus 0.32 L/s), which reflected a reduction in the amount of infiltration from adit water. The concentration of zinc at site WM104 (11,000 µg/L) was the same for both years, and so the zinc load decreased from 310 g/day to 130 g/day. However, the copper concentration increased from 106 to 1700 µg/L. Additionally, the sum of the 2003 loads at sites WM103 and WM104 for

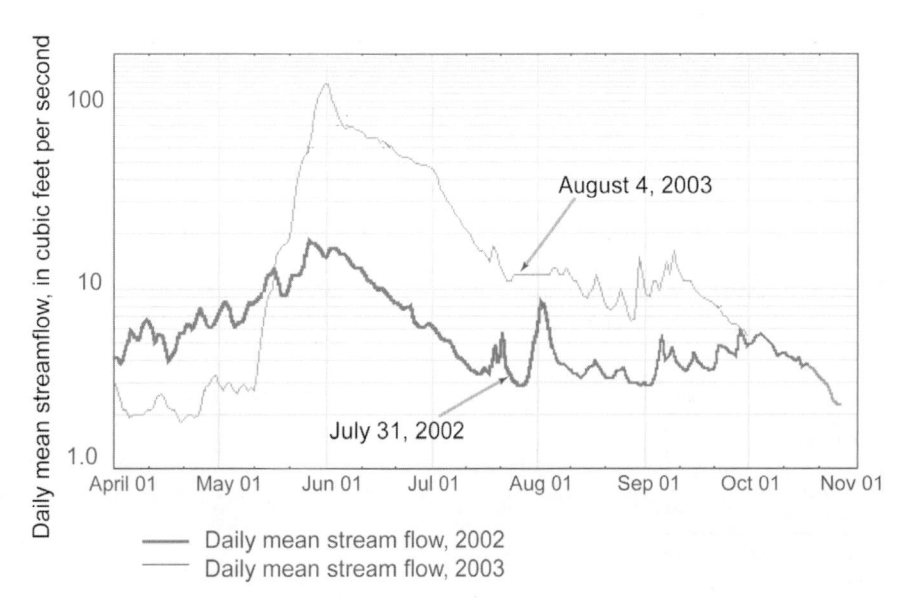

Figure 5. Discharge at U.S. Geological Survey gauging station (number 06714800) at the mouth of Leavenworth Creek near Georgetown, Colorado, for water, April through October of 2002 and 2003.

TABLE 1. FIELD NUMBERS, DISTANCE FROM TRACER INJECTION SITE, SAMPLE LOCALITIES, SITE DESCRIPTIONS, DISCHARGE, SPECIFIC CONDUCTANCE, pH, TEMPERATURE, DISSOLVED OXYGEN, AND SELECTED RECHARGE DATES, WALDORF MINE AND LEAVENWORTH CREEK WATER SAMPLES, 2002 AND 2003

Field Number	Distance in m	Latitude	Longitude	Additional Sample Attributes or Comments	Collection Date	Discharge L/s	Specific Conductance µS/cm	pH	Temperature °C	Dissolved Oxygen mg/L	Year of recharge*
2002 synoptic samples											
WM-104 (2003)											
2002-LC-000	0	39.63458	105.76170	Leavenworth Creek, just upstream of injection site	7/31/02	13.8	158	7.10	15.4	7.4	
2002-LC-018	18	39.63470	105.76155	Leavenworth Creek	7/31/02	13.8	153	7.38	15.3	8.5	
2002-TR-042	42	39.63485	105.76142	Seep, Left Bank (LB)	7/31/02	0	110	6.87	12.7	<4.5	
2002-LC-052	53	39.63494	105.76135	Leavenworth Creek	7/31/02	13.8	153	6.88	15.1	7.7	
2002-TR-062	62	39.63502	105.76127	Seep, LB	7/31/02	0.4	73	6.98	13.3	<5.4	
2002-LC-080	80	39.63514	105.76123	Leavenworth Creek	7/31/02	14.3	150	6.94	15.1	7.5	
2002-TR-095	95	39.63527	105.76120	Seep, LB	7/31/02	0.8	72	7.08	14.6	6.5	
2002-LC-150	150	39.63580	105.76095	Leavenworth Creek	7/31/02	15.1	148	7.02	15.0	7.7	
2002-TR-153	153	39.63580	105.76095	Seep, LB	7/31/02	2.4	74	7.11	15.2	7.9	
2002-LC-219	219	39.63613	105.76078	No visible inflows in this reach	7/31/02	17.5	143	7.22	15.0	7.5	
2002-TR-262	262	39.63648	105.76055	Small pocket of water on right bank (RB)	7/31/02	0.5	76	7.29	13.3	<2.9	
2002-LC-293	293	39.63648	105.76055	Leavenworth Creek	7/31/02	18.0	143	7.31	14.9	7.7	
2002-TR-293	293	39.63673	105.76060	Tributary inflow, LB	7/31/02	0.5	443	6.74	12.3	7.1	
2002-TR-295	295	39.63673	105.76060	Diffuse inflows from marshy area on LB	7/31/02	0.5	565	4.96	13.1	5.4	
2002-TR-297	297	39.63673	105.76060	Diffuse inflows from marshy area on LB	7/31/02	0.5	555	4.85	14.7	<4.0	
2002-LC-366	366	39.63728	105.76020	More diffuse inflows from marshy area on LB	7/31/02	20.8	186	5.78	14.5	8.3	
2002-LC-476	476	39.63768	105.75917	Leavenworth Creek	7/31/02	22.4	187	6.32	14.4	8.7	
2002-TR-476	476	39.63768	105.75917	Tributary inflow, LB	7/31/02	0.9	370	5.85	11.6	3.9	
2002-LC-500	500	39.63788	105.75892	Leavenworth Creek	7/31/02	23.4	187	6.54	14.7	7.4	
2002-LC-700	700	39.63897	105.75718	Leavenworth Creek	7/31/02	27.6	187	7.48	14.8	7.4	
2002-LC-800	800	39.63897	105.75718	Leavenworth Creek	7/31/02	28.3	187	7.35	14.5	7.3	
2002-LC-900	900	39.64005	105.75578	Leavenworth Creek	7/31/02	31.1	187	6.63	14.4	7.1	
2002-TR-902	902	39.64005	105.75578	Tributary inflow, LB; drains large tailing deposit	7/31/02	3.1	198	6.43	10.4	7.4	
2002-LC-930	930	39.64030	105.75545	Site of Autosampler, end of study reach	7/31/02	34.2	187	6.61	13.8	7.8	

(*continued*)

TABLE 1. (*continued*)

Field Number	Distance in m	Latitude	Longitude	Additional Sample Attributes or Comments	Collection Date	Discharge L/s	Specific Conductance μS/cm	pH	Temperature °C	Dissolved Oxygen mg/L	Year of recharge*
2003 synoptic samples											
2003-LC-000	0	39.63458	105.76170	Just upstream of injection site	8/6/03	168	61	7.02	10.5	7.5	1990 ± 0.5
2003-LC-200	200	39.63612	105.76075	Leavenworth Creek	8/6/03	168	60	6.92	11.0	7.6	
2003-TR-293	293	39.63673	105.76060	Tributary inflow, LB	8/6/03	5.7	307	6.44	12.9	6.7	
2003-TR-295	295	39.63673	105.76060	Diffuse inflows from marshy area on LB	8/6/03	5.7	463	4.85	13.6	3.1	
2003-LC-325	325	39.63697	105.76057	Leavenworth Creek	8/6/03	180	85	7.18	12.7	6.8	
2003-LC-605	605	39.63845	105.75822	Leavenworth Creek	8/6/03	120	89	6.82	10.4	7.6	
2003-TR-609	609	39.63845	105.75822	Tributary inflow, LB	8/6/03	0.9	263	5.85	7.3	6.0	
2003-TR-902	902	39.64005	105.75578	Tributary inflow, LB; drains large tailing deposit	8/6/03	3.8	1470	5.92	9.0	7.3	
2003-LC-930	930	39.64030	105.75545	Site of Autosampler, end of study reach	8/6/03	122	92.3	6.39	11.1	7.3	
Adit, pond, wetland samples											
Adit (2002)		39.63787	105.76647	Adit water directly out of collapsed colluvium	7/31/02	4.8	621	7.51	4.0	8.3	
Adit (2003)		39.63787	105.76647	Adit water directly out of collapsed colluvium	8/5/03	7.5	644	6.13	4.0	1.5	1981 ± 3
Pond (2003)		39.63692	105.76663	Pond southwest of Waldorf dump	8/5/03	7.5	624	7.01	9.8	8.4	
Fen (2003)		39.63673	105.76687	Wet slope southwest of Waldorf dump	8/5/03	0.0	31	5.50	8.5	5.7	1985.5 ± 1
WM-101 (2002)		39.63755	105.76645	Adit surface flow on dump; infiltrates	7/31/02	0.25	478	7.28	21.9	6.7	
WM-102 (2002)		39.63703	105.76638	Seep at base of dump	7/31/02	0.013	485	4.62	12.1	3.7	
WM-104 (2002)		39.63710	105.76587	Seep at base of dump	7/31/02	0.32	561	5.30	13.2	7.8	
WM-105 (2002)		39.63723	105.76568	Seep at base of dump	7/31/02	0.80	634	6.57	6.4	8.7	
WM-106 (2002)		39.63780	105.76560	Adit surface flow on dump; leaves dump	7/31/02	2.7	693	7.04	15.2	7.1	
WM-103 (2003)		39.63710	105.76618	Seep at base of dump	8/5/03	0.03	359	3.69	10.2	1.9	1990 ± 1
WM-104 (2003)		39.63710	105.76587	Seep at base of dump	8/5/03	0.14	670	3.68	5.4	3.7	1988.5 ± 1

* Calculated using CFC

TABLE 2. FIELD NUMBERS, DISTANCE FROM TRACER-INJECTION SITE, ALKALINITY, TOTAL-RECOVERABLE CONCENTRATIONS FOR Ca, Mg, Sr, SO$_4$, Al, Mn, Fe, Cu, Pb, Zn, Cd, AND Ni, WALDORF MINE AND LEAVENWORTH CREEK WATER SAMPLES, 2002 AND 2003

Field Number	Distance in m	Alkalinity mg/L CaCO$_3$	Ca mg/L ICP-MS	Mg mg/L ICP-MS	Sr µg/L ICP-AES	SO$_4$ mg/L ICP-MS	Al µg/L ICP-MS	Mn µg/L ICP-MS	Fe µg/L ICP-AES	Cu µg/L ICP-MS	Pb µg/L ICP-MS	Zn µg/L ICP-MS	Cd µg/L ICP-MS	Ni µg/L ICP-MS
2002 synoptic samples														
2002-LC-000	0	27	17.7	5.62	260	39	19	73	400	0.6	0.4	31.4	0.11	0.9
2002-LC-018	18	27	16.6	3.45	250	36	18	68	380	0.6	0.4	28.1	0.11	0.8
2002-TR-042	42	20	10.2	2.26	130	16	16	206	440	0.8	0.5	45.9	0.13	0.8
2002-LC-052	53	22	16.8	4.03	260	31	17	62	350	1.0	0.4	28.8	0.11	0.8
2002-TR-062	62	22	8.40	1.50	100	9.0	18	187	1600	0.7	0.7	64.1	0.13	0.7
2002-LC-080	80	24	17.0	3.52	260	37	18	59	300	<0.5	0.3	26.7	0.09	0.9
2002-TR-095	95	21	8.60	1.48	100	10	25	191	1900	0.5	0.7	55.2	0.14	0.7
2002-LC-150	150	24	16.3	3.63	250	41	9	44	300	0.6	0.4	29.3	0.12	0.8
2002-TR-153	153	18	8.50	1.42	110	11	31	85	690	1.4	0.6	80.1	0.22	0.8
2002-LC-219	219	21	15.4	3.80	240	27	14	55	370	0.5	0.4	30.2	0.12	0.7
2002-TR-262	262	17	4.50	2.19	110	12	44	514	1000	1.8	1.1	111	0.34	1.7
2002-LC-293	293	23	17.5	4.38	240	36	13	53	300	4.2	0.4	32.4	0.12	1.0
2002-TR-293	293	3.4	56.1	12.4	720	207	701	390	99	57	8.6	2950	7.6	16
2002-TR-295	295	1.0	71.8	17.0	950	250	1710	1150	320	184	82	3700	15	29
2002-TR-297	297	0.0	79.6	17.4	940	305	1890	1230	350	183	78	3920	13	27
2002-LC-366	366	22	21.4	4.99	310	52	80	109	250	7.4	2.5	309	1.0	2.6
2002-LC-476	476	21	20.8	4.85	310	51	76	98	220	7.8	2.1	302	1.1	2.5
2002-TR-476	476	7.3	29.7	7.37	380	73	181	171	230	34	55	2700	12	10
2002-LC-500	500	20	22.4	5.42	310	54	124	110	330	10	3.6	342	1.2	2.7
2002-LC-700	700	21	23.0	5.55	300	55	78	87	200	7.5	2.2	324	1.2	2.7
2002-LC-800	800	21	22.3	5.26	310	52	96	87	250	8.1	3.2	325	1.2	2.6
2002-LC-900	900	20	24.2	5.76	310	57	75	80	180	7.7	2.1	336	1.2	2.8
2002-TR-902	902	10	20.5	4.27	250	68	219	301	380	25	39	5140	15	19
2002-LC-930	930	22	21.5	5.30	300	52	128	96	380	10	9.8	514	1.8	3.3
2003 synoptic samples														
2003-LC-000	0	17	5.9	1.70	72	14	22	17	160	0.63	0.33	26	0.1	0.21
2003-LC-200	200	16	5.9	1.60	73	14	21	16	160	0.56	0.3	24	0.1	0.23
2003-TR-293	293	0.6	31.0	6.40	515	180	230	120	100	23	7.2	1600	4.4	6.8
2003-TR-295	295	0	49.0	9.60	776	300	1000	920	550	160	79	2400	13	18
2003-LC-325	325	20	8.2	2.10	110	29	52	35	180	3.8	1.7	180	0.5	1.0
2003-LC-605	605	16	8.6	2.10	124	30	59	43	180	5.2	3.5	200	0.6	1.1
2003-TR-609	609	8.2	24.0	5.60	310	150	480	29	1300	12	15	4600	17	16
2003-TR-902	902	3.1	12.0	2.50	170	70	210	370	330	51	65	2900	10	9.9
2003-LC-930	930	15	8.8	2.20	130	34	77	49	230	7.4	11	310	1.0	1.4

(continued)

TABLE 2. (continued)

Field Number	Distance in m	Alkalinity mg/L CaCO₃	Ca mg/L ICP-MS	Mg mg/L ICP-MS	Sr µg/L ICP-AES	SO₄ mg/L ICP-MS	Al µg/L ICP-MS	Mn µg/L ICP-MS	Fe µg/L ICP-AES	Cu µg/L ICP-MS	Pb µg/L ICP-MS	Zn µg/L ICP-MS	Cd µg/L ICP-MS	Ni µg/L ICP-MS
Adit, pond, wetland samples														
Adit, 2002	—	72	78	15	960	220	127	8830	920	122	4.4	5330	9.8	24
Adit, 2003	—	61	76	13	1040	370	180	12000	920	240	7.6	7100	23	29
Pond, 2003	—	54	74	13	1070	360	0.29	11000	<20	2	0.13	3400	15	25
Fen, 2003	—	2.0	1.8	0.35	20	6.3	81	26	68	0.97	0.46	100	0.46	0.86
Seep and surface samples from dump														
WM101 (2002)	surface	77	83	15.5	960	234	187	9000	1200	112	47	5260	9.7	25
WM102 (2002)	seep	2.9	62	16.1	680	207	692	5370	2600	350	193	3260	9.7	17
WM104 (2002)	seep	0.0	81	21.6	880	324	670	4300	1200	106	137	11300	30.5	36
WM105 (2002)	seep	21	88	16.9	1100	262	579	945	3600	111	193	2430	4.9	9.5
WM106 (2002)	surface	49	83	15.2	960	235	162	8530	1000	93.2	30	4990	9.3	24
WM103 (2003)	seep	0	39	11	608	280	730	4000	2000	560	110	5400	23	18
WM104 (2003)	seep	0	36	14	521	360	3000	10000	540	1700	250	11000	56	44

copper and zinc (22 g/day copper, 148 g/day zinc) following the remediation were relatively small compared to the loads reaching Leavenworth Creek through other pathways (discussed below).

The age of recharge, as determined by CFC analyses in 2003, was 1981 ± 3 years for the adit and 1985.5 ± 1 year for the adjacent wetland area. The adit water is apparently the same or slightly older than the wetland water, with a similar ground-water residence time of ~20 years. In comparison, the composite age of recharge for base flow in Leavenworth Creek at the beginning of the tracer reach (site LC-000) was 1990 ± 0.5, or about 5 years younger, a composite residence time for the catchment area of 13 years.

Spatial Changes in Stream Water Chemistry Before Remediation

In the following sections, spatial changes in the stream chemistry of Leavenworth Creek within the study reach will be attributed to mixing with tributary inflows and ground water and to geochemical processes such as dissolution, precipitation, or sorption. These sections present a discussion of spatial variations in field parameters and selected constituents with distance, based on the more detailed synoptic sampling that occurred prior to remediation in 2002. Total recoverable concentrations of Ca, Sr, SO_4, Al, Mn, Fe, Cu, and Zn are used for this discussion.

Field Parameters (Specific Conductance and pH)

Specific conductance and pH were useful indicators of the degree to which water had been impacted by mine waste. Specific conductance is the ability of a substance to conduct an electrical current, which in dilute solutions is directly related to the concentration of dissolved salts (Hem, 1992). The pH is a convenient method for reporting the hydrogen ion concentration, or acidity, of a solution.

The first four sampled inflows to Leavenworth Creek were largely unimpacted by mining and had a specific conductance of less than 80 µS/cm (Fig. 6 and Table 1). These tributaries had a combined discharge of ~4.1 L/s and decreased the specific conductance of Leavenworth Creek from 153 to 143 µS/cm. In contrast, sampled tributaries at downstream sites TR-293, TR-295, and TR-297 were clearly mine impacted, as indicated by their higher specific conductivities ranging from 450 to 550 µS/cm. Diffuse ground-water inflow also enters the creek in this area. Curiously, the specific conductance of Leavenworth Creek remained constant at 187 µS/cm between the mine-impacted inflows and the end of the study reach. We attribute this observation to a lack of change in the major constituents Ca and SO_4 concentrations in this reach (Table 2).

Differences in pH were also related to the degree of contact with mine waste. The pH of Leavenworth Creek ranged from 5.8 at site LC-366 to 7.5 at site LC-700. The largest drop in pH occurred between site LC-293 and site LC-366, where the pH decreased from 7.3 to 5.8 in response to the low-pH waters of the inflows of sites TR-295 and TR-297. The pH of sites TR-295 and TR-297 was

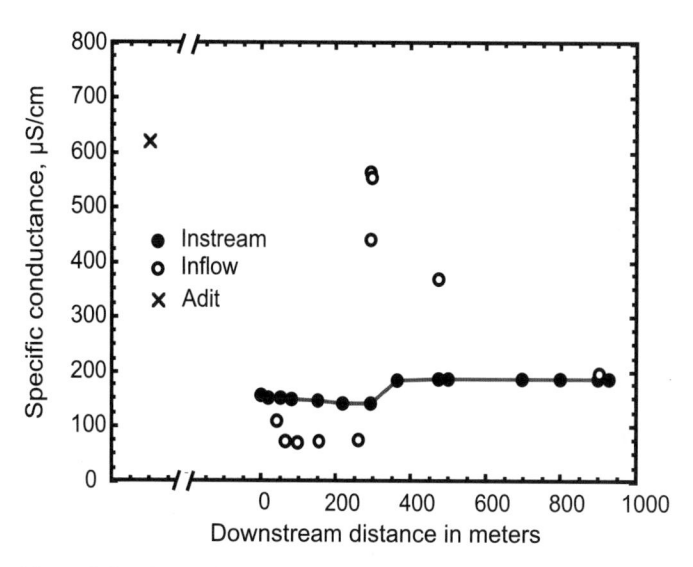

Figure 6. Graph showing specific conductance of Leavenworth Creek and inflow water samples from 2002 study. Plotted point for adit is at an arbitrary distance upstream from LC-000, as indicated by the break in the x-axis.

~5, compared to a pH of 7.5 for the collapsed adit drainage (Fig. 7). The likely source acidity was an extensive apron of eroded mine waste extending down-gradient (southeast and east) from the dump (Fig. 1). Both the adit water (diverted or not) and a relatively small but unknown fraction of wetland water flow through the waste apron. Some of this water probably infiltrates to become shallow ground water and becomes more acidic by oxidation of pyrite in the waste material or through dissolution of acid-storing salts (also originally produced as a result of pyrite weathering). Thus, adit water became acidic after traveling through the mine-impacted apron and before entering Leavenworth Creek.

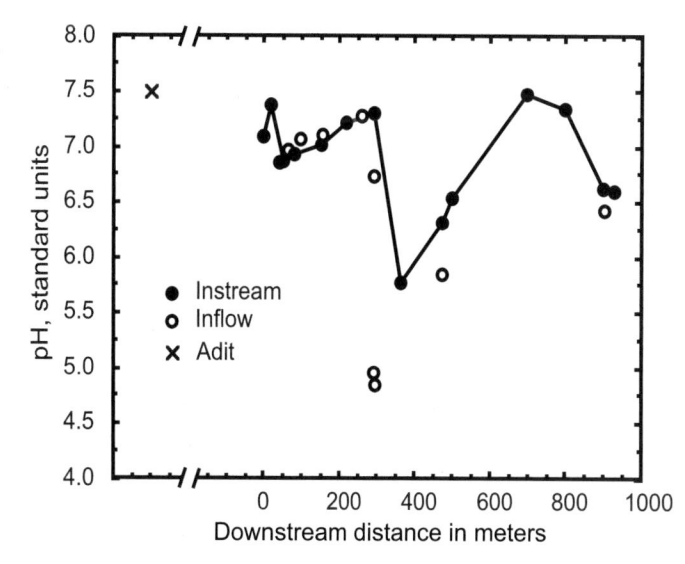

Figure 7. Graph showing pH of Leavenworth Creek and inflow water samples from 2002 study. Plotted point for adit is at an arbitrary distance upstream from LC-000, as indicated by the break in the x-axis.

Background Conditions and Mine Adit Inflows to Leavenworth Creek

Background conditions in the beginning of the tracer reach are characterized by low concentrations of ore-deposit–related constituents: Cu, <1 µg/L; Zn, 30 µg/L; Pb, <1 µg/L; Cd, <0.2 µg/L. Major cations in upper Leavenworth Creek were Ca (18 mg/L) and Mg (5.6 mg/L), and the major anions were SO_4 (39 mg/L) and bicarbonate alkalinity (22 mg/L $CaCO_3$). Several mining-impacted surface and ground-water inflows related to the adit enter Leavenworth Creek below site LC-293. Geochemical analyses indicate surface water from sites TR-295 and TR-297 was almost entirely adit water. Concentrations of Ca, Mg, and Sr, which in this environmental setting appear to be chemically conservative (nonreactive) elements, varied little between the adit and Leavenworth Creek in these two inflows, a distance of about one-half kilometer (Table 2). Concentrations of Ca, Mg, and Sr analyzed from site TR-293 were about three-fourths of those for the adit discharge, suggesting mixing with a nonimpacted source. Saturated groundwater-fed areas on the slopes between Leavenworth Creek and the Argentine Pass road (Forest Service Road 248) contained very low concentrations of these three elements (fen, Table 2). The water at site TR-293 thus appeared to be diluted by about one-fourth with fen water.

Conservative Constituents (Ca, Sr, and SO_4)

Calcium, Sr, and SO_4 appear to behave conservatively in this environmental setting, as indicated by the close agreement of instream and cumulative instream loads (Figs. 8 and 9). The sums of tributary inflow loads contributed by sites TR-293–TR-297 were essentially equal to the loads from the adit. The instream and cumulative inflow loads exceed the cumulative inflow loads below TR-293, indicating that there was significant inflow from unsampled diffuse groundwater.

Reactive Constituents (Al, Fe, Mn)

The behavior of aluminum, iron, and manganese was nonconservative (reactive), and is of particular interest because these elements are capable of scavenging, or removing, other metals from the water column. Concentrations of Al, Fe, and Mn (Fig. 10) and loads (Fig. 11) were affected by inflow contributions and by instream attenuation such as precipitation, or colloid or oxide coating formation.

At the beginning of the study reach, the instream Al load was 30 g/day, which increased by unsampled inflow to 45 g/day by site LC-052. However, this instream load decreased to ~1 g/day by site LC-150 through attenuation by precipitation/colloid formation and settling. The mining-impacted inflows from TR-293–297 added a total of 184 g/day of Al, but the full contribution was not fully evident downstream at site LC-366 (Fig. 11A). The instream Al load at that site had only increased by 87 g/day, so instream attenuation was clearly evident. White Al-oxide coatings have been observed on the streambed and rocks in this reach (R. Wanty, USGS, pers. comm., 2005). Farther downstream, the instream Al load continued to increase owing to diffuse inflow and then decreased due to

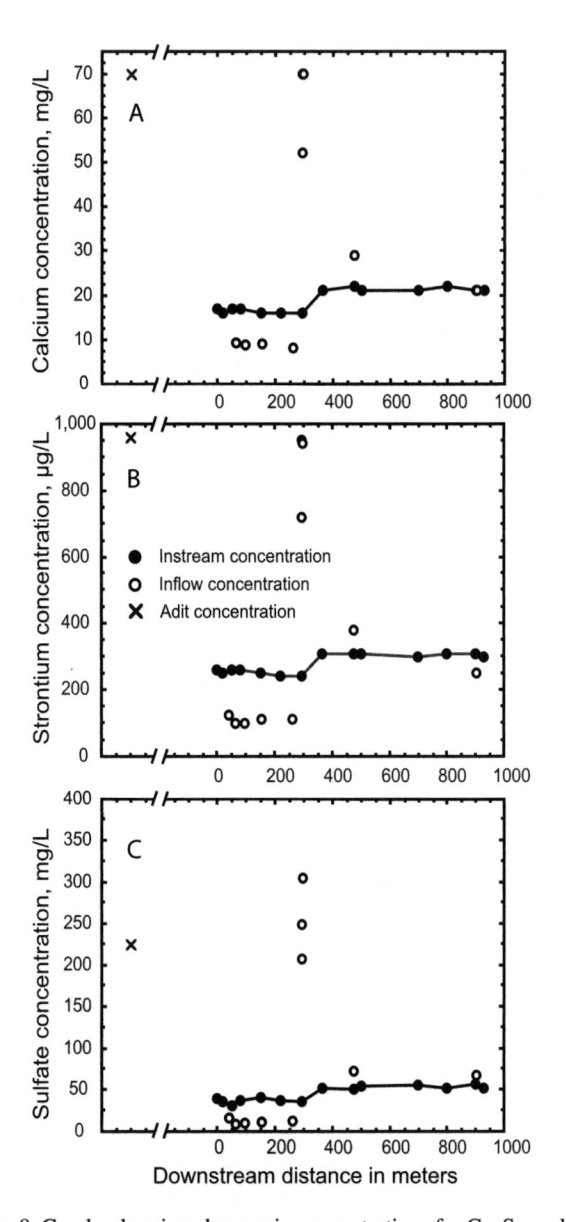

Figure 8. Graphs showing changes in concentrations for Ca, Sr, and SO_4 for Leavenworth Creek and inflow water samples from 2002 tracer-dilution study. Plotted point for adit is at an arbitrary distance upstream from LC-000, as indicated by the break in the x-axis. Note how high Sr concentration (B) from adit flow serves as a natural tracer.

precipitation over the remainder of the study reach. The fairly large inflow load at site TR-902 was only partially evident at site LC-930, again as a result of instream attenuation.

Iron concentrations in Leavenworth Creek samples and tributaries could not always be linked to mining-impacted inflows. Unexpectedly, non-mining-impacted tributaries upstream from site TR-293 contributed more than twice as much Fe as all the mining-impacted tributaries below site LC-293. At the beginning of the tracer reach, concentrations of total-recoverable Fe decreased from 450 to 300 µg/L between sites LC-000 and LC-150. Although tributaries TR-062 and TR-095 contributed 194 g/day Fe, the in-

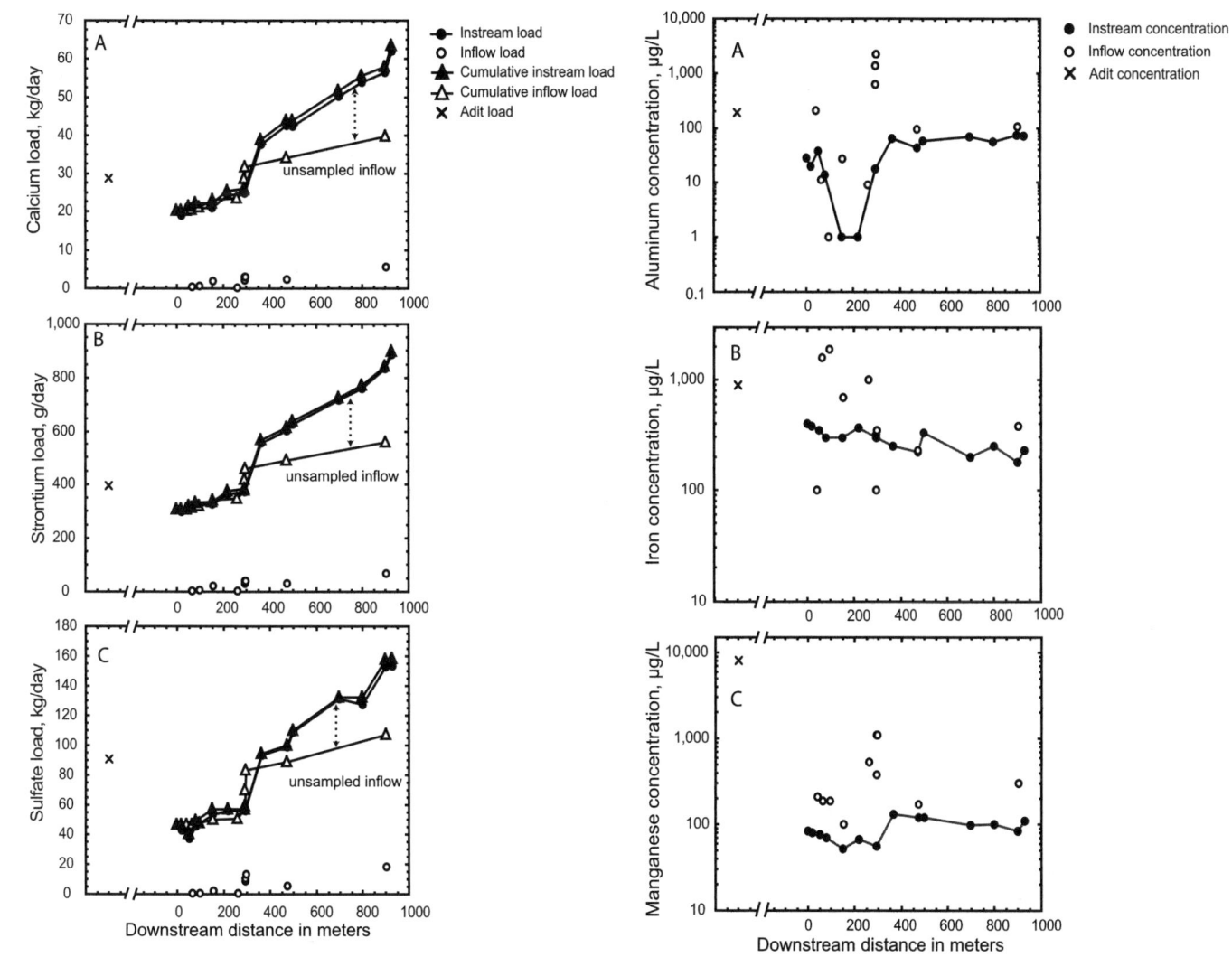

Figure 9. Graphs showing instream, inflow, cumulative instream, and cumulative inflow loads for Ca, Sr, and SO₄ for tracer-dilution study of 2002. Plotted point for adit is at an arbitrary distance upstream from LC-000, as indicated by the break in the *x*-axis. Unsampled inflow is indicated by vertical dashed lines with arrowheads, showing the difference between the cumulative instream loads and the cumulative inflow loads. Closeness of fit between lines for instream load and cumulative instream load indicates conservative behavior of these elements (A, B, C).

Figure 10. Graphs showing changes in concentrations for Al, Fe, and Mn for Leavenworth Creek and inflow water samples from 2002 tracer-dilution study. Plotted point for adit is at an arbitrary distance upstream from LC-000, as indicated by the break in the *x*-axis. Note that the Al concentrations in the adit inflows at ~300 m on *x*-axis (A) are much higher than in the adit discharge, and that the Mn concentrations (C) in the adit inflows are greatly depleted from the adit discharge.

stream load decreased slightly from 417 to 391 g/day. Surprisingly, the single largest inflow load to the tracer reach of 144 g/day occurred at site TR-153, which showed little if any evidence of contamination from mining. The Waldorf adit contributed a comparatively small load of Fe, 372 g/day, which was slightly lower than the instream load of Leavenworth Creek at the beginning of the tracer reach. In contrast to the upper reach, the reach between site LC-376 and site LC-902 received substantial loads of Fe from unsampled inflow (see sites LC-500 and LC-800, Fig. 11B).

The behavior of a reactive element is well illustrated by manganese (Fig. 11C). Concentrations of Mn in the Waldorf adit discharge were highly elevated (8100 μg/L in 2002). Ore veins in the Waldorf mine contain rhodocrosite and manganese wad (Lovering, 1935). Manganese oxide forms quickly upon discharging to oxygenated surface water, coating rocks and soils close to the mouth of the adit. Dark stains on the surface of the dump remained visible in 2003, 1 year after the adit discharge had been diverted around the dump. In 2002, the Mn load between the adit and Leavenworth Creek decreased from 3300 g/day to 111 g/day. Sorption to Mn oxides was probably responsible, in part, for reducing the copper and zinc loads in the adit flows (which is discussed in greater detail in the next two sections).

At the beginning of the tracer study reach, the instream Mn load was ~100 g/day and fluctuated slightly until reaching the

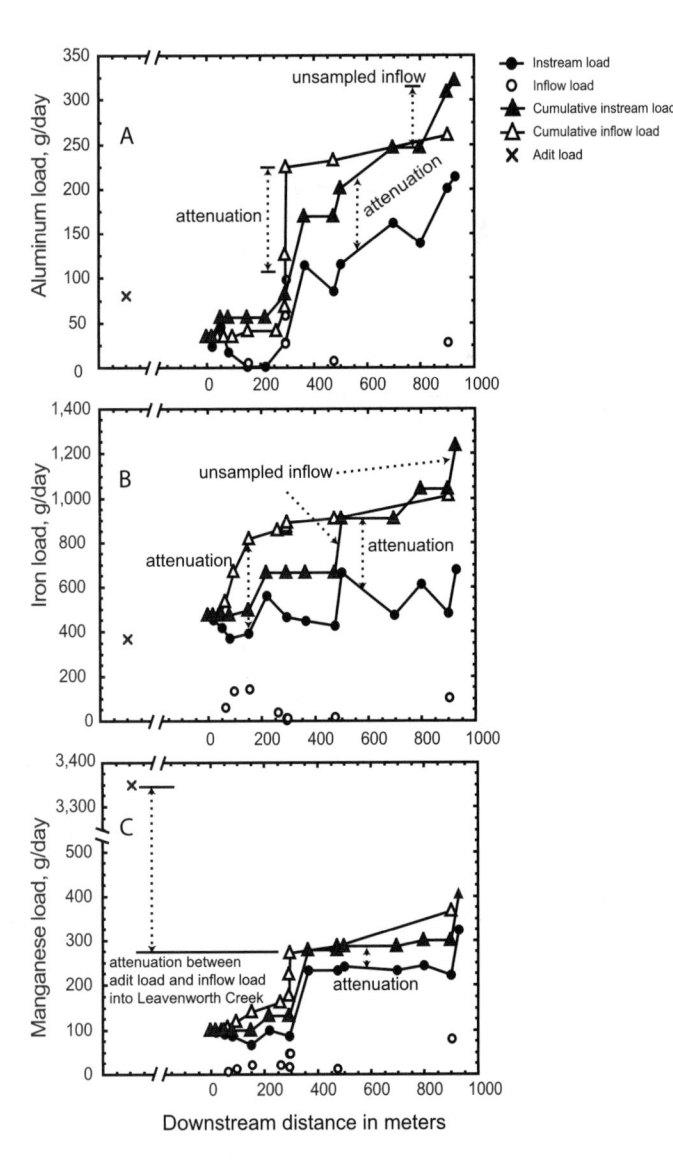

Sources of Zinc Loads in Leavenworth Creek

Attenuation by sorption to hydrous iron (Smith, 1999; Kimball et al., 2000) or Mn oxides (Hem, 1992) or organic matter (Drever, 1988) between the adit and Leavenworth Creek is thought to remove much of the zinc before it reaches the creek. The total Zn load in 2002 at the mouth of the collapsed adit was 2206 g/day, compared with a combined load of 457 g/day for sites TR-293, TR-295, and TR-297 (Fig. 12). By subtraction, 1257 g/day of the zinc load from the adit apparently did not contribute to these three tributary inflows and was being sequestered in material between the adit and the creek. The zinc is probably trapped in the mine-waste dump apron as well as in the wet organic-rich soils that extend between the dump apron and the creek.

Downstream from the most obvious adit inflows, site TR-476 was interpreted as a mix of adit water and background water, as indicated by a relatively low strontium concentration of 380 μg/L. The high zinc content of 2700 μg/L (with a corresponding load of 214 g/day) is attributed to a mixture of adit water and leachate from mill tailings rather than leaching of the waste apron mentioned in the paragraph above (see white area near TR-476 in Fig. 1). Between LC-476 and LC-900, diffuse inflows added 7.7 L/s in discharge and 210 g/day zinc. Inflows downstream from

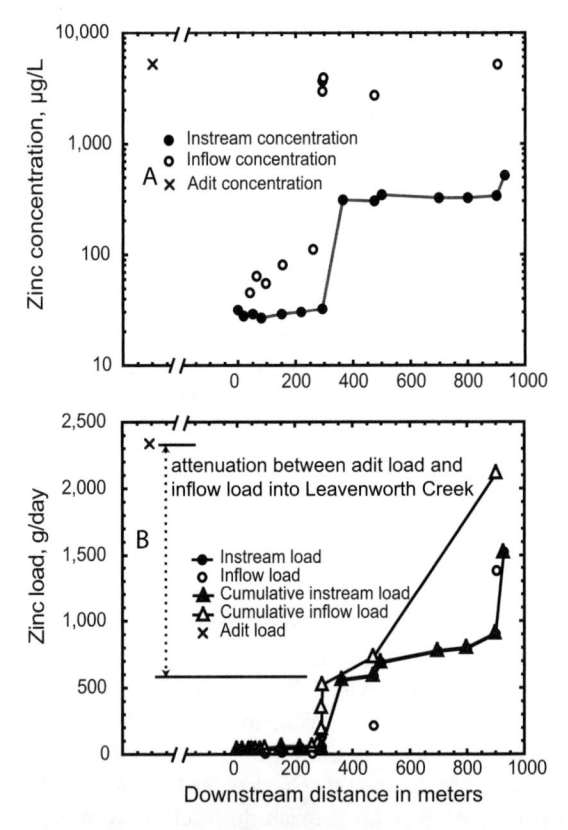

Figure 11. Graphs showing instream, inflow, cumulative instream, and cumulative inflow loads for Al, Fe, and Mn for Leavenworth Creek and inflow water samples from 2002 study. Plotted point for adit is at an arbitrary distance upstream from LC-000, as indicated by the break in the x-axis. Unsampled inflow for Al and Fe (A and B) is indicated by labeled vertical dashed lines with arrowheads, showing the difference between the cumulative instream loads and the cumulative inflow loads. Attenuation by colloid or oxide coating formation is indicated by labeled dashed lines, showing the difference between calculated cumulative instream and instream loads or the difference between cumulative inflow and instream loads.

Figure 12. Graphs showing concentration and loads for Zn from 2002 tracer-dilution study. Plotted point for adit is at an arbitrary distance upstream from LC-000, as indicated by the break in the x-axis. Note attenuation of Zn from adit as shown by the difference between the adit load and the cumulative inflow load at LC-297 (vertical dashed line with arrowheads).

inflows from the mine. There, 111 g/day was added by discrete inflow and 37 g/day by unsampled inflow (Fig. 11C). Between site LC-376 and the end of the study reach, additional Mn was derived from both discrete inflows and unsampled diffuse inflows; however, the process of Mn-oxide formation also removed Mn from the water column. Inflow at site TR-902 contributed the single largest load of 80 g/day.

LC-476 appear to contain little adit water but were derived from local recharge through the dispersed mill tailings.

Downstream from site TR-476, west bank inflows to Leavenworth Creek are impacted predominantly by mill-tailings deposits. Tributary TR-902 drains several hundred meters of the old abandoned ditch and the dispersed mill tailings. The zinc concentration for site TR-902 (5140 µg/L) is about the same as the adit (5330 µg/L). However, the Sr concentration was only one-fourth that of adit water and was the same as Leavenworth Creek water above the study reach (250 µg/L Sr), which suggests that no contribution from the adit is required to account for the Sr concentration at site TR-902. Therefore, the water chemistry of this site is interpreted as shallow ground water that has leached its zinc from the dispersed mill tailings.

In summary, only ~22% of the cumulative inflow load for total zinc at the bottom of the tracer reach (site LC-930) is attributed *directly* to the adit, from inflows TR-293 through TR-297. At *least* 50% of the cumulative inflow load for zinc is attributed to leaching of tailings drained by TR-902. The remaining 28% of the cumulative inflow load is attributed to *a mix* of zinc from the adit flow (that reached Leavenworth Creek by paths other than tributaries TR-293–297) and zinc leached from the tailings by a flow path other than TR-902 (diffuse inflow between LC-475 and LC-902).

Given the amount of spatial and temporal variation in the concentrations of strontium in Leavenworth Creek from 2002 to 2003 (Table 2), these percentages should be considered rough estimates. Based on the ratio of strontium in the inflows to strontium in the adit water, as much as 35% of the cumulative inflow load for zinc to Leavenworth Creek can be attributed to the adit, and as much as 65% can be attributed to leaching of mill tailings, but the actual proportions are probably closer to 25% for the adit, 10% a mix of adit water and waste apron leachate, and 65% tailings leachate from TR-902. Figure 13 shows likely flow paths for mine-related contaminants, including surface water paths from the dump to the creek and ground-water flow through the mill tailings. Included in the figure are proposed small-scale "basin divide" boundaries, which depict probable topographic constraints on the surface-water flow.

Sources of Copper Loads in Leavenworth Creek

The major source of the Cu load entering Leavenworth Creek at sites TR-293–TR-297 (18 g/day) (Fig. 14) is adit inflows that have picked up copper from the waste apron and wetland area, as opposed to the adit discharge where it first emerges. The Cu concentration of flow discharging from the adit in 2003 was 240 µg/L, whereas the concentration in the nearby pond, fed by the adit and less than 100 m away, was only 2 µg/L. At the circum-neutral pH of the adit and pond water, nearly 100% of the Cu can be expected to be sorbed onto hydrous iron oxides or manganese oxides (Smith, 1999) in the pond sediment. Between sites LC-366 and LC-930, small increases in the Cu load are attributed to diffuse inflow. Tributary TR-902 contributed 7 g/day Cu, leached from the dispersed mill tailings. The cumulative inflow load for the

Figure 13. Oblique photograph of slopes below Waldorf mine area taken from the southwest on Argentine Pass road, showing surface and near-surface flowpaths (dashed lines) and conceptual small-scale basin divides (solid lines). The Waldorf mine and dump are just to the left of photograph.

entire study reach was 33 g/day, compared with the instream load at site LC-930 of 30 g/day, indicating that the copper was in the solid-phase (colloidal) rather than the dissolved phase.

Zinc and Copper Loads after Remediation

Synoptic sampling at selected sites was repeated in 2003 to determine whether the remedial rerouting of adit discharge around the Waldorf dump had improved water quality and reduced metal loads of Zn and Cu in Leavenworth Creek. Interpretation of the before-remediation and after-remediation data sets was complicated by the large differences in base flow between 2002 and 2003. Both data sets were collected after the prolonged spring melt-off of the snow pack and before the main onset of summer monsoon thunderstorms. However, base flow discharge was at least five times greater in 2003 than in 2002 (Figs. 4 and 5). The water year 2002 was one of the driest on record with the average snow pack in Colorado for the winter ~52% of normal (Pielke et al., 2004), whereas water year 2003 was closer to normal.

Consequently, differences in zinc and copper loads between 2002 (a dry year) and 2003 (a comparatively wetter year) are largely attributed to changes in discharge, as opposed to remediation. The greater precipitation and snow pack in 2003 nearly doubled discharge from the adit draining the mine workings of McClellan Mountain. Correspondingly, the adit load for zinc in

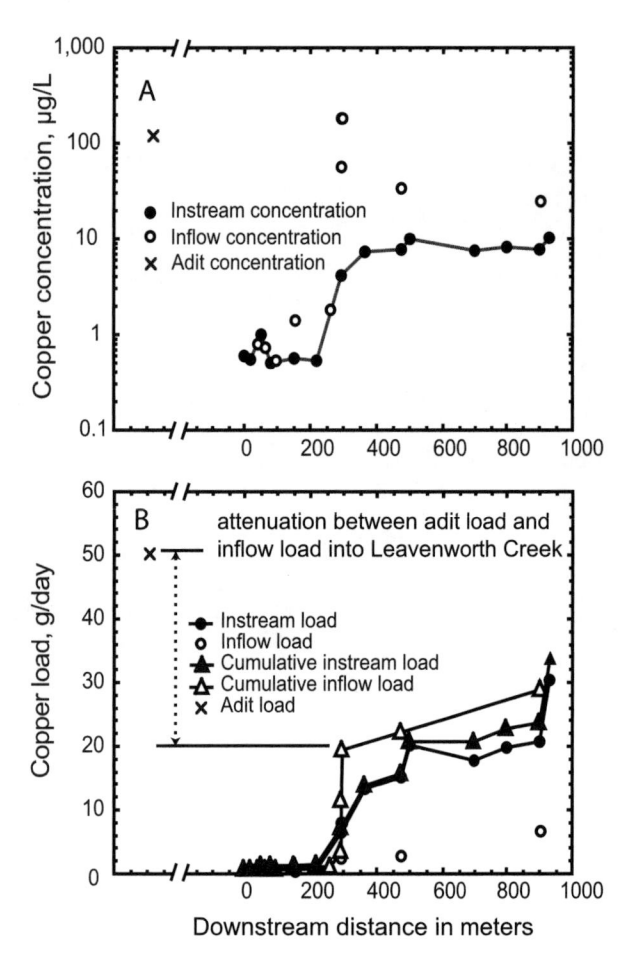

Figure 14. Graphs showing concentration and loads for Cu from 2002 tracer-dilution study. Plotted point for adit is at an arbitrary distance upstream from LC-000, as indicated by the break in the x-axis. Note the attenuation of Cu from the adit as shown by the difference between the adit load and the cumulative inflow load at LC-297 (vertical dashed line with arrowheads).

2003 was roughly twice that in 2002 (4586 g/day versus 2206 g/day), and the adit copper load in 2003 was three times that in 2002 (155 g/day versus 50 g/day). Disproportionately, discharge of the tributaries TR-293–TR-297 was 10 times higher in 2003 than in 2002. Zinc and copper loadings of these tributaries also increased greatly (zinc from 457 to 1900 g/day; copper from 18 to 87 g/day). The 2003 results indicate that significant amounts of zinc and copper were derived from the adit discharge after it had leached through the tailings apron and that any benefit from diverting the adit discharge around the dump in 2002 was masked by increased loading from the tailings apron.

Discharge from tributary TR-902, a small catchment draining the dispersed mill tailings, was only slightly higher in 2003 (3.8 L/s) than in 2002 (3.1 L/s). Some 2003 loads were lower (Ca, Sr, Zn), whereas others were slightly higher (Fe, Mn, Cu, Al). One possible explanation is that during the spring runoff period, more water would have leached through the dispersed mill tailings in 2003 than in 2002. More runoff in 2003 earlier in the season may have "flushed" and subsequently diluted the concentrations of the relatively more conservative elements prior to sampling and at the same time may have changed pH and redox conditions, making greater quantities of the more reactive constituents such as Zn and Cu more mobile. This would result in the lower loads evident at the time of sampling.

Loading trends for Zn and Cu from 2002 to 2003 attributed to the leaching of mill tailings (as measured at site TR-902) were inverse to one another, with the Zn load decreasing and the Cu load increasing (Fig. 15 and Table 3). The Zn load contributed by TR-902 in 2003 (958 g/day) was smaller than in 2002 (1378 g/day). This amounted to a smaller proportion of the cumulative inflow load than in 2002 (28% versus 65%), but still implicates the dispersed mill tailings as a major source of zinc. The Cu load contributed by TR-902 in 2003 was 17 g/day, compared to 7 g/day in 2002. The proportion of the cumulative inflow load of Cu derived from the mill tailings in 2003 was less than in 2002, 16% versus 25%, but the actual amount was greater. The overall cumulative inflow load at LC-930 for copper in 2003 was 105 g/day, whereas in 2002 it was 33 g/day.

Acute and Chronic Zinc and Copper Concentrations

Both zinc and copper can be toxic to aquatic life, depending on concentration (Horn, 2001). Their toxicity is partially ameliorated by calcium and magnesium ions in the water (hardness), and so hardness-corrected standards can be calculated and compared with instream dissolved concentrations. The formulas (Colorado Department of Public Health and Environment [CDPHE], 2005) for the calculations are:

$$Cu_{acute} = e^{(0.9422[\ln(hardness)] - 1.7408)}$$

$$Cu_{chronic} = e^{(0.8545[\ln(hardness)] - 1.7428)}$$

$$Zn_{acute} = e^{(0.8473[\ln(hardness)] + 0.8618)}$$

$$Zn_{chronic} = e^{(0.8473[\ln(hardness)] + 0.8699)}$$

where hardness = [(Ca concentration in mg/L)/40.08 + (Mg concentration in mg/L)/24.31] × 100.1.

Hardness-corrected aquatic life standards were calculated using concentrations of dissolved Zn and Cu from the Leavenworth Creek samples collected in 2003. These samples were processed with conventional 0.45-μm syringe-mounted filters and were not affected by contamination issues, as were the 2002 filtered samples. The data from the 2003 samples indicate that the standards were exceeded by factors of ~6 for Zn and slightly over 1 for Cu (Table 4) at site LC-930. Instream concentrations of Zn exceeded water-quality standards consistently downstream from site LC-293. In comparison, Cu equaled the acute standard and was 1.4 times the chronic standard. Remediation efforts should therefore be focused on reducing Zn over Cu concentrations. This implies that Zn loads not only from the adit but also from the tailings apron below the mine-waste dump and from the mill-tailings area east and north of the dump would all have to be reduced significantly to improve instream Zn concentrations.

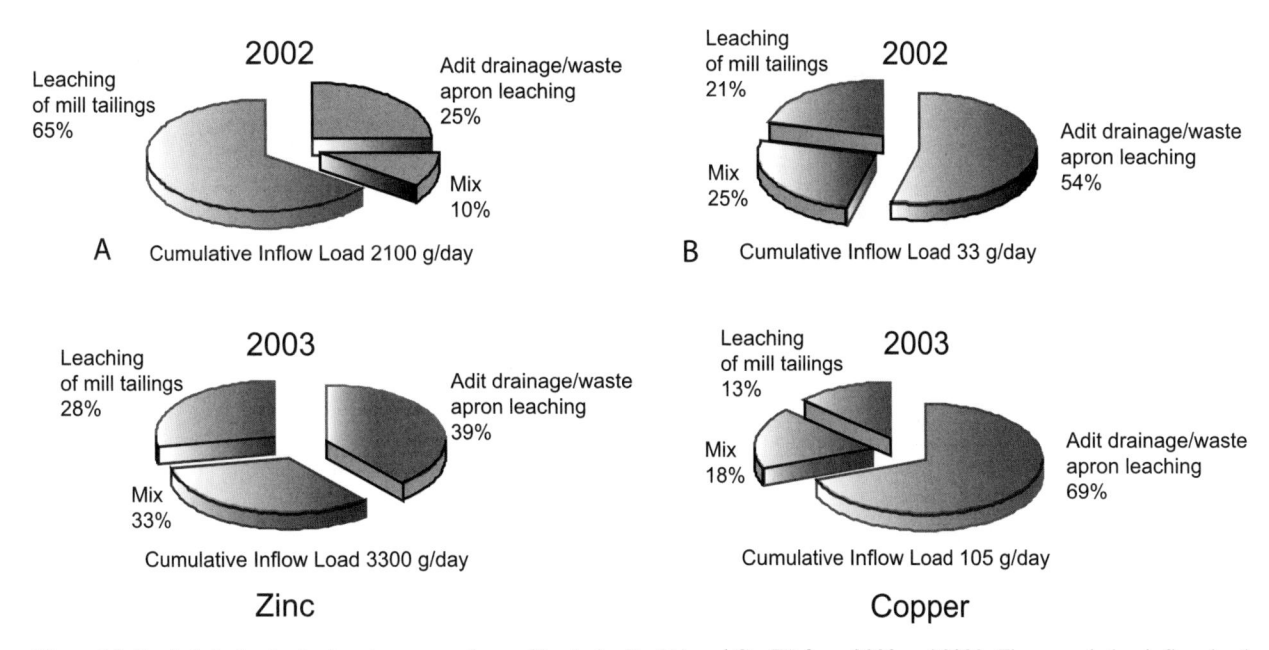

Figure 15. Exploded pie charts showing proportions of loads for Zn (A) and Cu (B) from 2002 and 2003. The cumulative inflow load is calculated for site LC-930. In 2002, the dominant loading for Zn was from the mill tailings, whereas in 2003 it was from adit discharge and shallow groundwater leaching of the waste apron below the dump (A). In both 2002 and 2003, the main source of Cu was from the adit discharge/waste apron leaching (B). Note the greatly increased cumulative load for both elements between 2002 and 2003.

SUMMARY AND CONCLUSIONS

Many mine remediation activities often proceed with little or no scientific study to identify the sources and pathways of metal transport. Consequently, these "remedial" strategies may not be successful in improving water-quality in the receiving stream. Assessment of sources and pathways to gain better understanding of the system is a necessary investment in the outcome of any successful remediation strategy. Tracer-injection and synoptic sampling studies were conducted in 2002 to determine mine-related sources of metals to Leavenworth Creek downslope from the Waldorf mine site. Follow-up synoptic sampling with current-meter measurements was conducted in 2003 to assess possible improvements in water quality resulting from the rerouting of adit flow away from the Waldorf dump. The studies show three main sources of contaminants to Leavenworth Creek: (1) adit discharge; (2) leaching of mine-waste material by adit and/or wetland water; and (3) leaching of mill tailings, which predominantly occurs through leaching with local ground-water recharge. Infiltration through the mine-waste dump by itself did not appear to be a major cause of Cu and Zn loading to Leavenworth Creek. Adit discharge initially accounted for up to 25% of the Zn; however, subsequent leaching of mine-waste material in the wetland accounted for as much as an additional 10% of the Zn load. In 2002, leaching of mill tailings below the mine and mill by shallow ground water was the largest source of contamination, which accounted for up to 65% of the Zn load.

Higher discharge for both the adit and Leavenworth Creek in 2003 masked any possible improvement resulting from remediation over the timeframe of these studies. Loads generally corresponded with discharge, especially the adit-related inflows near sites TR-295–TR-297. Unexpectedly, a higher proportion of the total Zn load was contributed by adit inflows in 2003 than in 2002 despite the fact that this flow had been rerouted away from the

TABLE 3. Zn AND Cu LOADS AT THE WALDORF ADIT, THE BEGINNING AND END OF TRACER REACH ON LEAVENWORTH CREEK, AND TWO INFLOWS FOR 2002 AND 2003 SYNOPTIC STUDIES

Field No.	Zn site load 2002 (g/day)	Zn site load 2003 (g/day)	Cu site load 2002 (g/day)	Cu site load 2003 (g/day)
Adit	2210	4590	50	155
LC-000	37	230	1	6
TR-293-297	460	1900	18	87
TR-902	1380	960	7	17
LC-930	1520	4820	30	115

TABLE 4. SITE-SPECIFIC CONCENTRATIONS OF COPPER AND ZINC IN FILTERED WATER SAMPLES COLLECTED IN 2003 AS A PERCENTAGE OF HARDNESS-BASED TOXICITY STANDARDS FOR AQUATIC LIFE

Field Number	Stream	Downstream Distance (m)	Site Copper/ Chronic Standard (Percent)	Site Copper/ Acute Standard (Percent)	Site Zinc/ Chronic Standard (Percent)	Site Zinc/ Acute Standard (Percent)
2003-LC-000	Creek	0	12	17	70	70
2003-LC-200	Creek	200	10	15	86	85
2003-TR-293	Tributary	293	210	140	1300	1300
2003-TR-295	Tributary	295	1040	660	130	1300
2003-LC-325	Creek	325	102	75	390	380
2003-LC-605	Creek	605	104	76	380	380
2003-TR-609	Tributary	609	43	29	4100	4000
2003-TR-902	Tributary	902	1080	770	4600	4600
2003-LC-930	Creek	930	140	100	580	580

waste dump. In contrast, the Zn load introduced by the mill-tailings deposits in 2003 was lower than in 2002, possibly due to more complete flushing of soluble salts by that year's higher spring runoff of the mill tailings prior to sampling. The overall cumulative inflow load for Cu was higher in 2003 than in 2002, as was the specific load from TR-902. The Cu load in 2003 at TR-902 was 2.4 times higher than in 2002. Follow-up sampling is recommended in future years to better define any water-quality trends that may be occurring.

Despite the reduction in stream concentrations and tailings loads in 2003 compared to 2002, Cu and Zn were still present in concentrations near or exceeding cold-water acute and chronic aquatic life standards. In the study area, removal of the mill tailings is recommended as the single most effective strategy for reducing metal loads to Leavenworth Creek, followed by removal of the tailings apron below the dump, followed by point-source treatment of the adit discharge. Removal of the mine-waste dump would not directly address the major sources of contamination during base-flow conditions but could have some utility regarding high flow or rainfall runoff. Plugging the adit also would not necessarily produce the desired result because McClellan Mountain has many underground workings and tunnels, and metal-impacted water would probably exit the ground through other available paths to the surface with similar results.

ACKNOWLEDGMENTS

The following U.S. Geological Survey employees assisted with the tracer study and synoptic sampling: Robert McDougal, Thomas Chapin, Rhonda Driscoll, Pete Theodorakas, LaDonna Choate, and Paul Wigton. Fairda Malem of the Colorado School of Mines provided background information and assisted with fieldwork. The CFC dating analyses were performed by Alan Rigby at the Department of Geology and Geophysics Dissolved and Noble Gas Laboratory of the University of Utah in Salt Lake City. Thanks are also extended to Kathy Smith and Rob Runkel of the U.S. Geological Survey for their helpful reviews. Finally, we gratefully acknowledge Historic Georgetown for site access and permission to conduct research studies. Funding for this study was through the U.S. Geological Survey Mineral Resources Program.

REFERENCES CITED

Bencala, K.E., McKnight, D.M., and Zellweger, G.W., 1990, Characterization of transport in an acidic and metal-rich mountain stream based on a lithium tracer injection and simulations of transient storage: Water Resources Research, v. 26, no. 5, p. 989–1000, doi: 10.1029/89WR03235.

Briggs, P.H., 2002, The determination of twenty-seven elements in aqueous samples by inductively coupled plasma-atomic emission spectrometry, in J. E. Taggart Jr., ed., Analytical methods for chemical analysis of geologic and other materials: U.S. Geological Survey Open-File Report 02-223, p. F1–F13. (Available online at http://pubs.usgs.gov/of/2002/ofr-02-223/.)

Briggs, P.H., and Meier, A.L. 2002, The determination of forty-two elements in geological materials by inductively coupled plasma-mass spectrometry; in Taggart, J.E., ed. Analytical methods for chemical analysis of geologic and other materials, U.S. Geological Survey Open-File Report 02-223, ch. I, 14 p.

Broshears, R.E., Bencala, K.E., Kimball, B.A., and McKnight, D.M., 1993, Tracer-dilution experiments and solute-transport simulations for a mountain stream, Saint Kevin Gulch, Colorado: U.S. Geological Survey Water-Resources Investigations Report 92-4081, 18 p.

Colorado Department of Public Health and Environment (CDPHE), Water Quality Control Commission, 2005, Regulation No. 31, The Basic Standards and Methodologies for Surface Water (5 CCR 1002-31). Accessed January 18, 2007, at http://cdphe.state.co.us/regulations/wqccregs/index.html.

Drever, J.I., 1988, The geochemistry of natural waters (2nd edition): Prentice Hall: Englewood Cliffs, New Jersey, 437 p.

Hem, J.D., 1992, Study and interpretation of the chemical characteristics of natural water: U.S. Geological Survey Water-Supply Paper 2254, 263 p.

Horn, B., 2001, Biotoxicity report: Animas River aquatic toxicity profiles with existing water quality conditions, Appendix 6c in Simon, W., Butler, P., and Owen, J.R., eds., Use attainability analysis for the Animas River watershed: Animas River Stakeholders Group, 240 p., appendices, CD-ROM.

Kimball, B.A., 1997, Use of tracer injections and synoptic sampling to measure metal loading from acid mine drainage: U.S. Geological Survey Fact Sheet 245-96, 4 p.

Kimball, B.A., Bencala, K.E., and Runkel, R.L., 2000, Quantifying effects of metal loading from mine discharge: ICARD 2000: Proceedings from the Fifth International Conference on Acid Rock Drainage, v. 2, p. 1381–1390.

Kimball, B.A., Runkel, R.L., Walton-Day, K., and Bencala, K.E., 2002, Assessment of metal loads in watersheds affected by acid mine drainage by using tracer injection and synoptic sampling: Cement Creek, Colorado, USA: Applied Geochemistry, v. 17, no. 9, p. 1183–1207, doi: 10.1016/S0883-2927(02)00017-3.

Lamothe, P.J., Meier, A.L., and Wilson, S.A., 1999, The determination of forty-four elements in aqueous samples by Inductively Coupled Plasma-Mass Spectrometry: U.S. Geological Survey Open File Report 99-115, 14 p.

Long, H.K., and Farrar, J.W., 1995, Report on the U.S. Geological Survey's evaluation program for standard reference samples distributed in May 1995: T-135 (trace constituents), M-134 (major constituents), N-45 (nutrients), N-46 (nutrients), P-24 (low ionic strength), Hg-20 (mercury), and SED-5 (bed material): U.S. Geological Survey Open-File Report; 95-395, 135 p.

Lovering, T.S., 1935, Geology and ore deposits of the Montezuma quadrangle, Colorado: U.S. Geological Survey Professional Paper 178, 119 p.

McDougal, R.R., and Wirt, L., 2007, this volume, Characterizing infiltration through a mine waste dump using electrical geophysical and tracer injection methods, Clear Creek County, Colorado, in De Graff, J.V., ed., Understanding and Responding to Hazardous Substances at Mine Sites in the Western United States: Geological Society of America Reviews in Engineering Geology, v. XVII, doi: 10.1130/2007.4017(02).

Pielke, R.A. Sr., Doesken, O.B., Green, T., Chaffin, C., Salas, J.D., Woodhouse, C.A., Lukas, J.J., and Wolter, K., 2004, Drought 2002 in Colorado: An unprecedented drought or a routine drought? Accessed April 5, 2005, at http://ccc.atmos.colostate.edu/pdfs/PAGEOPH_2002DroughtArticle.pdf.

Plummer, L.N., and Busenberg, E., 1996, Chlorofluorocarbons (CFCs) as tracers and age-dating tools for young ground water: Selected field examples, in Stevens, P.R., and Nichols, T.J., eds., Joint U.S. Geological Survey, U.S. Nuclear Regulatory Commission Workshop on "Research related to Low-level radioactive waste disposal," May 4–6, 1993, Reston, Virginia: U.S. Geological Survey Report 95-4045, p. 65–71.

Shelton, L.R., 1994, Field guide for collection and processing stream-water samples for the National Water-Quality Assessment Program: U.S. Geological Survey Open-File Report 94-455, 42 p.

Smith, K.S., 1999, Metal sorption of mineral surfaces: An overview with examples relating to mineral deposits, in G.S. Plumlee and M.J. Logsdon, editors, Reviews in economic geology, vol. 61, The environmental geochemistry of mineral deposits, p. 161–182.

Wilde, F.D., and Radtke, D.B., 1998, National field manual for the collection of water-quality data: U.S. Geological Survey Techniques of Water-Resources Investigations, book 9, chap. A6, sections 6.0, 6.0.1, 6.0.2, 6.0.2.A, and 6.0.2B.

Wirt, L., 2004, Sources of base flow in the upper Verde River based on multiple lines of geochemical evidence and the tracer-dilution method, in Wirt, L., DeWitt, E., and Langenheim, V.E., eds., Geologic Framework of Aquifer Units and Ground-Water Flowpaths, Verde River Headwaters, North-Central Arizona, U.S. Geological Survey Open-File Report 2004-1411F.

MANUSCRIPT ACCEPTED BY THE SOCIETY 28 NOVEMBER 2006.

On-site repository construction and restoration of the abandoned Silver Crescent lead and zinc mill site, Shoshone County, Idaho

Jeff K. Johnson
Idaho Panhandle National Forests, 3815 Schreiber Way, Coeur d'Alene, Idaho 83815, USA

ABSTRACT

From the early 1900s through the 1950s the Silver Crescent mine and mill processed lead, zinc, and silver from ore found in the Precambrian metasedimentary rocks of the Belt Supergroup. Approximately 150,000 cubic yards of tailings and waste rock were deposited in the floodplain of Moon Creek less than 2 miles upstream of what is now a residential area. The actively eroding tailings impoundments were a source of heavy metal contamination to the surface and groundwater flowing through the site. The U.S. Forest Service began a CERCLA non-time-critical removal action at the Silver Crescent mine in 1998. Removal action goals included reduction of particulate and dissolved metal loading into Moon Creek and local groundwater. These goals were successfully achieved in part by incorporating the tailings and waste rock dumps into an on-site capped repository. The nearly $2 million Silver Crescent removal action construction phase was completed in late 2000 with the final habitat restoration phase scheduled for completion in 2007.

Keywords: reclamation, repository, mine, lead

INTRODUCTION

Beginning in the early 1990s the USDA Forest Service (FS) has placed an increased emphasis on addressing environmental problems at abandoned mine sites on National Forest System lands across the nation. The FS has presidential authority to address problems at these sites through the Comprehensive Environmental Response, Compensation, and Liability Act (CERCLA; 42 USC 9604; and Executive Order 12580, 52 Federal Register 2923–26; January 23, 1987). Through a competitive funding process, efforts have been made by the FS to address the most severe problems first. Nearly $2 million have been spent at the abandoned Silver Crescent mine and mill site, hereafter referred to as the site, near Kellogg, Idaho. Since 1992 the FS and several partners have worked to achieve a comprehensive remedy at the site with the final habitat restoration phase scheduled for completion in 2007.

LOCATION AND SETTING

The site is located on National Forest System lands in the Coeur d'Alene Mining District of northern Idaho. This area is ~80 miles east of Spokane, Washington, and just 3 miles east of Kellogg, Idaho. The site is located at an elevation of 2700 feet and is situated along the East Fork of Moon Creek, ~2.8 miles upstream from its confluence with the South Fork of the Coeur d'Alene River near U.S. Interstate 90 (Fig. 1). The area consists of deeply incised, well-vegetated mountain valleys, and precipitation at the site ranges from 30 to 40 inches per year. Surface water base flow conditions at the site are ~0.5 cfs, with a peak flow in the spring of 1998 measured at 8.8 cfs.

Previous to FS CERCLA activities, the abandoned site features included a mill site, several buildings, waste rock dumps, tailings impoundments, several portals, and one shaft (Figs. 2 and 3).

Johnson, J.K., 2007, On-site repository construction and restoration of the abandoned Silver Crescent lead and zinc mill site, Shoshone County, Idaho, *in* DeGraff, J.V., ed., Understanding and Responding to Hazardous Substances at Mine Sites in the Western United States: Geological Society of America Reviews in Engineering Geology, v. XVII, p. 105–113, doi: 10.1130/2007.4017(06). For permission to copy, contact editing@geosociety.org. ©2007 The Geological Society of America. All rights reserved.

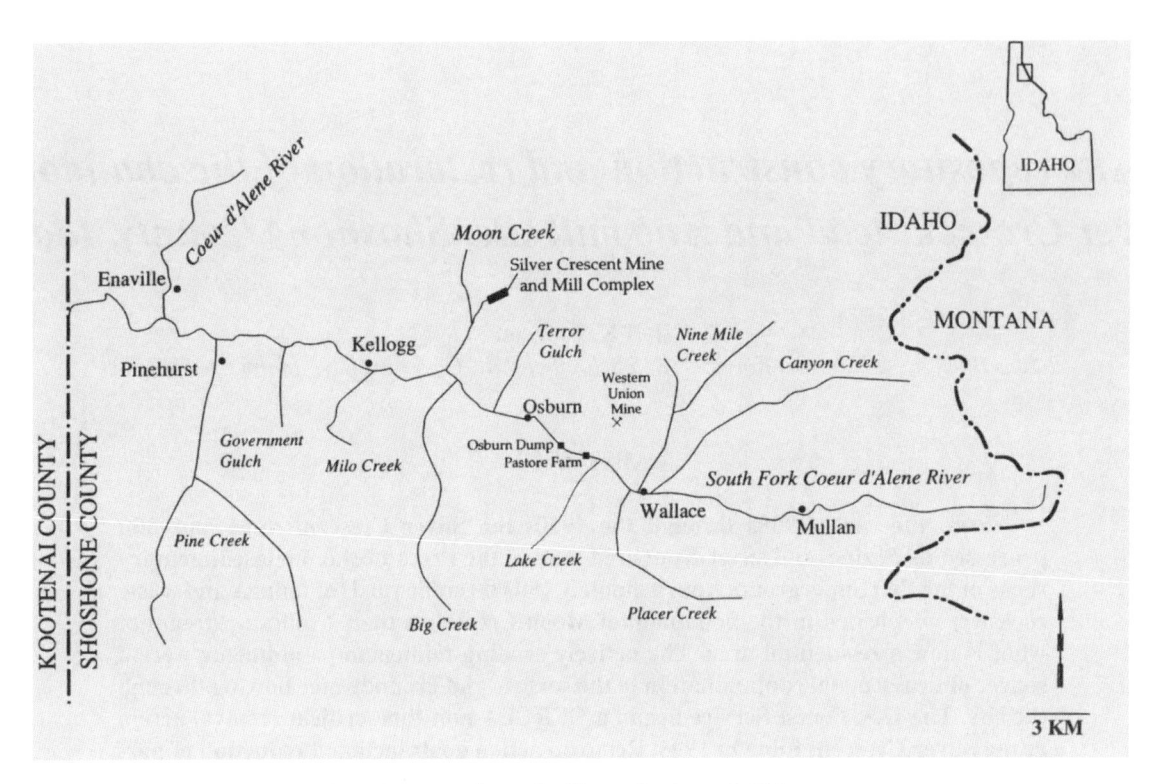

Figure 1. Location map of the Silver Crescent mine and mill complex in north Idaho.

Access to the site was unrestricted, and the flat tailings impoundments were being used as all-terrain-vehicle recreation areas.

Between 20 and 30 homesites are located in the downstream valley bottom adjacent to the East Fork of Moon Creek between the site and the South Fork of the Coeur d'Alene River. Shallow alluvial wells are the primary drinking water source at these homesites, with some residents using surface water from Moon Creek for irrigation.

GEOLOGY AND MINERALIZATION

The rocks in the area consist of the Precambrian Prichard Formation of the metasedimentary Belt Supergroup. Thick bedded gray quartzose argillite is the dominant rock type at the site, with graded bedding and ripple marks present (Hobbs et al., 1965). Ore minerals including galena and sphalerite are typically found in vein deposits throughout the area. Reported ore grades at the site were 3.48 oz/ton silver, 7.98% lead, 0.86% zinc, and 0.34% copper (Fryklund, 1964; Mitchell and Bennett, 1983). The valley bottom of Moon Creek is occupied by 15–30 feet of alluvium consisting mainly of sand and gravel.

SITE HISTORY

The earliest record of mining and milling at the site is from 1902 (USBM, 1995). Two separate properties existed at the site in the early 1900s, the Charles Dickens mine and the Silver Crescent mine. These properties were combined in 1935, and the site was active through the 1950s.

Early activity at the site involved a concentrator and jig mill at the upstream end. Later, in the 1920s and 1930s, a 150-ton-per-day flotation mill operated at the site, and several thousand tons of ore were processed. This activity involved several mine portals, as well as at least one vertical shaft. The majority of processing and tailings generation at the site occurred while the mill operated as a custom mill in the 1940s and 1950s. During this time the mill reprocessed older ore grade tailings from other sites in the Coeur d'Alene mining district (USBM, 1995) resulting in three large tailings impoundments at the site. The site was inactive from the 1960s through the 1990s with the exception of a few brief exploration projects.

SITE CHARACTERIZATION

In May 1992, the U.S. Bureau of Mines (USBM) in partnership with the FS began a comprehensive site characterization effort. This characterization study was focused on evaluation of the site features and identification of any hazards associated with the mine waste. The characterization study was completed in 1995 and provided to the FS as part of the CERCLA process but was never formally published. Widespread water, sediment, and soil contamination was shown at the site involving lead, zinc, cop-

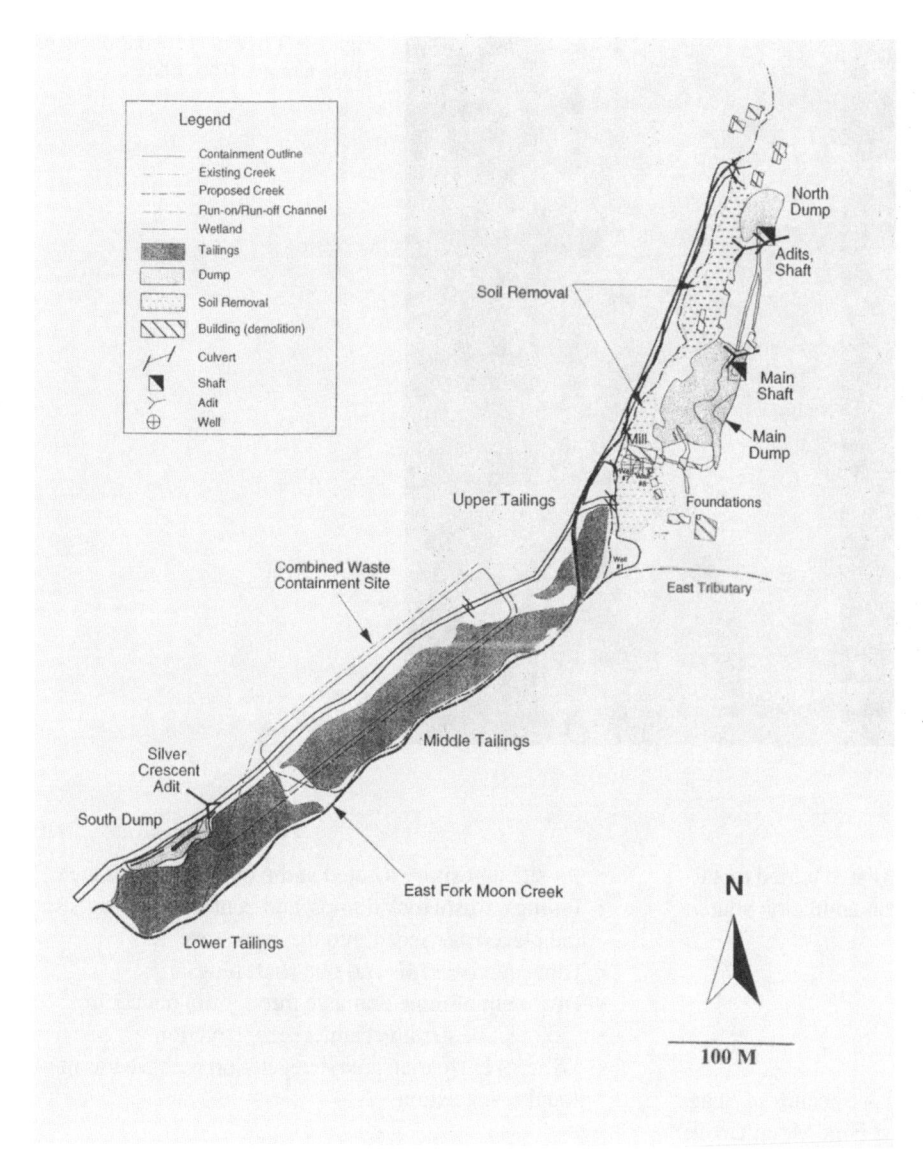

Figure 2. October 1996 preremoval map of the Silver Crescent site including the combined waste containment (repository) area.

per, and cadmium. The following summary outlines results from the USBM characterization study:

- Both mill tailings and mine drainage appeared to contaminate Moon Creek with metals.
- Mill tailings appeared to contribute sediment to Moon Creek.
- Approximately 35,000 cubic yards of mill tailings from 3 to 15 feet thick were present at the site.
- Tailings occurred as two distinct layers in four subponds.
- The basal layer of each pond was a blue gray "slime" (heavy-metal rich, silt-sized tailings); the upper layer of each pond was an orange and brown sandy material.
- More than 90% of the dissolved cadmium and 50% to 80% of the zinc in Moon Creek came from a seep along the creek bed near the mill site.
- Copper and lead were removed from the stream as it passed through the tailings.
- Zinc and lead concentrations were higher in pore water from the sandy upper layer of tailings material than in the lower slimes.
- Concentrations of lead were ~50% higher in the sandy material (top of tailings).
- Concentrations of zinc were ~50% higher in the slimes (bottom of tailings).
- Biologically, the benthic community was impacted immediately below the mine and tailings pile on Moon Creek.

In 1996, additional characterization work was accomplished by FS personnel in order to determine the nature of the metal-rich seep in Moon Creek near the mill. The origin of the seep was found to be influent-contaminated groundwater. The saturated

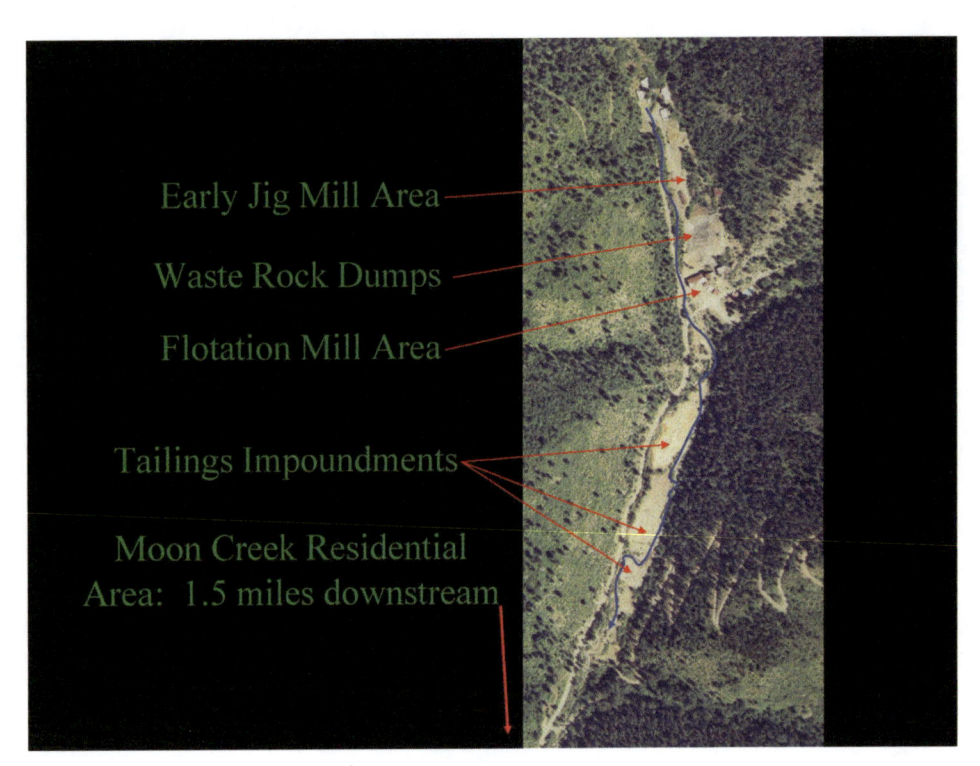

Figure 3. Preremoval aerial photo of the Silver Crescent site, 1996.

zone up-gradient of the seep was found to consist of buried metal-rich jig tailings that were identified as the contamination source for the groundwater.

CLEAN-UP ALTERNATIVES

In October 1996, Ridolfi Engineers and Associates of Seattle, Washington (Ridolfi), completed the East Fork Moon Creek Engineering Evaluation and Cost Analysis (EECA) for the FS as part of the CERCLA process. Site conditions, nature of contamination, and human/environmental health risk evaluations were used to develop and select a removal action alternative.

Focus areas for the alternatives included flotation and jig tailings, waste rock dumps, contaminated soils, buildings, mine openings, mine drainage, stream channel, revegetation, erosion control, and use of a repository. The three alternatives analyzed in the EECA were no action, on-site stabilization with containment, and off-site disposal. Off-site disposal would appear to be a logical choice for a large removal in such a narrow stream valley. The selection of the off-site alternative was hindered due to the added risk and complication associated with moving the wastes through the downstream neighborhood combined with the possible consequences and added cost of introducing the wastes to another site.

On-site stabilization with containment was shown to be economical as well as effective and was selected as the removal action alternative for the site. Components of the selected alternative included the following:

- On-site repository located at the middle tailings area
- Tailings, waste rock dumps, and contaminated soils sorted and placed as needed into the repository
- Three adit backfills and one shaft plug
- Treatment of mine drainage through source control
- 2700 feet of stream channel reconstruction
- 4.6 acres of riparian area revegetation with 5 acres of additional revegetation

The middle tailings area was selected as the location for the repository after careful analysis of the contamination processes and pathways at the site. Although the middle tailings depth extended into the saturated zone, the potential for long-term leaching and loading of dissolved metals from this area into the ground and surface water systems was low (USBM, 1995).

REPOSITORY CONSTRUCTION

Description of the repository construction is primarily derived from the author's on-site observations as well as the project design drawings, specifications, and engineer's report prepared for the FS by Ridolfi (Ridolfi Engineers, 1997 and 2000).

Base Preparation

The in-place fine-grained mill tailings of the middle tailings area required extensive preparation work in order to serve as a

Figure 4. Preparation of the repository base in the middle tailings area.

suitable base for the repository structure. Initial excavation of the middle tailings area was accomplished, and excess tailings were stockpiled outside of the repository footprint. The repository base was also built into the valley wall by excavating 6-foot-high, 10-foot-wide stair-step notches into the bedrock (Fig. 4).

A 5-foot-deep, point-down triangular compacted gravel drain was installed along the perimeter of the in-place tailings base. The deepest 12 inches of the compacted gravel drain consisted of compacted limestone gravel with a 4-inch perforated PVC pipe at the bottom. This drain was connected to a subsurface outlet with monitoring ports at the downstream end of the repository area. Intermediate drains were also constructed across the width of the repository base. These drains are located on 88-foot centers along the length of the repository and are 130 feet long, 5 feet deep, and 1 foot wide (Fig. 5). The intermediate drains are lined with geofabric, filled with gravel, and slope at 2% toward their perpendicular intersection with the perimeter drain. The drainage network within the repository base serves to provide a conduit for water flowing from and around the base area. The intermediate drains also served to relieve hydrostatic pressure during construction associated with loading heavy, wet material on top of the possibly saturated fine-grained base materials.

The repository base preparation next involved covering the area with geofabric and construction of a geocell mattress (Fig. 6). This mattress is comprised of numerous perforated plastic cells in a honeycomb-like orientation. The dimension of each cell is

Figure 5. Intermediate drain construction in the repository base.

Figure 6. Installation of the geocell mattress system on the repository base.

~1.5 feet by 1.5 feet and 8 inches deep. The mattress sections were laid out and fastened together to form a seamless horizontal base layer containing thousands of empty cells. The empty cells were then overfilled with crushed waste rock and compacted to form a solid flat surface (Fig. 7). The geocell mattress layer is an important soft foundation engineering design component of the repository providing a stable and strong, yet porous, reinforcing layer on which to load waste material. Without this reinforcing layer, repository base stability would be questionable, and unacceptable settling would likely occur over time.

Repository Waste Placement

Wet tailings at the site were excavated and stockpiled for several days in order to dry the materials as much as possible. Initially, depth of excavation to clean substrate was determined using sampling and laboratory metals analysis data. Visual field indicators of contamination, such as material color and grain size, were rapidly recognized and used during the majority of the removal to guide excavation, with additional periodic sample analysis used for confirmation. Geochemical analysis of the waste rock, tailings, and

Figure 7. Filling of the geocell mattress cells with crushed waste rock.

soils indicated that the materials on site were compatible and could be placed together in one repository. Tailings were added to the repository area in 6-inch lifts. After each lift of tailings, a much more coarse-grained mixture of contaminated soil and waste rock was placed on top in another 6-inch lift. The 12-inch layer of tailings and waste rock/soil was then mixed with a disc and compacted. This process was repeated until the repository was full. This mixing and compacting method was incorporated to create a solid and stable fill absent of discrete fine-grained layers that could potentially act as failure planes.

Repository Cover

Results of synthetic leach tests on the material at the site were used with the hydraulic evaluation of landfill performance (HELP) computer model to predict leachate generation and develop the overall containment design. This analysis indicated that a bottom liner would not be necessary for this repository because a sloped water balance capillary break cover would provide adequate protection from water infiltration. By eliminating water flow through the wastes, contaminated leachate generation would cease. An even more conservative approach was preferred by the FS, and they chose a water balance/low-hydraulic-conductivity hybrid design that would provide an extra level of security. The 3.5-foot thick cover design (Fig. 8) incorporated eight specific layers. The bottom five layers of the cover consist of a geosynthetic clay liner (GCL) with associated cushion, geofabric, and coarse gravel drainage layers. The premanufactured GCL is comprised of bentonite sandwiched between two pieces of woven geotextile and has a permeability at 5 pounds/inch^2 of 5×10^{-9} cm/second. Because of this extremely low permeability layer, precipitation such as snowmelt that passes down through the top three soil and vegetation layers of the cover will preferentially flow downslope above the GCL in the gravel drainage layer. This multilayer cover effectively keeps water from infiltrating the contents of the repository. The cover design emphasizes low maintenance requirements, with the thickness and choice of materials allowing for future rooting of plants and trees. Concurrent with completion of the cover, an armored surface water diversion channel was built between the upper repository perimeter and the hillside in order to prevent future precipitation events from flowing across and eroding the cover surface. Figures 9, 10, 11, and 12 illustrate stages of the cover placement and early revegetation.

In total, 131,700 cubic yards of tailings, waste rock, and contaminated soils were moved into the repository. The completed repository measures 807 feet in length, 220 feet in width, and 44 feet in height.

POSTCONSTRUCTION MONITORING AND FUTURE RESTORATION

The FS has been monitoring the site since the end of construction in 2000. The repository has been stable, with no significant settling cracks or failures. All-terrain-vehicle use at the site has ceased, because the area is now closed to all nonauthorized motor vehicles.

Analytical results from groundwater and surface water samples strongly indicate an increase in water quality due to a reduction in heavy-metal contaminants. With the exception of dissolved zinc concentrations measured at 0.2 ppm in low-flow surface water conditions, all other metal concentrations measured in surface waters through the site have been below human health and chronic aquatic life criteria, representing a large improvement.

With an increase in water, soil, and sediment quality and all human health concerns addressed at the site, the East Fork of Moon Creek boasts potential for fisheries, wildlife, and terrestrial restoration. The entire site is in the winter range for area big game animals such as elk, and the East Fork of Moon Creek provides valuable spawning and rearing habitat for Westslope Cutthroat trout in the South Fork of the Coeur d'Alene River drainage.

During the removal action construction process, the stream channel, floodplain, and flood-prone area were greatly altered. Over 3000 feet of the East Fork of Moon Creek's channel needed to be moved several times in order to remove tailings, waste rock, and contaminated soils. Although contaminants at the site are now controlled and the disturbed area is relatively stable, the channel and floodplain areas are still in need of restoration work in order to achieve premining habitat quantity and quality. The existence of the repository adds many unique hydrological considerations when adapting stream restoration methods for the site. The stream reach next to the repository will be further enhanced to not only protect the repository from erosion but also provide for fish passage and habitat opportunities. In addition, the repository itself

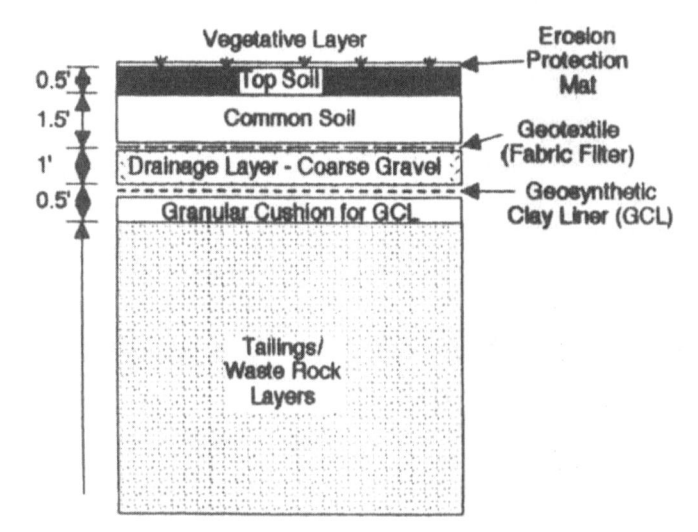

Figure 8. Cross-sectional representation of the engineered repository cover.

Figure 9. Cover construction: granular cushion layer installation.

will be planted with native shallow-rooted brush species to further prevent erosion, reduce future maintenance, and provide beneficial wildlife habitat. The eventual brush field on top of the cover will dominate the growing area, delaying the natural introduction of deeper rooted trees.

Fisheries, wildlife, and terrestrial restoration work at the site is currently in the final planning stages. Funding has been acquired from multiple cooperators, and the site restoration work is being considered a demonstration for future restoration of similar sites in the Coeur d'Alene watershed.

Figure 10. Cover construction: common soil layer on top of geotextile and drainage layers.

Figure 11. Cover construction: final placement of organic erosion protection mat.

Figure 12. July 2003 photo of the repository area 3 years after construction.

REFERENCES CITED

Fryklund, V.C. Jr., 1964, Ore deposits of the Coeur d'Alene Mining District, Shoshone County, Idaho: U.S. Geological Survey Professional Paper 445.

Hobbs, S.W., Griggs, A.B., and Wallace, R.E., 1965, Geology of the Coeur d'Alene Mining District, Shoshone County, Idaho: U.S. Geological Survey Professional Paper 478.

Mitchell, V.E., and Bennett, E.H., 1983, Production statistics for the Coeur d'Alene Mining District, Shoshone County Idaho, 1884–1980: Idaho Bureau of Mines and Geology Technical Report 83-3.

Ridolfi Engineers, 1996, East Fork Moon Creek engineering evaluation and cost analysis: Produced for the U.S. Forest Service, unpublished.

Ridolfi Engineers, 1997, Construction plans for East Fork Moon Creek Reclamation Project: Produced for the U.S. Forest Service, unpublished.

Ridolfi Engineers, 2000, Moon Creek Reclamation Project engineer's post-reclamation report: Produced for the U.S. Forest Service, unpublished.

USBM Western Field Operations Center Staff, 1995, Preliminary characterization Silver Crescent Mill tailings and mine site, East Fork Moon Creek Watershed, Shoshone County, Idaho: Produced for the U.S. Forest Service, unpublished.

Manuscript Accepted by the Society 28 November 2006

Approaches to contamination at mercury mill sites: Examples from California and Idaho

Jerome V. DeGraff
U.S. Department of Agriculture Forest Service, Sierra National Forest, 1600 Tollhouse Road, Clovis, California 93611, USA

Michelle Rogow
U.S. Environmental Protection Agency, Emergency Response Section, 75 Hawthorne St., San Francisco, California 94105, USA

Pat Trainor
U.S. Department of Agriculture Forest Service, Payette National Forest, 800 West Lakeside, McCall, Idaho 83638, USA

ABSTRACT

Abandoned or inactive mercury mines are found throughout the western United States. Mercury contamination from these mines has migrated into a variety of different media in varying forms. Cleanups and mitigation projects have been undertaken by various agencies and private entities at a number of these mines, although many remain to be addressed. Although each cleanup has similar objectives, such as source control, the methods employed in each area of the site may differ. By having an understanding of mercury and its effects and assessing different methods used at mercury-mine cleanups, future actions can be more effective at addressing the variety of issues posed by mercury contamination at former extraction and processing sites. This paper provides background on mercury, its occurrences, its health effects, and the mercury mining process. Four cleanup sites that utilized different methods for addressing mercury contamination illustrate how different sources at abandoned mercury mill sites may be addressed to mitigate impacts.

Keywords: mercury mill, calcine, retort, western United States

INTRODUCTION

History

Mercury mining in the western United States began almost 200 years ago, had periods of great prosperity and decline, and now has almost entirely ceased. Any examination of mercury mining in the western United States is intertwined with the 1849 California Gold Rush and the many gold mining booms than followed. Ironically, it was mercury rather than gold that was first discovered in California. In the 1820s, Mexican settlers discovered the red deposits used by local Native Americans living in the vicinity of modern-day San Jose. However, it took Andres Castillero, a native of Spain and a Mexican Army officer, to recognize that the ore represented a source of mercury. In 1845, he and others established a company to exploit the deposits. By 1846, the rudimentary mercury mine and mill site was sold to an English industrial firm that named the site New Almaden after the famous Almaden quicksilver mine in Spain.

DeGraff, J.V., Rogow, M., and Trainor, P., 2007, Approaches to contamination at mercury mill sites: Examples from California and Idaho, *in* DeGraff, J.V., ed., Understanding and Responding to Hazardous Substances at Mine Sites in the Western United States: Geological Society of America Reviews in Engineering Geology, v. XVII, p. 115–134, doi: 10.1130/2007.4017(07). For permission to copy, contact editing@geosociety.org. ©2007 The Geological Society of America. All rights reserved.

The California Gold Rush caused a dramatic increase in the demand for mercury to use in amalgamation, the principal means for gold recovery at that time (Davis and Bailey, 1966). Another peak in demand came in the 1860s with the introduction of hydraulic mining in the Sierra Nevada. Consequently, there was both an expansion of production at existing mines like New Almaden and an intensive exploration and development of deposits throughout the Coast Ranges. The greatest mercury-mining activity in California took place in the 1870s with production continuing at high levels until 1883 (Bureau of Mines, 1965). To one degree or another, this pattern of mercury and gold discovery and mining was repeated throughout the western United States.

Around 1900, changes in gold mining and the introduction of cyanide recovery methods reduced demand for mercury with a corresponding effect on production. Peaks in domestic mercury production were linked to worldwide availability of mercury, price fluctuations, and, especially, increased demands during wartime from World War I through the Korean War (Bureau of Mines, 1965). In 1991, large-scale mercury mining ended in the United States. This was a consequence of the easy availability of mercury from recycling and increasing environmental concerns making continued mining uneconomic (Sznopek and Goonan, 2000). As a result, mercury mines across the western United States have ceased production, and many are abandoned. Because of the fluctuating demand for mercury, mitigation measures at mines were not often under-taken, for many operators hoped to restart production when mercury was in higher demand. Profits made during earlier mining are often long gone, and the legacy of these historical mercury mines is one of contamination migrating into surrounding soils and waterways.

Mercury Occurrence

In assessing contamination resulting from mercury mining, it is notable that mercury occurs within a number of mineralized belts in North America including those within the Coast Ranges of California, the Pacific Northwest, and the Great Basin. Rytuba (2003) has documented that mercury mining from these areas since the 1800s amounts to ~130,000 tons. Mercury deposits are present in areas where extensive volcanic and tectonic activity has occurred. The mercury deposits form by deposition of ore minerals from low-temperature aqueous solution (Davis and Bailey, 1966). The primary ore mineral is cinnabar, or mercury sulfide (HgS). The cinnabar fills fractures and voids or replaces the host rock. In a number of instances, the replaced host rock is a silica-carbonate. This mode of emplacement results in small, irregular ore bodies. Deposits associated with hot springs are consistent with general emplacement at a shallow depth. Based on these characteristics, mercury deposits are classified as epithermal deposits. While cinnabar is most prevalent, elemental mercury (Hg^0) may also be present in small but important amounts (Gray, 2003a). Because cinnabar is located within certain areas and in particular strata, a conglomeration of mercury mines located near each other is frequently found, and these mines may impact the same watershed or water bodies within close proximity. In prioritizing mercury mines for cleanup, it is important to look at not only each mine but also the types of mercury that are present and any hydrologic connection of the mines to local and regional water bodies.

Recovering Mercury from Ore

The ore quality and the type and age of processing systems have a substantial impact on the quantity, types, and locations of mercury that may be present at a mercury mine. Understanding the processing system and the ore present at the mine can provide clues for source identification and potential migration pathways. A common feature of the mercury ore development was the presence of processing facilities at or near the mines. Rytuba (2003) notes that co-location of mining and processing was especially true of Coast Range mercury mines. This co-location was facilitated by the extraction process being both relatively simple and inexpensive (Gray, 2003a).

Once the ore was separated from the host rock, the first step in recovering the mercury was a simple crushing of the ore. The crushed ore was roasted at a temperature above 600 °F, greater than the stability of the mercury mineral, causing the release of elemental mercury vapor. Mercury vapor was cooled and collected in liquid form. The quality and quantity of mercury was partially dependent upon the means used for roasting the ore. Roasting of a few tons of ore per day could take place using a retort, a small, externally heated, oven-like furnace (Gray, 2003b) (Fig. 1). Processing of up to 100 tons of ore per day required employing a large rotary furnace. As its name implies, the internally heated furnace was rotated as ore was introduced. The elemental mercury released from cinnabar ore as a vapor. The mercury vapor was contained and piped or conducted through a condenser assembly. Water was often utilized to assist in cooling the vapors so that the mercury became liquid. The liquid mercury accumulated in troughs from which it was collected and placed in flasks for shipment. Flasks are cylinders of cast iron or steel that hold 76 pounds of mercury, a unit of trade recognized worldwide (Davis and Bailey, 1966). So, recovering mercury at a mine only required enough space to build a retort or furnace, convenient sources for heating fuel and cooling water, and rudimentary access to transportation. This resulted in processing areas being located near mining operations.

The mercury recovery process of heating ore to vaporize the mineralized mercury is called calcination. The resulting mine wastes are termed calcines. Calcines have a characteristic red brown color due to oxidation of iron sulfides and some remaining amounts of cinnabar (Rytuba, 2003). Because the processing involves both crushing and heating the ore, a significant portion of the calcines is relatively fine-grained. This waste was usually placed near the processing area, sometimes forming a large mass adjacent to the mill where tracks permitted ore carts to dump the material. In other instances, the location of the mill on a slope or adjacent to a stream resulted in either a large fan-like deposit of ma-

Figure 1. Typical retort of the type used for roasting ore before rotary ovens came into widespread use.

terial on the slope or extending into the stream channel where some volume would be periodically swept away during flood events.

Mercury Forms, Transport, and Effects

Understanding the history of the mine and its mineral processing offers key indications into the location and forms of mercury-contaminated wastes and potential migration pathways. It is also important to comprehend the different forms of mercury that exist, their potential exposure pathways and relative risk. This understanding can provide the basis for selection of the appropriate removal action and mitigation of harm to human health and the environment. From solid or liquid forms, mercury may be released into the air as a gas or a particulate. Depositional forces encourage settling of airborne mercury onto land and surface waters. Mercury on land may be transported via rainfall, erosion, and leaching into streams and larger waterways. This mercury migration has the potential to cause exponential harm due to the mercury's changing forms and providing opportunities for bioaccumulation.

The human health effects related to mercury exposure are dependent on the chemical form, dose, age and health of an exposed individual, and duration and route of exposure (EPA, 2005). There are three forms of mercury: inorganic, elemental, and organic. These three forms have differing primary routes of exposure, elements of toxicity, and impact on humans and the environment. In addition, the three routes of potential mercury exposure—ingestion, inhalation and absorption—may cause varying degrees of health effects. By having knowledge of the forms of mercury that exist and the locations of these wastes, relative risk can be assessed and mitigated.

Inorganic mercury, such as mercury sulfide, is a prevalent contaminant in the environment, most especially at mercury mine sites. Inorganic mercury, or mercury salts, generally occur as white powders or crystals except for mercury sulfide, which is red (ATSDR, 1994). Although inorganic mercury is often the primary chemical of concern at mercury mine sites, it is less toxic to humans than other forms of mercury, and ingestion is the least common or likely exposure pathway. Inorganic mercury is not readily absorbed by the digestive system and has a low bioavailability. In large doses, inorganic mercury can cause gastrointestinal and kidney distress. Dermal and inhalation exposure can result in skin rashes/dermatitis, memory loss, muscle weakness, pulmonary edema, and elevated blood pressure.

Elemental mercury is liquid heavy metal, metallic silver in color. Although it is a metal that is liquid at room temperature and pressure, it has the tendency to volatilize if not properly contained. As a gas, mercury is colorless and odorless. Although elemental mercury is rare in the natural environment, anthropogenic alteration, such as mining and thermal processing, of mercury sulfide can form elemental mercury. Both inorganic and elemental mercury can be converted into organic mercury, such as methylmercury (CH_3Hg), under certain biological and chemical conditions.

Elemental and organic mercury are more detrimental because these forms are more likely to impact the brain. Inhalation of elemental mercury vapor is the most significant exposure pathway because mercury is readily absorbed into the bloodstream from the lungs and travels to the other organs, including the brain and kidneys. Inhalation of elemental mercury can cause lung damage, eye irritation, tremors, headaches, nausea, emotional impacts, insomnia, and gingivitis. At higher concentrations, exposures may result in kidney damage and possibly death. Dermal exposure through direct contact can cause allergic reactions such as dermatitis. When ingested, elemental mercury will pass through the digestive system with minimal absorption.

Organic mercury compounds are formed when mercury binds with carbon. Organic mercury is readily absorbed by the body and passes into the bloodstream, targeting the brain and central nervous system. Unborn fetuses and young children experience the greatest risk from exposure. Impacts to children may result in effects to neurological development (ATSDR, 1994). Organic mercury can be rapidly absorbed through the skin, although dermal absorption of methylmercury is minimal.

Although there are a number of organic mercury compounds, methylmercury is the most common. Methylmercury may be formed in the aquatic environment depending on a number of factors including mercury type and concentration, water temperature, pH, and the presence of organic matter, dissolved solids, and microorganisms. Other compounds such as sulfur may increase concentrations of methylmercury (EPA, 2005; U.S. Geological Survey, 2000). Methylmercury is soluble in water, and it can be absorbed by aquatic biota and bioaccumulate in the tissue of fish and other aquatic organisms. It also may biomagnify in the food chain to result in greater concentration with increasing position in the food chain (Gray, 2003a). Human exposure to methylmercury is primarily through ingestion of mercury-contaminated fish and seafood. Methylmercury impacts the immune, genetic, enzyme, and nervous systems. Because it is a neurotoxin, it is believed to be one of the most harmful forms of mercury to humans and wildlife. Adverse effects of mercury on wildlife have been documented and may result in death, reduced reproductive success, and impaired growth and development (NEWMOA, 1999).

Although the predominant source of mercury in most mercury-mining sites is inorganic, other forms of mercury pose a greater risk to humans and the environment. Effective mercury mine cleanups require a thorough determination of the forms of mercury waste, concentrations of contaminants, and environmental conditions on- and off-site. Mercury mine tailings in and around streams have the potential to be transported into larger waterways and environments that may encourage methylation. Therefore, migration of contamination off-site and the propensity for methylation of mercury must be considered. This information is essential to selection of the option(s) to address the site contamination. Following the pathways of potential exposure and knowing the receptors, fate and transport mechanisms, and forms of mercury is imperative to mitigation of the threat posed by the mercury contamination at a mercury mine site.

Contaminated Media and Receptors

Possible media impacted by a mercury release include surface water, air, soil, and groundwater. Surface water includes overland flow during precipitation events, drainage pathways, streams, lakes, and oceans. Contamination of surface water may occur when rain causes materials to be transported in solution or as sediment from the site into the stream network. Due to the generally insoluble nature and density of elemental mercury, it does not tend to travel readily in solution. Rather, inorganic mercury, especially in fine-grained sediment, has a tendency to be transported via rainfall into surface waters. This occurrence can be far more detrimental because the mercury-rich sediment can be deposited where action is favorable to microbial formation of methylmercury. At mercury mill sites, avoiding formation of this toxic and soluble form of mercury is a significant reason for preventing migration of mercury-contaminated sediment into the downstream environment.

Mercury contamination in air has two contributing factors, elemental mercury vapor and airborne particulates from mercury-laden sediments and soils. Elemental mercury may be in and around process areas and may be trapped within parts of the equipment resulting in localized high concentrations. The presence and levels of elemental mercury vapors released are dependent on the amount of elemental mercury at the surface, disturbance, and temperature. Temperature is proportional to the release of elemental mercury vapor; therefore, release of mercury vapors on-site increases at higher temperatures. Other air pathway exposure results from disturbance of mercury-contaminated soils and tailings. Simply walking on a mercury-contaminated site may cause an increase in airborne particulates yielding an increased exposure. Both elemental and inorganic mercury may be inhaled by persons coming into contact with the wastes.

Soil is another medium by which people or wildlife may to come in physical contact with elevated concentrations of mercury. Mercury-contaminated soil can be transferred to food and ingested. Dermal exposure to mercury can occur via direct contact with soils. Soils may be transported and deposited in other locations.

Groundwater impact occurs when contaminants are leached or released into soils or groundwater. When contaminants are soluble in water they are likely to travel with rainfall in its path of groundwater recharge. At mercury mine sites this is not a major medium of concern due to the relatively insoluble nature of mercury and depths to groundwater. At mercury mines where mining activities have intercepted the groundwater table, evaluation of potential methylmercury impact to groundwater may be necessary.

Receptors are the people and the environment that will be impacted by the contaminant. The people affected may include nearby residents, recreational users (hikers, campers, ATV riders), and workers who may be on the site during the course of their job. Environmental receptors would include wildlife exposed on-site and those subject to mercury transported from the site to their location. Introduction of methylmercury into areas downstream is an example of the latter circumstance.

CLEANUP OF MERCURY MILL SITES

Authority

The actions taken to address a release or potential release of contaminated materials from a mercury mill site are authorized under the Comprehensive Environmental Response, Compensation, and Liability Act (CERCLA; 42 U.S.C. § 9601, et seq.). The specifics of the response are developed pursuant to the guidelines of the National Oil and Hazardous Substances Contingency Plan (NCP), 40 C.F.R. 300, et seq. (1995).

At the four sites, removal actions were implemented to address contaminated materials and the physical situation in which they were located. Removal actions do not necessarily leave the site free of contamination. In a removal action, the response mitigates a release or threat of release of materials that would impact humans and the environment. Mercury-contaminated materials at mill facilities include but are not limited to: tailings, condenser soot, soil and structures impacted by emission deposition, as well as residual mercury within the retorts and other processing structures. These materials are often uncontained and have the potential to migrate.

Regulatory Constraints on Mercury Disposal

One important factor to consider in the selection of any removal action is the disposition of the wastes that are generated by the response. Mercury is a hazardous substance according to CERCLA and under certain circumstances may be a Resource Conservation and Recovery Act (RCRA; 42 U.S.C. § 6901 et seq.) hazardous waste. RCRA is the federal law that regulates the storage, treatment, recycling, and disposal of waste. Mercury is both a listed waste and characteristic waste under RCRA. The former means that certain types of mercury wastes are specifically cited in RCRA as hazardous. The latter establishes a certain level of contamination by which mercury is designated as a hazardous waste. The characteristic that mercury exhibits is toxicity. The toxicity test for RCRA is the toxicity characteristic leaching procedure (TCLP), and mercury contaminated waste is deemed hazardous if the TCLP test reveals concentrations exceeding 0.2 mg/L.

If the mercury waste is determined to be hazardous in accordance with federal regulations, the RCRA land disposal regulations (LDRs) apply to wastes that are disposed rather than recycled. Depending on the total mercury concentration, wastes may need to be stabilized or treated prior to disposal. States may also have hazardous waste regulations that need to be considered. For example, California law specifies that a total mercury concentration equal to or greater than 20 mg/kg determines it to be California hazardous.

For high mercury concentration hazardous waste, the only approved treatments at this time are incineration or utilization of a retort (e.g., thermal treatment of the waste). These processes may be possible but are often impractical and cost prohibitive for the large quantities of waste found at mercury mine sites. Other disposal alternatives are also possible for this type of waste, through request for an alternative treatment standard or a variance to the land disposal regulations. Both of these options require receptive federal, state, and disposal entities and take a substantial amount of time to approve.

Remedy Alternatives

The limited number of treatment and disposal options impacts potential remedies at mercury mine sites. Off-site disposal of mercury mine wastes requires adequate characterization to assist in determination of applicable regulations, potential disposal facilities, and response actions necessary. On-site repositories may be utilized, but concerns regarding location, maintenance of the repository, and leachability of the mercury need to be considered in assessment of this alternative. Treatment of mercury wastes has been accomplished with some success, although an analysis of the waste and pilot testing of the treatment protocol are essential to the success of this alternative. Traditional stabilization and treatment methods are often not sufficient for mercury wastes with high TCLP levels.

Because of the complexity of laws and regulations applicable to mercury waste, it is imperative to evaluate the on-site wastes in a manner that provides the best information for remedy selection. For example, although information exists regarding the relative total mercury concentration and potentially corresponding TCLP estimate, often this information is misleading, and it is better to run TCLP tests for the specific waste at the site. At one site, Rinconada mine, the predominant waste was calcines. Composite analytical sampling indicated total mercury concentrations throughout the site that were high, but TCLP results were under the federal hazardous level for mercury. This was due to the relative concentrations of inorganic, organic, and elemental mercury in the waste stream. Quantities of inorganic mercury versus elemental mercury and coarser materials versus fine materials are factors in the correlation between total and TCLP mercury. This characterization sampling at Rinconada mine revealed that the wastes could be treated as a federal solid waste and California hazardous waste, which drastically impacted the cost associated with disposal as an alternative.

MERCURY CONTAMINATION AT ABANDONED MILL SITES

Contamination Sources

Although mercury in our environment originates from a variety of sources, Gray (2003b) identifies closed and inactive mercury mines as locations where some of the highest concentrations of mercury can be found. These sites not only have high concentrations but also large quantities of mercury-contaminated wastes that are often uncontained, resulting in transportation of mercury off-site. Many of these abandoned mercury mines are located on public lands, and some are accessible to the public. At closed or abandoned mines, ventilation systems are either no longer operating or altered by collapsed workings, closed portals, or other circumstances. This increases the likelihood of mercury vapor concentrations within the mine. Elemental mercury vapor in the underground portions of the mine represents an inhalation exposure risk. Studies at mine sites in California and Nevada have documented the elevated mercury vapor present (Gustin et al., 2000).

Other than the mercury vapor that may be encountered in mine adits and openings, mercury mine works have been found to have relatively minimal mercury contamination compared to other areas of the mine. Most mercury mines have already extracted the cinnabar from the host rock with not much left behind. The host rock, which is often silica carbonate, has not been demonstrated to contain significant quantities of mercury in the absence of the cinnabar. Rock that has been removed from the mine adits but does not have value as ore is referred to as muck, waste rock, or overburden. Piles of waste rock are often found at the portals of mine adits and shafts and have varying mercury content depending on the grade and quantity of the mined ore. Generally, at mercury mines where there are higher-grade ore bodies or elemental mercury is present, there is a tendency to discard lower-grade material, resulting in higher mercury concentrations in waste rock. Conversely, at mercury mines that have a smaller quantity or lower grade of ore, most of the cinnabar has been processed, resulting in waste rock that is very low in mercury content. It is essential in examination of the mercury mine site to determine the concentration of mercury and other contaminants that may exist in waste rock piles. At mercury mine sites, waste rock may also be responsible for the significant problem of acid mine drainage (AMD). This is an important factor not only for stream health but also with respect to mercury methylation because increased methylation has been noted in lower pH environments (EPA, 2005). In addition, mine shafts and adits may have been used as repositories for mine wastes and present a risk due to concentrations and pathways for migration of contaminants.

Rather than the actual mine workings, it is the associated mill facility that is often responsible for the highest and most mobile concentrations of mercury at the mine site. As noted earlier, the inexpensive and simple nature of ore processing allowed most mercury mine sites to have a processing facility nearby. Concentrated levels of mercury in and around the processing area may result in exposure to humans who may traverse a contaminated mine site. Processing equipment may contain residual elemental mercury in addition to processed and unprocessed cinnabar. Any retorts, kilns, condensers, tanks, and troughs have a high likelihood of containing some elemental mercury.

Surrounding the processing area, mine wastes—or calcines—are often abundant and represent the largest source of mercury remaining at a mercury mine site. Like most ore recovery processes, calcination is not completely efficient. Calcines at sites in California were found to have concentrations of total mercury ranging from 10 to 1500 mg/kg (Rytuba, 2003). The calcines can contain not only residual cinnabar but also elemental mercury, metacinnabar, and various sulfate, chloride, and oxychloride compounds of mercury (Kim et al., 2000). Because of this, one of the main sources of mercury contamination and exposure present at a mercury mine site is the calcines. The concentration and location of these calcines is important in determining the relative risk and potential impact of these materials on humans and the environment. Although inorganic mercury poses the least amount of risk resulting from direct contact, it is the amount of fine-grained particles, elemental mercury present in the calcines, and potential mobility of these wastes that poses a threat through transport mechanisms and other exposure pathways.

A source closely related to calcines is condenser soot. Condenser soot is a result of the fine-grained particles of cinnabar traveling through the processing system with the mercury vapor. In the condenser, where vapor is cooled into liquid, the fine-grained particulates drop out of the vapor stream and are collected. This condenser soot was collected in various locations in the condenser system as well as in cooling water tanks associated with the condenser system. Because the particulates were in contact with the elemental mercury vapor stream, this mercury-rich by-product was often collected from the condensers and reprocessed in small retorts. Depending on the efficiencies or inefficiencies of the processing operations, the amount and quality of the condenser soot varied. Condenser soot often ended up on the ground around the condensers or was disposed of with the calcines. Because this material is fine grained in nature, it is subject to erosion and is likely to be transported greater distances than other, coarser-grained wastes on-site. In addition, former locations of cooling water tanks or ponds have been found to have some of the highest mercury concentrations found on-site, due to the presence of remaining elemental mercury and/or fine-grained inorganic mercury particles.

Elemental mercury generated during processing would be present in the retort or rotary oven and within the condenser system. Eventually, the air emissions were released through a stack (Fig. 2). When cooling of mercury vapor in the condensers was insufficient, mercury vapor and particulates were released through the stack and deposited on nearby vegetation and surface soil. Over time, the near surface soil around the stack areas could accumulate significant concentrations of mercury. Any structures in the vicinity of the stack had the potential for accumulation of depositional mercury. Porous surfaces such as wood buildings

Figure 2. Overview of the mill at Rinconada mine taken on January 3, 1954. Note the dust and gases venting from the chimney and coating the rocks on the adjacent slope. (U.S. Geological Survey photo)

may have accumulated mercury as well. The mill buildings often were not fully enclosed structures, making it easy for emissions to enter and result in contamination inside the mill itself. During processing operations, mercury vapor would also be present around the retorts, furnaces, and condensers. These vapors would penetrate any porous material such as bricks in the retorts and ovens. Some vapor would cool within the interior of the retort, furnace, or condenser leaving a mercury residue. Similarly, minor amounts of elemental mercury or mercury minerals from the ore could remain within the interior of the retort or furnace.

Each area of potential mercury contamination needs to be assessed when addressing a mercury mine site in order to identify contributing sources and forms of mercury that exist. In summary, the mill facility at a mercury mine may have varying forms and elevated concentrations of mercury present in a number of locations. A primary location is the on-the-ground disposal area(s) where the accumulated calcines and, sometimes, the condenser soot was placed. These are often referred to as the tailings piles or areas. A second location would be near the stack where emission deposition may be present in the near-surface soil. Emission deposition may also be found on porous structures near the stack. Finally, processing equipment such as retorts, furnaces, condenser structures, tanks, and associated piping systems represent a fourth location where mercury concentrations may be elevated. This would include porous elements of this equipment where vapor deposition would occur, their interior areas, and any waste material that has been disposed on the ground in the vicinity of this equipment.

FOUR MERCURY MILL SITES IN CALIFORNIA AND IDAHO

Response actions taken at three mill sites in California and one in Idaho illustrate some of the ways identified mercury contamination can be addressed to mitigate threats to human health and the environment posed by abandoned mercury mines (Fig. 3). These sites represent some of the varied site conditions encountered and the typical mill features that may need to be addressed. Three sites are situated within the Coast Range of California:

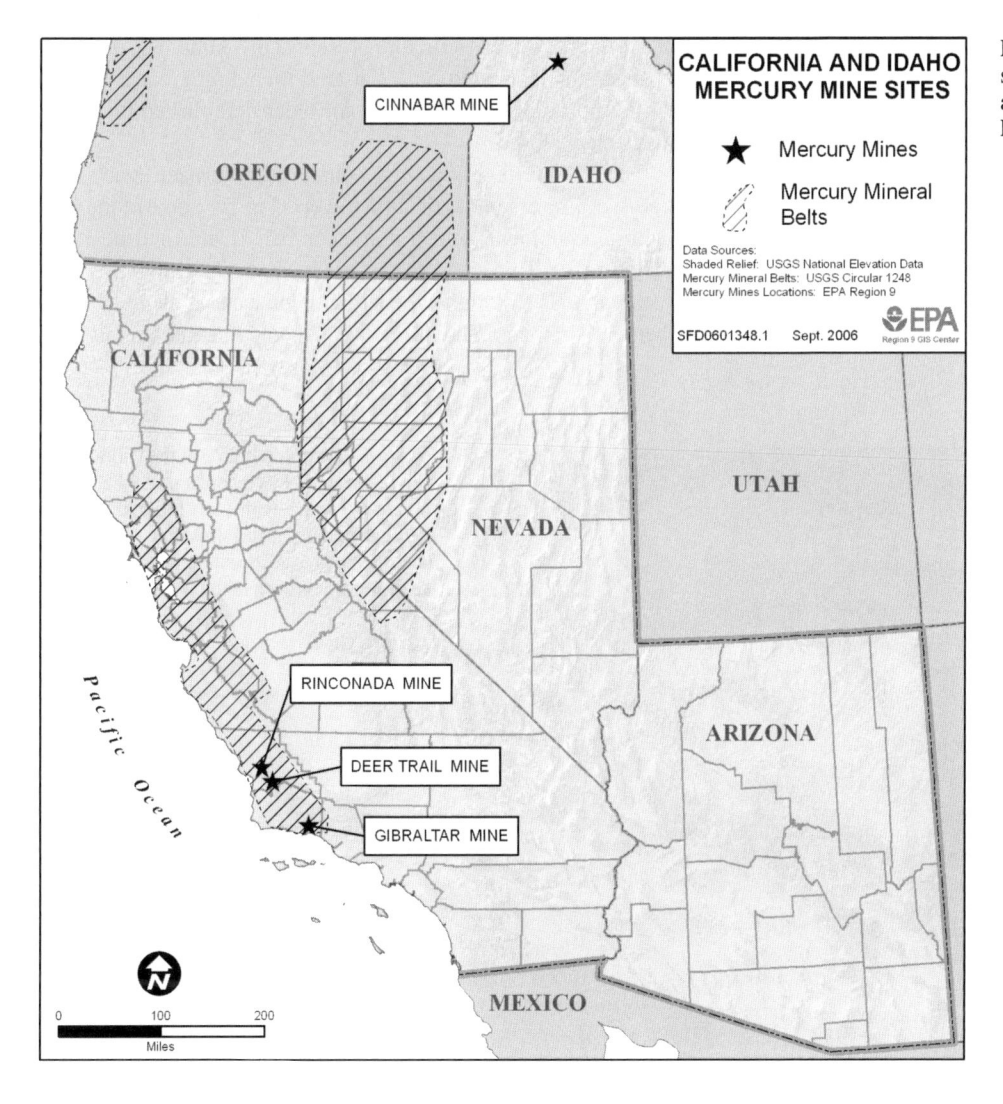

Figure 3. Map of western states showing some of the major mercury mineral areas and the location of Gibraltar, Deer Trail, Rinconada, and Cinnabar mines.

Gibraltar mine, Deer Trail mine, and Rinconada mine. The fourth site, Cinnabar mine, is located in central Idaho. All of these sites are located on or have impacted National Forest System land. Portions of the Rinconada mine and its associated wastes are located on adjacent private land, and federal land administered by the U.S. Department of the Interior, Bureau of Land Management (BLM). The Cinnabar mine is also located on both National Forest System and private lands.

Gibraltar Mine

The southernmost of these mine sites in the Coast Range of California is Gibraltar mine. The mill site is situated adjacent to the mine on the Santa Barbara Ranger District of the Los Padres National Forest. This location is ~8 miles north of Santa Barbara, California, on the south side of Gibraltar Reservoir near its easternmost end within the Santa Ynez River Valley.

At the site, the two-story mill building has an extensive, heterogeneous mass including calcines from the mill and waste rock from the mine. This mixture of mill tailings and mine waste rock extends from the building into the reservoir (Fig. 4). From the lower mine haulage way, ore was carted to a crusher in the uppermost part of the mill. On the lower main mill floor, ore was processed in two 30-inch-diameter, 40-foot-long rotary ovens with their respective condenser tower assemblies. The resulting calcines from the ovens emptied into a collection area where the material was transported using tracked carts moved to the tailings pile. The USDA Forest Service addressed the mercury contamination at the mill site, undertaking a removal action completed in 1999. Sediment and water monitoring continued until 2004.

Like many mines, the Gilbraltar mine's history of operations involves brief windows of activity separated by long idle periods. The Gibraltar mine traces its history to the 1860 discovery of mercury deposits in the Santa Ynez Valley by Jose Moraga. However, significant production did not occur until 1874. The major production period for the present mill site began around 1937. It was about this time that the two rotary ovens were installed. Activity ceased around 1942, resumed in 1956, and continued through 1963. Some sporadic activity occurred between 1965 and 1971 before the site was abandoned. The primary mercury-contamination issues addressed at Gibraltar mine are a consequence of the 1937–1963 production activity.

Deer Trail Mine

North of Gibraltar mine, in the vicinity of Santa Maria, California, is the Deer Trail mine. The Deer Trail mine site is located ~20 miles due east of Arroyo Grande, California, in San Luis Obispo County. The mill and associated mine are on the Santa Lucia Ranger District of the Los Padres National Forest. The mill is the principal surface structure present at this mine. The processing apparatus is located on a flat area including a concrete pad near the portal of the main adit. An older demolished mill may have been enclosed.

The mill consists of a single small rotary furnace attached to a free-standing assembly of eight towers (Fig. 5). Some of the ore-crushing equipment with a conveyor belt is attached near the rotary furnace. Two small retorts are located adjacent to this processing unit. The facility is constructed on a cut at the head of an ephemeral drainage. The mine access road passes across this area

Figure 4. The mill building at the Gibraltar mine. One of the two banks of condenser towers is visible with men removing the mercury-contaminated residue present. The calcine produced, extending to Gibraltar Reservoir, is seen in the upper right.

Figure 5. The mill at Deer Trail mine with a small rotary oven visible in the foreground. The condenser towers and ore hopper feeding the oven are seen in the background. Note the man for scale.

next to the mill. In 2004, the USDA Forest Service completed a removal action that addressed mercury contamination at the mine.

The Deer Trail mine was first located in 1915. By 1916, a 12-pipe retort was built and over a 3-month period produced 70 flasks of mercury. Disagreement among the owners resulted in the site's being idled in 1917 (Bradley, 1918). The records indicate that renewed production occurred from 1931 through 1935, yielding 106 flasks of mercury (Franke, 1935). The 12-pipe retort was torn down and ultimately replaced with a 3-pipe furnace. This was supplemented with a two-pipe retort. Bureau of Mines (1965) indicates that mining and processing occurred intermittently from 1928 through 1940. It was also active in 1951. The total production at the Deer Trail mine appears to be 200 flasks of mercury.

Rinconada Mine

Rinconada is the northernmost mine of the three California Coastal Range mine sites discussed in this paper. It is located on federal and private lands within and adjacent to the Santa Lucia District of the Los Padres National Forest, 11 miles southeast of Santa Margarita, San Luis Obispo County, California. The mine is adjacent to the trailhead for the Rinconada Trail, a hiking and equestrian recreational area. An ephemeral drainage, aka Mine Creek, flows through the main processing area mine tailings and discharges to Pozo Creek, which leads to the Salinas River. Although the primary processing area is ~0.5 miles from the trailhead, the other, older processing areas of the mine are located in closer proximity to recreational areas.

The Rinconada mine consists of ~7000 feet of underground workings (Bureau of Mines, 1965) on at least four levels through serpentine, sandstone, and shale (Eckel et al., 1941). Portions of the mine are still open and accessible, although the federal agencies are working cooperatively to restrict public access. The main mill processing area and the surrounding vicinity are the primary area of concern at the site. The processing area was comprised of a 54-foot-long rotary furnace, retorts, dust collectors, condensing units, settling tanks, and a variety of stacks. Formerly, the area contained a jaw crusher, fan, cast-iron condensers, and redwood settling tanks (California Division of Mines, 1939). Condenser dust, partially processed materials, and mill tailings were located in and around the processing area and in and around Mine Creek. Mill tailings and condenser soot have been transported from the site to Pozo Creek and beyond (Olsen and Rytuba, 2003).

Two other areas were identified as processing areas. The first, near the Rinconada Trailhead parking lot, contained a retort in a collapsed aluminum structure. The second, to the southwest of the main entrance road, contained the remnants of an old retort. Adjacent to the retort were the remnants of processing equipment, partially processed ore, condenser soot, and tailings. The mine's processing areas, tailings piles, and ~840 feet of impacted stream were addressed in two phases of a removal action during 2004–2005.

The mine was alleged to have been discovered during the period of Mexican sovereignty, but federal records show production began in 1872; an inefficient furnace was installed in 1876; and operations occurred intermittently for the following 11 years. Mine operations were believed to have been conducted in 1897 and from 1915 until 1917 (California Division of Mines, 1939). In 1920, the 4-foot by 54-foot rotary furnace was installed, and it operated intermittently for a decade with various entities making improvements to the plant. In 1930, a fire destroyed the plant, but it was rebuilt and operations continued. In 1943 and 1944, the mine was operated as part of the war effort and operations were documented from 1951 until 1961 (Fig. 2) and from 1965 until 1968. It has been estimated that the mine produced 3000 flasks of mercury (Bureau of Mines, 1965). Although the mill building was consumed by a wildfire in 1985, the foundation, retort, condenser structure, rotary oven, and related structures remained intact.

Cinnabar Mine

The Cinnabar mine site is a mixture of private and federal land in Valley County, Idaho. The federal land is administered by the Payette National Forest. Cinnabar mine is located ~15 miles east of the community of Yellow Pine, Idaho. The site included a rotary kiln, flotation circuit, assay lab, and a number of tailings piles (Fig. 6). The locations of two tailings piles on National Forest System land known as the south tailings and north tailings impoundments involved removal actions by the USDA Forest Service.

Because the mine impacted private and public lands, cleanup involved participation of different agencies. A removal action in 1992 diverted Cinnabar Creek from the south tailings to minimize mercury transport. In 1996 and 1998, two additional removals included disposal of the smelter roaster and other piles on private land near Cinnabar Creek. The most recent removal action, in 2003, addressed the north tailings impoundment.

Mining and processing for mercury took place intermittently during the twentieth century (Trainor, 2003). Although the mercury deposit was discovered ~1902, initial development at the Cinnabar mine (then called the Hermes mine) did not occur until 1921. Intermittent production through 1942 produced only 23 flasks of mercury, although a total of 6800 flasks of mercury were produced during the next 7 years. In 1952, 1244 flasks were produced, and another 633 flasks were produced from 1953 to 1956, when the mill building was destroyed by fire. The mill was rebuilt with a wet flotation and electro-separation process that produced 4299 flasks through 1959. Subsequent production was sporadic from 1963 to 1966, when operations ceased.

MERCURY MINE WASTE SOURCES AND SOLUTIONS

As discussed above, a thorough assessment of the mine site is necessary to identify the sources, forms, and potential exposure pathways of contamination. This information is essential to development of a clean-up action that effectively mitigates mercury-contamination and prevents migration. Selection of a particular response action should be based on the levels of mercury-contamination present, the types and quantities of mercury-contaminated materials, the physical environment at the mill site, exposure pathways, accessibility to the mill site or treatment and disposal areas, and costs involved with the action. The following discussion highlights a variety of mercury-contamination sources commonly found at mill sites and provides details on different remedy alternatives that were utilized at the four mine sites.

Processed Tailings and Condenser Soot

Tailings from processing of mercury ore create the largest volume of contaminated material at mill sites. Tailings consist of a variety of particle sizes, from coarse- to fine-grained materials. The tailings piles often contain a variety of processed materials, such as calcines and condenser soot. Mercury spills or discarded processing equipment may also be present in the tailings piles. Due to space constraints, tailings piles may be in numerous locations. Because concentrations of mercury in ore bodies and pro-

Figure 6. Cinnabar mine with the main mill building in the foreground. Various other structures and tailings piles are visible.

cessing itself were so variable during the operation of the mine, the concentrations of mercury present may vary greatly within this heterogeneous mass. For example, Gibraltar mine includes a large tailings pile. Although the initial sampling of the tailings found concentrations ranging from 13 to 130 mg/kg, concentrations up to 378 mg/kg were detected during a later tailings sampling event (Ecology & Environment, 1995).

At mercury mines, tailings were typically discarded adjacent to processing areas and are frequently on steep slopes, adjacent to stream channels, or at similar locations that facilitate an off-site release from these mill site features. Tailings may have been moved around the site or off-site or used for construction purposes and have been identified in roads in the area of these mine sites. Removal actions for tailings piles must be tailored to what may be difficult to reach locations or difficult working environments for heavy earth-moving equipment. The following examples illustrate some of the options that may be applied to address tailings at mercury mine sites.

Encapsulation

One strategy for addressing mercury mine tailings and condenser soot is to cap or encapsulate the material at its current location or nearby to prevent further release through the soil-contact, surface-water, or air pathways. This option was implemented at the north tailings impoundment of the Cinnabar mine. Development of the north tailings impoundment began in 1958 when the mine switched to a flotation process to recovery mercury. Cinnabar Creek passed along the west side of the tailings and tailings were migrating into the creek. Sampling of tailings yielded mercury concentrations as high as 1000 mg/kg as compared to background mercury concentrations of 15 mg/kg, and sampling in the creek identified mercury at higher levels downstream (91.3 mg/kg) from the tailings than from the upstream location (7.3 mg/kg).

When the removal action was implemented in 2003, an estimated 3000 cubic yards of mercury-contaminated tailings were contained in or relocated to this impoundment area for encapsulation (Trainor, 2003). Implementation required a 12-cubic-yard dump truck, excavator, front-end loader, and a small dozer. Tailings along Cinnabar Creek were excavated and removed to the tailings impoundment. The material in the impoundment was regraded to ensure that no water would pond on the pile surface, and the channel bank slope was reduced to a 3:1 grade. A layer of 8-ounce nonwoven geotextile was rolled onto the regraded surface of the impoundment and a geotextile was also placed on the channel slope and covered with riprap of 1-foot-diameter or greater adjacent to the creek. The riprap covered the toe to a vertical height of 3 feet above the current high-flow levels. A minimum of 18 inches of topsoil was then placed on the geotextiles to ensure free-draining surface flow across the encapsulated area. To prevent channelized erosion developing on the topsoil, the surface was mechanically roughed. Additionally, straw mulch and woody debris were placed after hydroseeding of the topsoil (Figs. 7A and B). Utilizing encapsulation, large quantities of tailings can be managed on-site to prevent erosion of mercury into Cinnabar Creek and beyond.

Selective Removal and Off-Site Disposal

Another method to address tailings and condenser soot at mercury mill sites is selective removal and off-site disposal. This option involves removal of highly contaminated tailings and transport to an approved off-site treatment, storage, and disposal facility. Excavations can be filled with nonimpacted soils, which also serve to cover any remaining subsurface tailings, preventing exposure pathways through soil, surface water, and air. Implementation is assisted by sampling to characterize the lateral and vertical extent of the mercury contamination. Confirmation sampling is necessary in some instances to demonstrate that the remaining tailings are below an established threshold value or to document what concentrations are left in place. Off-site disposal facilities require sampling or waste profiling to confirm that the concentrations of mercury in the material being sent for treatment and disposal are in compliance with the facility permit. Variations on this option for addressing contaminated tailings using selective removal and off-site disposal were applied at both the Rinconada and Deer Trail mines.

The main processing area at the Rinconada mine was built on a cut into the mountain, with the former mill site constructed on a concrete pad with a retaining wall. This configuration allowed for tailings piles to surround ~70% of the processing area upslope of Mine Creek. Throughout the duration of the operation and afterward, erosion caused migration of the tailings into the creek, which then washed tailings farther downstream. Sampling of the tailings piles around the processing area revealed concentrations of up to 7020 mg/kg (Ecology & Environment, 2004), whereas sampling of sediments in the creek downstream of tailings piles indicated concentrations of up to 319 mg/kg (Ecology & Environment, 2005a) as compared to 9.8 mg/kg (Rytuba, 2005) upstream of this area (Fig. 8).

During the implementation of the removal action in 2004, tailings were excavated from the west, north, and east slopes around the processing area in an attempt to reduce mercury concentrations in surface soils and minimize migration of remaining contamination.

An excavator, dozer, haul truck, and front-end loader were utilized in the operation. Excavated wastes were loaded into a haul truck and transferred to a stockpile area for consolidation and waste characterization sampling. The mercury-contaminated material was loaded into trucks for transportation and off-site disposal.

Excavated areas were backfilled with at least 18 inches of fill material taken from an unimpacted area of the site. The slope beneath the processing area was secured using subgrade key and fill to establish a slope of ~2:1. Topsoil was used to cover fill material to add some nutrient value, and disturbed areas were planted with native shrubs and hydroseeded. Erosion control measures, such as jute, coir logs, and straw wattles, were utilized to encourage temporary slope stability while vegetation established. Without the space for constructing a repository, by utilizing the selective removal and off-site disposal alternative, tailings that had the potential to migrate into Mine Creek were mitigated in a timely and cost-effective manner.

126   DeGraff et al.

Figure 7. North tailings of the Cinnabar mine showing conditions of the tailings material prior to encapsulation (A) and after encapsulation (B). The channel of Cinnabar Creek can be seen in the lower-right corner of these views.

Another site that utilized the selective removal and off-site disposal alternative was the Deer Trail mine. At this site, mine tailings had been dumped down a hillside. The steepness of the slope limited access except by the extended reach of an excavator (Fig. 9), and the road had proved too difficult for normal dump trucks to negotiate. Therefore, a long-arm excavator and an articulated dump truck established a working location on the mine access road above the tailings. Approximately 400 cubic yards of material were excavated from the road to near the bottom of the slope (Tetra Tech EMI, 2004). The tailings averaged 2–3 feet in thickness with a clear change in color when the underlying native soil was revealed.

Excavation of the tailings at the toe of the slope was facilitated by pioneering a temporary road with a backhoe. The backhoe excavated and placed material within reach of the excavator. As the backhoe was assisting in completing the removal of tailings

Figure 8. Excavation of mine tailings and condenser soot on the hillside below the Rinconada mine processing area and up-slope of the creek. A pipe in the concrete wall (seen below the rotary oven and stack) discharged processed material.

from the steep slope, it became apparent that an additional mass of tailings existed near the base of the slope. Remnants of an old condenser tower were found, with this material suggesting this was the approximate site of the original retort area. An additional 100 cubic yards of material were excavated, but an estimated several hundred cubic yards of tailings remained, due to inaccessibility.

The articulated dump truck transported the excavated tailings to roll-off bins at a staging area where they were loaded and taken to an off-site treatment, storage, and disposal facility. The remaining tailings were backfilled with clean soil to a 2-foot depth and mulched with rice straw. The surrounding slopes and the drainage bottom are fully vegetated with no riparian fringe and a poorly defined channel indicating that water rarely flows. With the capping and stabilization measures, erosion is unlikely to expose the underlying tailings at this geomorphically stable location.

Wastes Deposited in Creeks or Waterways

As discussed previously, aquatic environments are the main route of mercury exposure from mercury mill sites. Therefore, one important source area to address during mercury mine cleanups is impacted waterways. Methods of addressing impacted waterways include removal of tailings, diversion of waterways, and stabilization in place. Restoration and restabilization of the creek after response actions have been completed are essential to the effectiveness of the remedy as illustrated by the following example from the Rinconada mine.

Tailings Removal and Creek Restoration

During the 2004 phase of the removal action at the Rinconada mine, cemented and loose tailings were excavated from ~360 feet of creek around the processing area. Although mercury concentrations in the creek were lower than the tailings on land around the processing equipment, this mercury-laden material was being transported off-site and downstream (Lockheed Martin, 2004). Creek tailings were removed to native materials, although in some instances native materials were found to be contaminated with mercury, requiring excavation to be continued to bedrock. The excavated channel section slope averaged 20%, and the bottom width was 4–10 feet. Due to the grade of the channel segment and presence of tailings still remaining upslope of the creek, a fully lined riprap channel was installed to reestablish the creek. Riprap design parameters such as filter fabric, filter bed size range and depth, and riprap size range and depth were determined for a 100-year return period storm duration. Riprap was placed to an average depth of ~3 feet. As the creek rounded the processing area to the end of the 360-foot length, the bottom width increased, and the channel side slopes became more gentle. A step transition was installed at the interface between the excavated creek and its downstream portion. The riprap channel was designed to provide a structurally stable channel that maximized flow and minimized ponding, thereby discouraging mercury methylation and erosion of tailings that remained in hillside slopes (Figs. 10A and B).

In 2005, another segment of Mine Creek downstream of the Rinconada mine was excavated because of a large tailings deposit in the course of the creek. Due to the magnitude of tailings and the fine-grained nature of the material, mercury methylation in this area and beyond was likely; therefore, this area was prioritized for removal and restoration. During this portion of the removal action ~480 feet of creek were excavated, including 4747 tons of tailings, which were transported off-site for disposal (Ecology & Environment, 2005b.)

Figure 9. Excavation of calcines at the Deer Trail mine using the 40-foot long excavator arm to reach from the existing mine road. The temporary road below is used by a backhoe to move the farthest calcines upslope within reach of the excavator.

In this segment, the channel slope of the restored reach averaged 12%; the bottom width ranged from 8 to 18 feet; and the side slopes from 5:1 to 1:1. The restoration of the streambed in this area was facilitated with the installation of a step pool profile. The step pool was designed to withstand peak flow from a 25-year storm; however, after the system is fully silted in, it should withstand a 100-year return period event. A total of 24 step pools were installed. The length of the step pools ranged from 13 to 20 feet to accommodate the meander, channel slope, and bottom width variations. This design allowed for high flows in the stream channel while mirroring the natural stream segments up-gradient and down-gradient of the excavated channel section. The design included a transition step between the natural channel and the excavated segment (Johnson, 2006). In the restoration of both segments of Mine Creek, channel banks were secured with installation of coir logs and, in steeper areas, straw wattles. Jute netting was installed on the largest excavated bank, and slopes were planted with native shrubs and hydroseeded.

Retorts, Furnaces, and Condenser Structures

Processing equipment at mercury mill sites may not encompass a large footprint but often contains some of the highest mercury concentrations in a variety of forms. Retorts, furnaces, and condenser structures often contain fine-grained mercury-laden particles and elemental mercury. To select the best mitigation, each piece of equipment should be evaluated for contamination, risk, and potential route of exposure. Then a determination can be made regarding whether it is better to remove the structures or to mitigate the risk from the equipment without removal. Three of these alternatives—removal of residue and securing from public access, partial removal and sealing, and complete removal—were utilized at our removal sites and are discussed below.

Removal of Residues and Securing from Public Access

At Gibraltar mine, the highest mercury concentration detected was in residues present in a condenser trough within the mill building, suggesting that mercury posed a threat from both contact and vaporization. Consequently, the removal action involved collection and containment of all residues and soil in and around the condenser troughs. The two rotary ovens and the condenser tower interiors also contained mercury-laden residue. To assess airborne concentrations of mercury, the area was monitored with a Jerome mercury vapor analyzer. Prior to disturbing the residues and soil, mercury was detected in the air at concentrations from 0.003 to 0.188 mg/m^3. During the work to collect and contain the mercury-contaminated material, mercury concentrations doubled. The high-level contamination was addressed by collecting and disposing of the residue in the condenser troughs. Visual inspection and mercury vapor monitoring of the troughs was done to confirm the effectiveness of this action. A total of five 55-gallon drums of residue and soil was removed from the two condenser areas and sent for treatment and, if possible, recycling of mercury (Ecology & Environment, 1999). Although mercury vapor concentrations were reduced by these actions, elevated concentrations of mercury were still present.

The mill building was fenced to restrict the public from access, thereby preventing exposure to remaining concentrations of mercury. A total of 1200 feet of 6-foot high chain-link fence was used to enclose the mill building. Additionally, warning sings were placed on the fence at 60–70-foot intervals (Ecology & Environment, 1999). Monitoring of the security measures continues to ensure the area remains restricted.

Partial Removal and Sealing

At Rinconada mine, there was a substantial amount of infrastructure remaining from abandoned mill operations. Although the mill building had burned down in the Las Pilitas wildfire in 1985, the 54-foot rotary oven, condenser system, stack, retorts, tanks, and associated piping system were present until the time of the removal action in 2004. It was decided that each of the retorts would be fully removed due to the amount of contaminated material, porous nature of the retort materials, and lack of an effective way to prevent exposure and migration of the contaminants associated with each unit. Three retorts, associated contaminated tailings, and soils beneath were excavated and disposed off-site.

The rotary oven, condenser unit, and piping were made of materials that were nonporous (steel) and could be effectively sealed off to prevent migration and exposure to any remaining contents inside. To address risk from loose mill process equipment, tanks and any loose piping were disconnected from the system. If any mercury processing waste was present, it was removed, and equipment was sent to a scrap metal facility for recycling. Remaining process equipment, such as the rotary oven and condensers, were assessed for openings and access ways. Metal caps or covers were purchased or fabricated for each of the openings and installed to prevent any exposure from or migration of any remaining material inside the structures. The trough at the base of the condensers, the inlet to the rotary oven, and the cast iron bowl beneath the vapor piping were filled with concrete to prevent water from accumulating and encouraging migration of contaminants. In this manner, each piece of equipment was evaluated and effective solutions were implemented that allowed some of the historical process equipment to remain, without the risk to the public that had previously existed.

Complete Removal

Unlike Gibraltar or Rinconada mines, the rotary furnace, flue, condenser towers, and mercury collection troughs at the Deer Trail mine were completely removed (Fig. 11). There was no building enclosing the small ore processing facility, which simplified demolition and removal. The condenser towers were separated from the trough assembly, and the rotary furnace was cut from its supports with an oxyacetylene torch (Tetra Tech EMI, 2004). Bulkier items were cut or flattened by equipment to maximize the space in the 20-cubic-yard roll-off bins. Because the retorts were constructed primarily of bricks and mortar, it proved simple to reduce both retorts at the site into piles of rubble using a backhoe. When scrap areas were cleaned up, any extra condenser piles, brick potentially associated with retorts, or possible furnace components were placed into the bins with the rubble from the retorts and mill.

A total of six bins of material weighing a total of 26.5 tons was utilized to contain the debris for transportation (Tetra Tech EMI, 2004). Like the tailings excavated during this removal action, this debris was taken to a treatment, storage, and disposal facility.

Materials Contaminated by Mercury Emission Deposition

Besides the processing areas and associated waste materials, another source of mercury contamination at mill sites results from process emissions deposited on structures, soils, and vegetation. Unlike deposited tailings, or calcines, soil contaminated by aerosol mercury emissions is often identified near the surface. This form of contamination can be found near or within the mill structure where mercury vapor escaped and circulated, as well as at the point where the combustion and process gases exited the condenser system. Within the mill structure, mercury can be deposited on the surface of the wood or impregnated within it. Soil contamination near the stacks results from both vapor and particulate sources. Cooling vapors and particles drop to the ground surface or deposit on vegetation that later falls to the ground to become incorporated into the surface organic layer.

Emission deposition often reflected inefficient operations or temporary difficulties with the processing equipment. For example, it is reported that operational inefficiency at Rinconada mine reduced expected mercury recovery by over two-thirds (Collier and Pampeyan, 1954). Consequently, significant amounts of mercury vapor passed through the condensers and out the stack with other gases. It is important to identify areas that may have been impacted by mercury emissions. Once these areas have been identified, a number of solutions may be employed. Remedies at our study sites that are discussed below include: securing from public access, partial removal and capping, and complete removal.

Securing from Public Access

The mill at Gibraltar mine is large and includes two large rotary furnaces, each of which is connected to a 14-pipe condenser structure. A small brick retort is also present within the building. Clearly, a significant opportunity existed for mercury vapors to be emitted during the long operation of this facility. Samples of wood taken from the interior framing near the furnaces confirmed evidence of mercury contamination. Based on data gathered, the wood floor near the ovens and the retort would have similar contamination.

To prevent human and wildlife contact with the mercury-contaminated wood within the mill, it was surrounded by a 6-foot-high chain-link fence. Nearly 1200 feet of fencing were built to fully enclose the building and provide a safety area around it

Figure 10. (A) A large mass of eroded tailings in Mine Creek below the Rinconada mine processing area as seen prior to the removal work. (B) Mine Creek restored with a riprap-lined channel and stabilized revegetated slopes following removal of the tailings.

(Fig. 12). The safety area was primarily to protect the fence from future parts of the building collapsing on the fence because engineers had already established that the building was structurally unsound. Monitoring necessary to ensure the integrity of the fence identified a breach in 2002. The fence was repaired and remains intact.

Partial Removal and Capping

At the Deer Trail mine, mercury deposition was addressed with partial removal and capping of contamination. Most of the ore processing at Deer Trail mine utilized a small rotary oven fed by an ore crusher-conveyor-hopper apparatus, with mercury recovered from an attached condenser system consisting of 10 towers and ending in a small stack. Collection troughs were used at each tower. Similarly, a large and a small retort were located adjacent to the main ore processing facility. The retorts were likely used for processing condenser soot and ore evaluation. Both retorts exhausted combustion and ore-roasting gases into the air through an attached stack. It was expected that mercury emission deposition would be in the soil adjacent to the processing area and retorts. Sampling confirmed the presence of mercury up to 1100 mg/kg in soil samples (CDM, 2001).

As noted earlier, the main ore processing facility including the condenser towers and oven were removed from the Deer Trail mine as part of the removal action. Both retorts were also demolished and removed at the same time. To address the mercury emission deposition, the concrete slab and the soil extending out to the limits of the sampled transects were excavated. Soil was excavated within this footprint to a depth of ~18 inches (Tetra Tech EMI, 2004) and sent to the off-site treatment, storage, and

Figure 11. Removal of the rotary oven at the mill at Deer Trail mine took place after demolition of the condenser towers and ore hopper.

disposal facility along with the tailings material and mill and retort debris.

After excavation, confirmation samples were taken from the former ore processing and retort areas. Clean soil from a nearby borrow source was placed over the excavated area to cap any remaining contamination. Enough soil was placed to restore the original grade of the excavated area. Excavation equipment was used to compact the fill, restore the roadway, and ensure proper drainage.

Complete Removal

At the Rinconada mine, due to inefficiencies of the processing system and the numerous changes in mill operations, mercury deposition was extensive in one area, and a complete removal was utilized to effectively mitigate the contamination there. When the Rinconada Mill operated, the vapor stream from the 40-ton-capacity, 4-foot-diameter by 54-foot-long oil-fired rotary furnace flowed through asbestos-lined piping to a four-tower condenser unit system, settling tanks, and out a stack. At the time of the removal, the gas discharge was through a steel tank with four steel pipe stacks (Ecology & Environment, 2005a). This system was located adjacent to the condenser and settling tank system in relatively close proximity to the processing area.

From a review of historic documents and photographs, it was discovered that previous to the use of the steel tank and pipe stacks, a system of 22-inch terra cotta pipe, redwood tanks, and wooden pipe stack was utilized (Frank, 1935). In a later system, gas from the condenser passed through the clay pipe into concrete boxes prior to release out of the stack. The clay pipe was routed up the mountain on the southeast side of the processing area and was ~75 feet in length. Portions of the clay pipe, redwood tank, and concrete box systems were discovered during the removal assessment. Concern regarding contamination in the area of the clay pipes was raised upon further review of the historic documents, which recounted the inefficiencies of the recovery process (Yates, 1943), and old photographs, which revealed a plume emanating from the old stack (Fig. 2).

Sampling of this area revealed mercury concentrations up to 6530 mg/kg. The highest concentrations were found in the concrete settling box, in close proximity to the clay pipe and around the former stack.

Due to the porous nature and condition of the clay pipe and concrete boxes, as well as the indication that mercury contamination could have migrated around and beneath the clay pipe, this system was removed, along with mercury-contaminated soils in the immediate vicinity. Excavation of the area was difficult due to the slope and the location of the pipe and processing equipment. The pipe was crushed in place and removed with soils beginning at the bottom and climbing slowly to the top, excavating and building benches for the excavator. Excavation in the area exceeded 2 feet in depth; in the upper segment, bedrock was exposed. This area was backfilled with borrow soils to restore slope and grade. Topsoil was applied, and the area was planted and hydroseeded, and erosion control measures were installed. At the uppermost section of the pipe slope excavation, bedrock was left exposed. As a result of heavy rains in the winter of 2004–2005, the slope was repaired in the summer of 2005, and a riprap drainage was installed on the slope and at the toe to minimize erosion and increase stability.

PLANNING A MERCURY MINE SITE CLEANUP

Aside from site assessment and remedy selection, other factors also come into play when planning and implementing removal actions at mercury mine sites. Cost and available funding influence what remedy may be selected. Weather strongly influences scheduling because many of the sites are subject to the seasonal impact resulting from winter snows or heavy rains. Due to the high elevation and location of Cinnabar mine, snow limited the time of the year that the site was accessible for cleanup. Removal operations at Gibraltar mine were impacted not only by the rainy season but also by high flows on the Santa Ynez River, preventing use of the four low-water crossings necessary to access the site. When planning a mine cleanup, it is essential to consider these factors to ensure as smooth as an operation as possible.

Access Issues

Another issue to consider in the planning process is access. Access roads to the mines have often fallen into disrepair and may not have been designed for the type of earthmoving equipment or trucks used during a mine cleanup. For example, typical tractor-trailer trucks for hauling bins could only come within 3 miles of the Deer Trail mine and 6 miles of the Gibraltar mine. A smaller, single-bin truck was needed to ferry the bins to a staging area accessible by the larger trucks. These issues complicate matters by requiring added equipment, locations, security, and coordination of scheduling. At Deer Trail mine, the mine access road was built at a grade too steep for normal dump trucks; therefore, specialized equipment such as an articulated 18-cubic-yard dump truck was utilized.

Where the roads are in disrepair, there may be a need to remove roadside brush to improve vehicle access to the site. At Gibraltar mine, several locations along the last 2.3 miles of the road were undermined by small landslides. To ensure that the large and heavy trucks could safely pass to the site, small crib-block retaining walls were used to restore those locations. To complete the removal at Cinnabar mine, a temporary bridge was installed in order to get across Cinnabar Creek.

Access may differ from the time when the mine was in operation due to changes in land use or ownership. Reaching Deer Trail mine required getting permission from private land owners and negotiating an agreement for ensuring their private road was restored after operation. At Rinconada mine, a trailhead parking lot for recreation use had been built. For the safety of recreational visitors, site clean-up workers, and heavy equipment, and protection of the response measures, the parking lot had to be closed during the removal action. This avoided both accidental exposure to mercury contamination and potential accidents between heavy equipment, trucks, and visiting vehicles.

Sometimes, portions of a site are difficult to access, and special measures must be used to address these areas. At Rinconada mine, access to mine waste was complicated by the topography, mill configuration, and location of processing equipment. The

Figure 12. Part of the fence placed to surround the mill at Gibraltar mine following removal of contaminated residue. Warning signs were placed at intervals on the fence.

north tailings pile was extremely problematic due to the construction of the processing area on top of a tailings fill with a 15-foot retaining wall. In this area, excavation was conducted slowly with geophysical monitoring to ensure that the retaining wall was not destabilized during removal operations. Tailings over the cleanup standard were left in place once the slope became questionable for the stability of the wall (Fig. 8).

Unanticipated Sources

Sometimes during a removal action, unanticipated sources are encountered. At one location at Rinconada mine, a relatively small tailings pile associated with an older retort was excavated for removal. Whereas mercury concentrations at the surface were found up to 1168 mg/kg (Ecology & Environment, 2004), upon excavation concentrations up to 20,424 mg/kg were identified and led to the discovery of portions of a retort buried in the tailings waste (Ecology & Environment, 2005a).

Usually at mill sites, wastes are found immediately downslope from the existing mill site. At Deer Trail mine, although elevated concentrations of mercury were anticipated downslope of the mill, sampling indicated mercury contamination over background but inconsistent with typical calcine waste (CDM, 2001). The material mantling the slope downslope from the mill was also not the characteristic red brown of mercury calcines. Further investigation revealed a nearby exposure on the slope below the access road that appeared more consistent with calcines in color. Sampling there found mercury concentrations from 380 mg/kg to 1000 mg/kg. For reasons known only to the operators at that time, the calcines were dumped for disposal at this location rather than at the mill.

Because these mines operated for such a long duration and through many operators and technologies, it is important to know the history and look for visual clues to potential mercury sources to fully and appropriately address the sources at the site. The more information that can be gathered and supported with sampling data, the more effectively a clean-up action can address the highest risk areas at mercury mill sites.

SUMMARY

Abandoned mercury mines are scattered throughout the western United States, with damaging effects of impaired waterways and fish advisories. In order to adequately assess the areas of impact from mining operations, it is imperative to have an understanding of the varied forms of mercury, its transport and transformation mechanisms, media of potential impact, and related health effects. Identifying sources and types of contamination as well as pathways of potential exposure and routes of migration is essential. For mercury mine sites, at the same time that it is important to consider the direct exposure pathways, it is also essential to prevent migration of contamination into waterways, especially those that may facilitate methylation and bioaccumulation in the food chain.

In order to appropriately address contamination at these sites, operational history, types and locations of processing, waste materials, and impacted areas are key pieces of information. These factors, along with the planning process, including addressing regulatory requirements, access, and scheduling, provide the necessary background to determine appropriate remedies for a particular mercury mine site.

Although mercury mines may have similarities, every situation is unique to the location, equipment, process utilized, types and quantities of waste generated, topography, and exposure and migration pathways. In addition, it is important to realize that a number of different alternatives may be utilized on a site in order to best address the contamination and situation. The four mine sites in this paper illustrate various removal options to address mercury contamination at ore processing sites. These sites represent a range of production levels that influence the volume and complexity of site cleanup. Each site involved a number of contaminated elements, exposure pathways, and remedies. These examples emphasize the range of options for particular sources of contamination where removal actions were designed to be the best approach to dealing with each particular site. The valuable experience gained through the implementation of cleanups at mercury mine sites in the past leads the way for future work to address contamination at mercury mine sites.

REFERENCES CITED

ATSDR, 1994, Toxicological profile for mercury: U.S. Department of Health and Human Services, Public Health Service, Agency for Toxic Substances and Disease Registry, Atlanta, p. 1–268.

Bradley, W.W., 1918, Quicksilver resources of California, California State Mining Bulletin 78: California State Mining Bureau, San Francisco, p. 124–133.

Bureau of Mines, 1965, Information Circular 8252, Mercury potential of the United States: Washington, United States Department of Interior, Bureau of Mines, p. 165–166.

California State Division of Mines, 1939, California Journal of Mines and Geology: v. 35, no. 4, p. 445.

CDM, 2001, Site inspection report, Deertrail Mine Site, Los Padres National Forest: CDM Federal Programs Corporation, Walnut Creek, California, 23 p.

Collier, J.T., and Pampeyan, E.H., 1954, Rinconada Mine, Rinconada District, San Luis Obispo, California: United States Geological Survey, p. 1–2

Davis, F.F., and Bailey, E.H., 1966, Mercury: Bulletin 191, Mineral resources of California, California Division of Mines and Geology, San Francisco, California. p. 247–254.

Eckel, E.B., Granger, A.E., and Yates, R.G., 1941, United States Department of Interior Bulletin 922-R, Quicksilver deposits in San Luis Obispo County and southwestern Monterey County California: United States Government Printing Office, Washington, D.C., p. 566–568.

Ecology & Environment, 1995, Gibraltar mine site: Baseline human health and ecological risk assessment, Los Padres National Forest, California: Seattle, Washington, p. 1–103.

Ecology & Environment, 1999, Non-time-critical removal action, Gilbraltar mine site, Santa Barbara, California: San Francisco, California, p. 1–12 plus appendices.

Ecology & Environment, 2004, Removal assessment report, Rinconada mine, San Luis Obispo County, California: Region 9 START for United States Environmental Protection Agency, San Francisco, California, p. 1–64.

Ecology & Environment, 2005a, Rinconada removal action report: Region 9 START for United States Environmental Protection Agency, Long Beach, California, p. 1–31.

Ecology & Environment, 2005b, Rinconada mine creek restoration, CY5 summary report: Region 9 START for United States Environmental Protection Agency, Long Beach, California, p. 1–21.

EPA, 2005, Mercury human health: United States Environmental Protection Agency, Washington, p. 1–8.

Franke, H.A., 1935, California Journal of Mines and Geology: San Francisco, California: State Division of Mines, v. 31, no. 1, p. 449–453.

Gray, J.E., 2003a, U.S. Geological Survey Circular 1248, Geologic studies of mercury by the U.S. Geological Survey: 41 p.

Gray, J.E., 2003b, U.S. Geological Survey Bulletin 2210-C, Leaching, transport, and methylation of mercury in and around abandoned mercury mines in the Humboldt River basin and surrounding areas, Nevada: http://geology.cr.usgs.gov/pub/bulletins/b2210-c/, 15 p.

Gustin, M.S., Lindberg, S.E., Austin, K., Coolbaugh, M., Vette, A., and Zhang, H., 2000, Assessing the contribution of natural sources to regional atmospheric mercury budgets: The Science of the Total Environment, v. 259, p. 61–71, doi: 10.1016/S0048-9697(00)00556-8.

Johnson, T., 2006, Rinconada mine creek excavation and restoration: United States Environmental Protection Agency, Las Vegas, Nevada, p. 1–2.

Kim, C.S., Brown, G.E. Jr., and Rytuba, J.J., 2000, Characterization and speciation of mercury-bearing mine wastes using X-ray absorption spectroscopy (XAS): The Science of the Total Environment, v. 261, p. 157–168, doi: 10.1016/S0048-9697(00)00640-9.

Lockheed Martin, 2004, Sample results (updated), Rinconada mercury mine: Lockheed Martin REAC for United States Environmental Protection Agency, New Jersey, 1 p.

NEWMOA, 1999, Mercury P2Rx topic hub: Northeast Waste Management Officials' Association, Boston, Massachusetts.

Olsen, J., and Rytuba, J.J., 2003, Geochemistry of mine tailings and sediment samples downstream from the Rinconada mercury mine, California: U.S. Geological Survey and U.S. Bureau of Land Management, Menlo Park, California, p. 1–3.

Rytuba, J.J., 2003, Environmental impact of mercury mines in the Coast Ranges, California: U.S. Geological Survey Circular 1248, Geologic studies of mercury by the U.S. Geological Survey, p. 13–18.

Rytuba, J.J., 2005, Summary of chemistry of sediment and waters sampled in mine and Rinconada Creeks: U.S. Geological Survey, Menlo Park, California: p. 1–3. Internal report to Bureau of Land Management.

Sznopek, J.L., and Goonan, T.G., 2000. The materials flow of mercury in the economies of the United States and the world: U.S. Geological Survey Circular 1197, 28 p.

Tetra Tech, E.M.I., 2004, Deertrail mine and mill site removal action completion report, Tetra Tech, EMI, Rancho Cordova, California, 27 p.

Trainor, P., 2003, Removal report, north tailings impoundment, Cinnabar mine, Yellow Pine, Idaho: USDA Forest Service, Payette National Forest Internal Report, McCall, Idaho, 10 p. plus appendices.

U.S. Geological Survey, 2000, Mercury in the environment: United States Department of the Interior, U.S. Geological Survey, Fact Sheet 146-0, Washington, p. 1–6.

Yates, R.G., 1943, Notes on the Rinconada mine, San Luis Obispo County, California: United States Department of the Interior, p. 1–2.

MANUSCRIPT ACCEPTED BY THE SOCIETY 28 NOVEMBER 2006

Approaches to site characterization, reclamation of uranium mine overburden, and neutralization of a mine pond at the White King–Lucky Lass mines site near Lakeview, Oregon

Kent Bostick
Professional Project Services, 545 Oak Ridge Turnpike, Oak Ridge, Tennessee 37830, USA

Norm Day
U.S. Department of Agriculture–Forest Service

Bill Adams
U.S. Environmental Protection Agency

David B. Ward
Jacobs Engineering Inc., 4300 B Street, Suite 600, Anchorage, Alaska 99503-5922, USA

ABSTRACT

Remediation of uranium mine overburden and an acidic mine pond at the White King–Lucky Lass mines near Lakeview, Oregon was completed in November 2006. The site was remediated under Superfund due to risk from arsenic and radium-226 in overburden soils. Separate clean-up standards were developed for each mine site for arsenic and radium-226 due to differing ore-body geochemistry. Gamma surveys were used to identify overburden with elevated radium-226 activities and to provide confirmation of visual clean-up of materials. Because arsenic is collocated with radium-226 at the White King mine, gamma surveys reduced the number of arsenic confirmation samples. Secular equilibrium in the uranium-238 decay series was used to determine the extent of leaching of uranium-238 and daughter products from overburden to groundwater. Trilinear geochemical analysis distinguished mineralized groundwater within the ore bodies from regional groundwater and detected any influence from seepage from overburden piles. Remedial actions include neutralization of an acidic mine pond and consolidation of elevated-activity overburden into a pile with a soil/rock cover at White King mine. Ecological toxicity studies determined that neutralization of the pond would provide a benthic community supportive of aquatic wildlife. An overburden pile at the Lucky Lass mine and disturbed areas were covered with clean soil. The remedial actions comply with State of Oregon siting regulations, which required removal of radioactive overburden from the 500-year flood plain. Protection of human health is assured by institutional controls to prevent use of mineralized groundwater and by fencing to prevent site access.

Keywords: uranium, arsenic, thorium, radium, radon, acid mine drainage, open pit, overburden, groundwater, surface water, neutralization, fracture flow, sulfides

INTRODUCTION

This report describes remediation of uranium mine overburden material and neutralization of an acidic mine pond at the White King–Lucky Lass mines near Lakeview, Oregon that was completed in November 2006. The mines were remediated under Superfund due to risk from arsenic and radium-226 in overburden soils. This report provides information on the location, historical land use, enforcement history, site risks, geology, and hydrogeology. It also presents the process of determining background levels for soil clean-up in the presence of ore bodies and the use of gamma surveys and gamma/radium-226 correlation coefficients to identify overburden for removal and to confirm clean-up. Evaluations of secular equilibrium in overburden materials; trilinear geochemical analysis of groundwater; and the transport of uranium, sulfate, and arsenic in groundwater were used to determine the potential for future contamination of groundwater by leaching of chemical constituents from overburden materials. The last section of the report describes remedial actions at the mines.

SITE DESCRIPTION

Location

The White King–Lucky Lass mines are two former uranium mining areas located in south-central Oregon, ~29 km (18 miles) northwest of Lakeview. The mines are located in the mountains adjacent to the northern boundary of the Goose Lake Valley within the Lakeview Ranger District, Fremont National Forest, Lake County, Oregon. The mines are near the edge of upland meadows within the Augur Creek drainage at an elevation of ~1800 m (6000 feet). The White King mine is on both the Fremont National Forest, which is managed by the United States Department of Agriculture–Forest Service (Forest Service), and private lands owned by Fremont Lumber Company and a trust. The Lucky Lass mine is situated 1.6 km (1 mile) northwest of the White King mine on Forest Service property. Locations of the mines are shown in Figure 1.

Historical Land Use

Both mines have had several owners, operators, mineral claims holders, and leases since mining began in 1955. Underground mine shafts were developed to a depth of 95 m (312 ft) at the White King mine but abandoned in 1959 due to infiltrating water. Open-pit mining techniques were used at the White King mine until mining stopped around 1965. Mining at Lucky Lass was conducted using open-pit mining techniques from 1956 to 1958 and from 1961 to 1964. An extensive exploratory drilling program was carried out at both sites through 1979. Since then, little activity has taken place on these sites. Available records indicate the White King mine produced ~125,324 metric tons (138,146 tons) of ore and the Lucky Lass mine produced ~4944 metric tons (5450 tons) of ore during their periods of operation. A total of 56.7 ha (140 acres) have been disturbed by mining: 48.6 ha (120 acres) at the White King mine and 8.1 ha (20 acres) at the Lucky Lass mine. Both mines have pits that have filled with water. The White King mine had an overburden and a proto-ore (protore) pile that are referred to collectively as overburden piles. The Lucky Lass mine has one overburden pile. Photographs of the mines are shown in Figure 2.

Enforcement History and Site Risks

The White King–Lucky Lass mines site was added to the National Priorities List in April 1995. The U.S. Environmental Protection Agency (EPA) was the lead regulatory agency, and the Forest Service, the Oregon Office of Energy (OOE) and the Oregon Department of Environmental Quality was the respective federal and state support agencies. A record of decision (ROD) was issued in September 2001 (EPA, 2001). Because the site was remediated under Superfund, it was not reclaimed, as is typical for other mines under other regulations. A major regulatory concern of the OOE was to remove overburden materials from the flood plain of Augur Creek. The Superfund response action was necessary to protect public health and the environment from actual or threatened releases of hazardous substances into environment. It was determined that the White King mine posed elevated risks for workers, recreational users, and future residents from multiple exposure pathways: ingestion of arsenic in soil, surface water, groundwater, and pond sediments; external radiation from radium-226 in soil; and inhalation of radon emanating from shallow groundwater. The risks at the Lucky Lass mine were similar, except there was no risk from ingestion of arsenic in surface water or pond sediment. An ecological risk assessment determined that terrestrial and aquatic receptors could be potentially harmed by exposure to surface soils, sediments, and stockpiled soil (EPA, 2001).

Geology

The White King–Lucky Lass mines site is located within the northwest terminus of the Basin and Range province. This area is characterized by north-trending fault-block mountains and intermontane basins with internal drainage. Geologic units include a thick sequence of volcanic flows and volcaniclastic rocks that have been extensively faulted and fractured. Six geologic units have been identified in the surface and subsurface of the White King mine. They are, from oldest to youngest, volcaniclastic rocks, rhyolite intrusive and associated tuff breccia, basaltic flows, additional volcaniclastic rocks and pyroclastics, alluvium, and stockpile overburden. Three geological units occur near the Lucky Lass mine. From oldest to youngest, they are volcaniclastic rocks, alluvium, and stockpile overburden.

The White King–Lucky Lass mines site is part of the Lakeview Uranium District, which encompasses an area extending 35 km (22 miles) to the north of Oregon Highway 140 and 28 km (18 miles) west of Lakeview. This 1000 km^2 (400-square-mile) area is host to ~20 areas of uranium and arsenic mineralization and former mining operations (Weston, 1996). The entire district has relatively high geochemical regional background values in uranium,

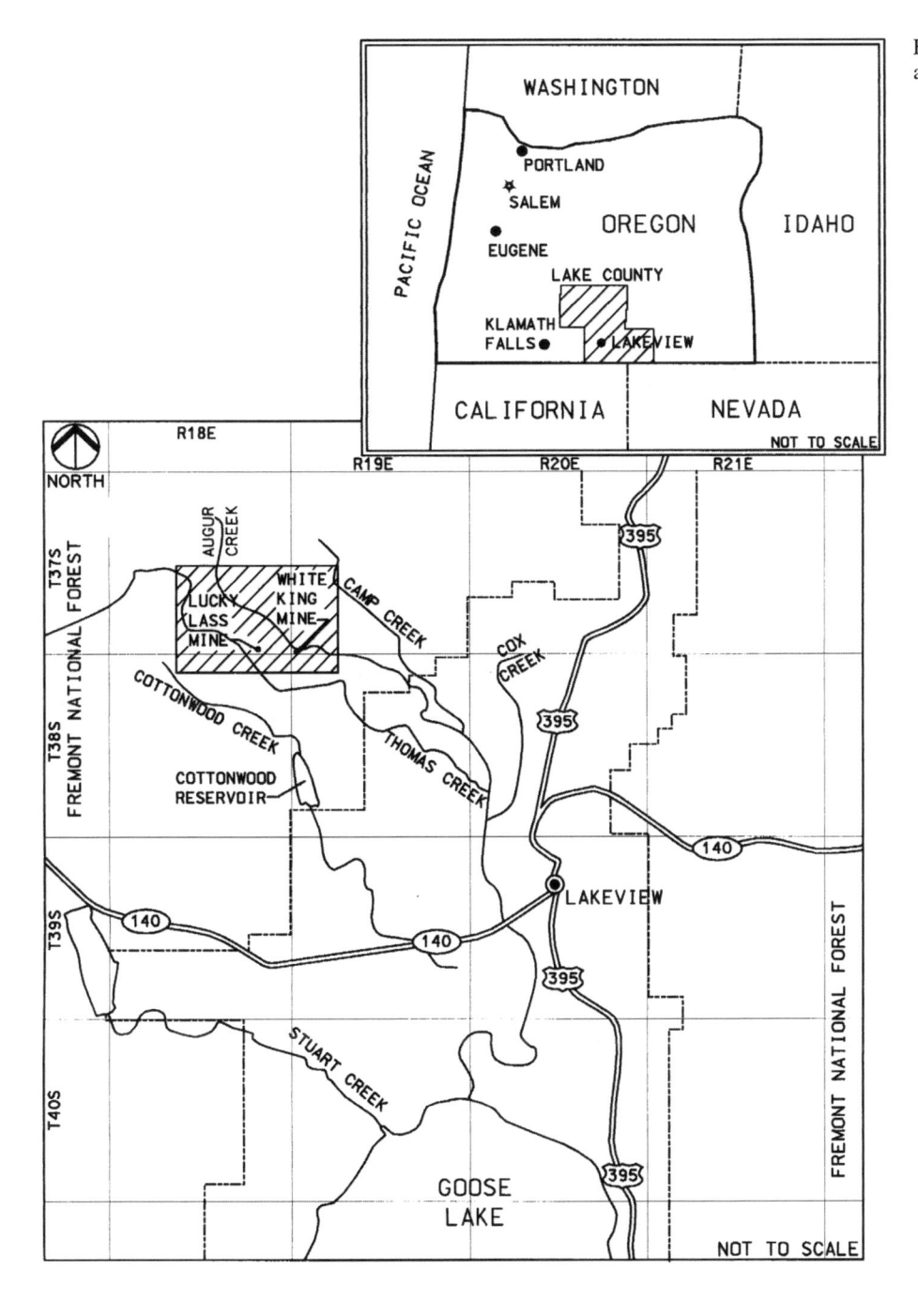

Figure 1. Locations of the White King and Lucky Lass mines.

arsenic, and other elements associated with the deposition of uranium ore. The extent of the ore bodies at each of the mine sites was limited to approximately the size of the mine ponds presently at the site. The ore bodies developed from precipitation of uranium from groundwater where geochemically reducing conditions developed from volcanic gases moving upward through fractures in the bedrock. Each ore body is surrounded by an area of mineralization with elevated background levels of uranium, arsenic, and other elements associated with the deposition of uranium ore.

Soils on valley side slopes are formed by weathering of basalt or tuff parent materials. Augur Creek meadow is covered by a thick mat of organic soil developed from the accumulation of partially decayed vegetation. Overburden piles at the mines were partially situated in the meadows.

Hydrogeology

Groundwater flow near the mines is influenced by both the local and regional topography and geology. Hydrogeologic units at the White King–Lucky Lass mines include the overburden piles, alluvium, shallow bedrock, and deep bedrock. The overburden piles consisted of mineralized soil with uranium- and metal-bearing

Figure 2. Photographs of the White King (A) and Lucky Lass (B) mines.

sulfides at the White King mine. Perched groundwater in the overburden piles was mounded on top of the underlying alluvial unit. A graph of White King mine groundwater elevation versus measuring-point elevation shows that water levels in the overburden piles constitute a distinct phreatic surface (Fig. 3). Recharge to the overburden piles is primarily from infiltration of precipitation. Discharge is by evaporation, downward flow into the underlying alluvial unit, or horizontal flow to seepage faces on the sides of the piles. The mean hydraulic conductivity for the White King mine overburden piles is ~1.7×10^{-6} cm/sec (4.8×10^{-3} ft per day).

Groundwater in the alluvial unit is unconfined. During the spring and early summer months, the alluvial unit is completely saturated. The water table in the alluvial unit reflects the local topography, with groundwater flowing down the Augur Creek valley. Groundwater elevations at the White King and Lucky Lass mines are shown in Figure 4. The alluvial unit is recharged by precipitation, by seeps and springs on bedrock outcroppings, and along some reaches of Augur Creek. Groundwater in the alluvial unit discharges to Augur Creek, shallow bedrock, and by evapotranspiration. The mean hydraulic conductivity of the White King mine alluvium is 4.7×10^{-7} cm/sec (1.3×10^{-3} ft per day) (EPA, 2001).

The shallow bedrock unit extends from the ground surface (except where it is overlain by the alluvial unit) to a depth of 30 m (100 ft) below ground surface. The depth to water in the shallow bedrock in the valley is less than 3 m (10 ft), whereas along ridges it can be more than 15 m (50 ft). Groundwater in the shallow bedrock unit flows in fractures and is unconfined. This unit is recharged by precipitation and by the overlying alluvium, where present. The mean hydraulic conductivity for the shallow bedrock unit at the White King mine is 1.8×10^{-6} cm/sec (5.1×10^{-3} ft per day) (EPA, 2001).

The deep bedrock unit is 30 m (100 ft) or greater below ground surface. Groundwater flow and storage in the deep bedrock unit also occurs in fractures. Deep groundwater occurs under semiconfined to confined conditions, suggesting that the deep fracture system is only weakly connected hydraulically to the shallow fracture system. The mean hydraulic conductivity of the deep bedrock at the White King mine is 1.4×10^{-6} cm/sec (4.0×10^{-3} ft per day) (EPA, 2001).

DETERMINATION OF BACKGROUND LEVELS FOR SOIL CLEAN-UP

Local background levels for arsenic and radium-226 in natural undisturbed mineralization at each site were used to distinguish overburden and native materials. These constituents were

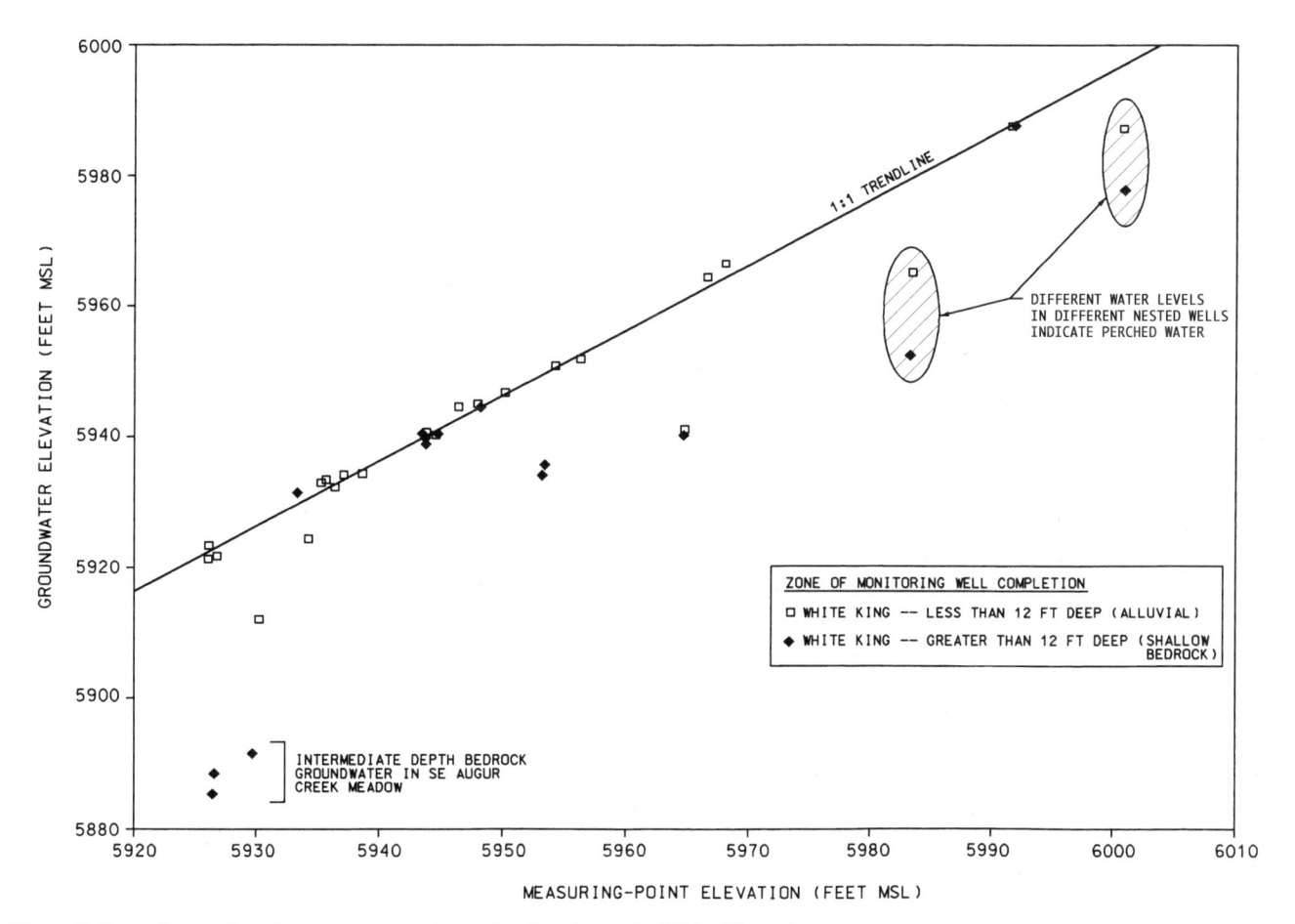

Figure 3. Groundwater elevation versus measuring-point elevation at the White King mine.

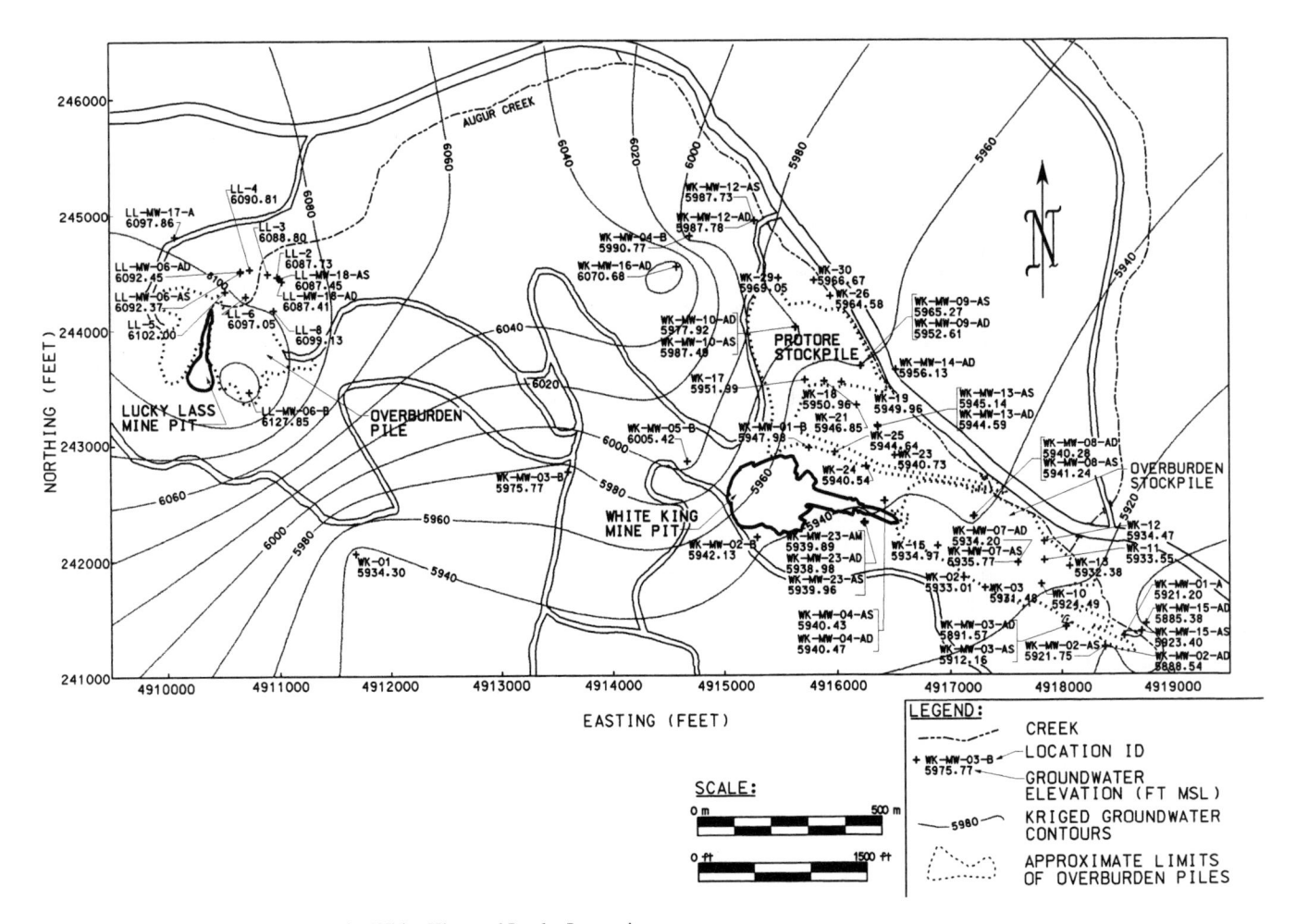

Figure 4. Groundwater elevations at the White King and Lucky Lass mines.

the major contributors to carcinogenic risk associated with soil exposure pathways. These background levels were essential to the development of clean-up goals that prescribe remediation of anthropogenic hazards while leaving naturally occurring material in place below and adjacent to the overburden piles. Although regional background levels were investigated, local background levels were found to be elevated by the presence of uranium ore in bedrock beneath the overburden piles. Because of different geochemical characteristics at the White King and Lucky Lass mines, separate local background levels were determined for each site. At the White King mine, these levels showed that a former organic meadow soil underlying the overburden piles was relatively unmineralized. The organic meadow soil was used as a marker bed for visual clean-up of overburden materials and then served as a natural cover for underlying bedrock mineralization.

Regional background soil concentrations for the soil samples collected outside the ore bodies are 6.4 mg per kilogram (mg/kg) for arsenic and 3.3 picoCuries per gram (pCi/g) for radium-226 (Jacobs Engineering, 1998). These were calculated based on the 95% upper tolerance limits (UTLs) for log-normal distributions of soil data. As site characterization proceeded, it became apparent that levels of mineralization in local native soil at the mines were substantially above regional background values. To evaluate local background levels, soil samples were grouped as native soil or overburden. Locations beneath or near the piles were omitted because they might have been affected by geochemical or physical processes that were not representative of local background conditions. In the Draft Final Remedial Investigation Report (DFRI), Weston (1997) proposed a local soil background of 1570 mg/kg for arsenic and 9.9 pCi/g for radium-226 based on maximum observed values at the White King mine. However, EPA selected the 95% UTLs because the maximum observed values were not sufficiently conservative for soil clean-up. In addition, EPA recommended in the ROD that local soil background levels be determined separately for each mine because of geochemically dissimilar ore bodies.

At the White King mine, 95% UTLs for subsurface soil arsenic and radium-226 levels were 442 mg/kg and 6.8 pCi/g, respectively. To check the validity of these local background numbers as preliminary remediation goals (PRGs), samples were ranked by radium activity according to material types and compared to maximum background and 95% UTLs. Figure 5 shows ranked arsenic concentrations at the White King mine. Ranked radium-226 activities and arsenic concentrations for the White King mine are presented in Figure 6. These figures indicate that

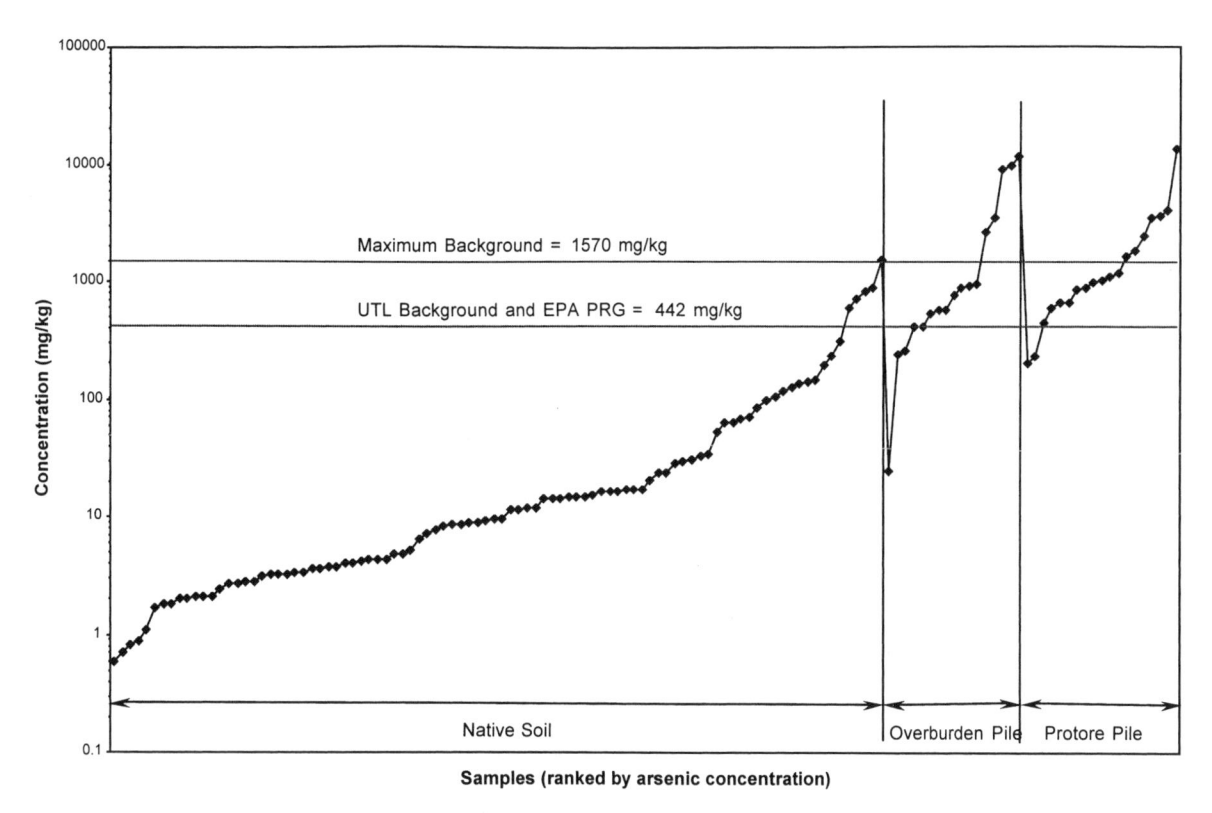

Figure 5. Ranked arsenic concentrations at the White King mine.

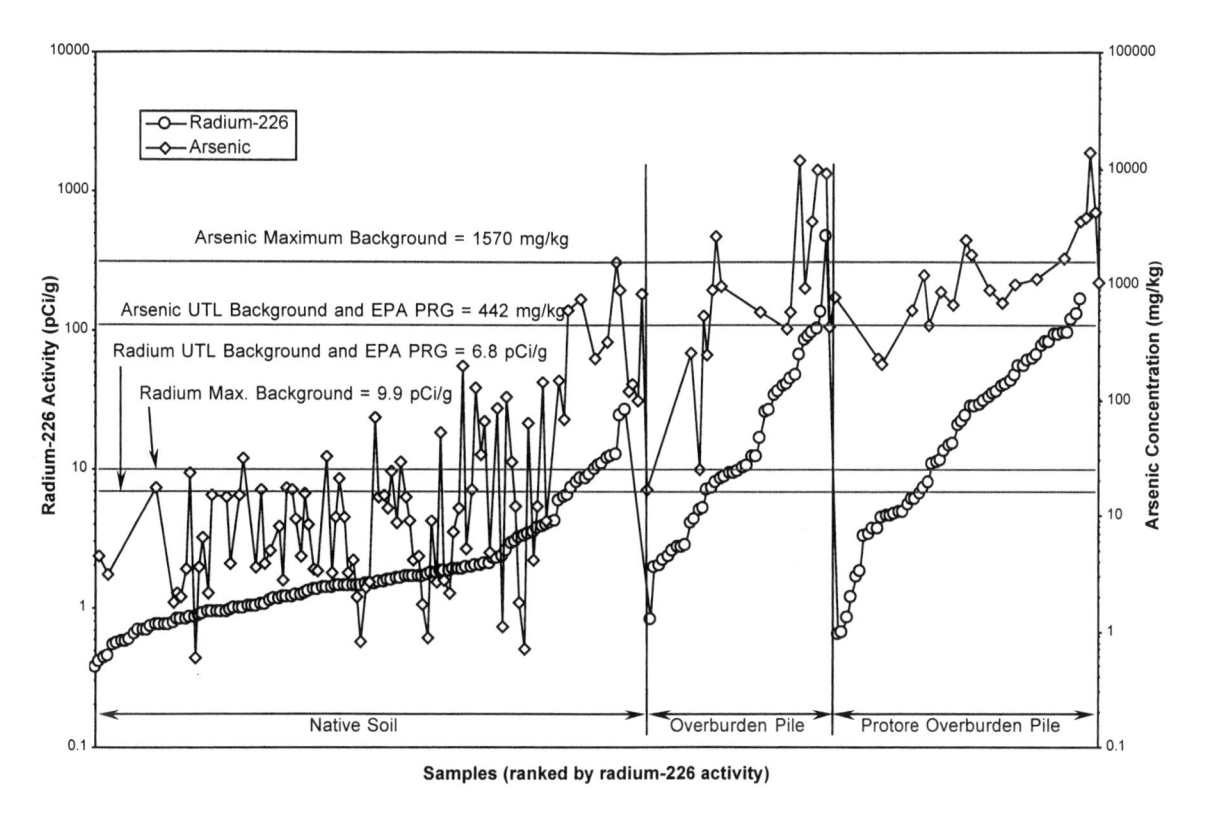

Figure 6. Ranked radium-226 activities and arsenic concentrations for the White King mine.

the PRGs set at the 95% UTLs would not result in clean-up of native soil, whereas use of the maximum background levels would not require clean-up of all overburden. Figure 6 also illustrates that elevated arsenic concentrations were generally collocated with radium-226. In most samples, arsenic values greater than 442 mg/kg (the arsenic PRG) were accompanied by radium-226 values greater than 6.8 pCi/g (the radium-226 PRG) and would require remediation. Thus, gamma surveys for radium-226 could be used for field verification of clean-up of both arsenic and radium.

At the Lucky Lass mine, the 95% UTLs for subsurface soils derived for local background within the ore body were 5.4 mg/kg for arsenic and 3.6 pCi/g for radium-226. Mineralized meadow and surface soil samples were not used in the calculation because surface mineralization in the meadow was probably transported there by physical means during mining and is not representative of local background. The sample population of meadow soils affected by mining processes is identified in Figure 7. This figure indicates that radium-226 activities in the meadow at the Lucky Lass mine are substantially higher than in the overburden pile. Excluding the meadow surface soil samples, local background radium-226 soil activity at the Lucky Lass mine is lower than at the White King mine. Figure 7 indicates that the local radium-226 maximum soil background (9.9 pCi/g) level would erroneously classify some of the mining-affected meadow samples as native soil. An arsenic PRG of 38 mg/kg (EPA Region IX Soil Clean-up Standards) was selected for the Lucky Lass mine because background is less than the clean-up standard. Based on these analyses, mining-affected meadow soils were included for clean-up with overburden at the Lucky Lass mine.

USE OF GAMMA SURVEYS TO VERIFY CLEAN-UP

Gamma surveys were used initially at the mines site as a field characterization technique to define areas of disturbed soil and overburden with elevated gamma activity. They also were used during site clean-up to verify that the radium-226 and arsenic materials were removed and clean-up levels were achieved in the remaining soil. Because arsenic and radium-226 were collocated, clean-up of soils to the radium-226 standard assured that disturbed soil with elevated arsenic levels was also remediated. Specifications for performing the surveys were submitted in a remedial design work plan (Golder Associates, Inc., 2003). The specifications outlined several areas where separate correlations were developed for radium-226 activity versus gamma counts detected in the field. Gamma correlations for the White King mine are presented in Figure 8, with two correlations for the two different meter systems used. The gamma surveys were combined with visual clean-up of overburden to assure that site clean-up standards were met. Use of gamma surveys reduced the requirements for laboratory analysis of confirmation samples and facilitated clean-up decisions in the field.

SECULAR EQUILIBRIUM IN SITE RADIOLOGICAL DATA FOR SOILS

Secular equilibrium in site radiological soil sample data was used to determine whether uranium, radium, and arsenic were

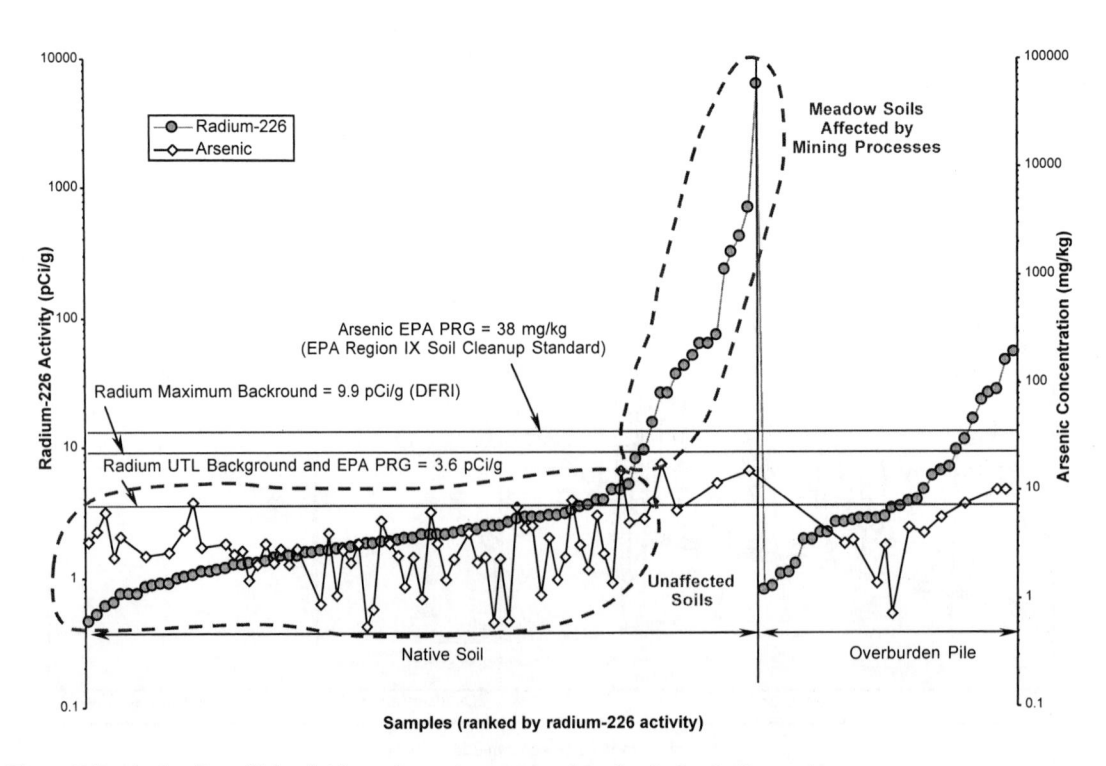

Figure 7. Ranked radium-226 activities and arsenic concentrations for the Lucky Lass mine.

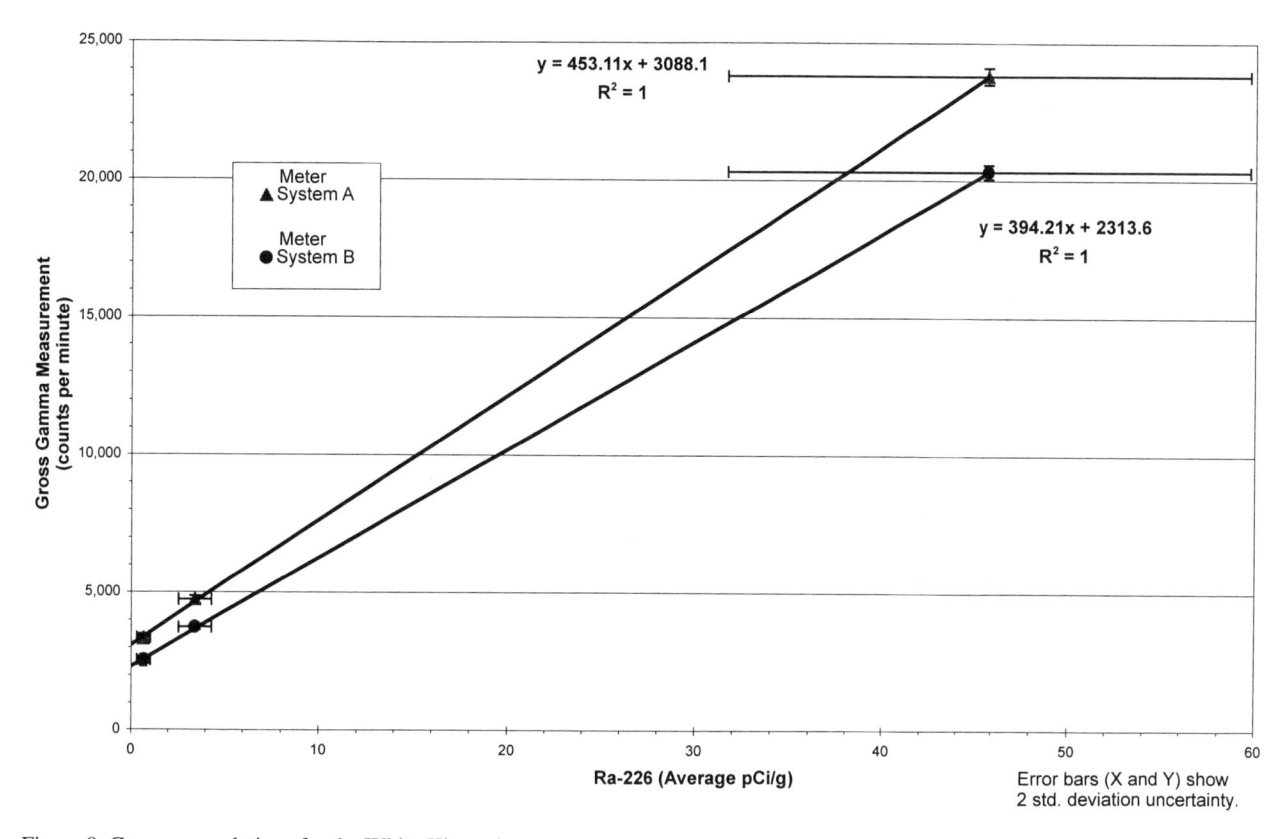

Figure 8. Gamma correlations for the White King mine.

leaching from overburden piles into underlying soil. Secular equilibrium also was used to assess whether radiological data collected prior to the remedial investigation (pre-RI data) were valid for use in calculation of local background levels. Pre-RI radiological data were needed to augment site characterization. Because the pre-RI radiological data were collected without an EPA-approved work plan, their usability had to be demonstrated. This was achieved qualitatively via evaluations of secular equilibrium.

Radiological analytical data for a soil sample are in secular equilibrium when all daughter nuclides in a decay series exhibit the same activity as the parent nuclide. At White King–Lucky Lass mines, samples were analyzed for uranium-238 and three daughter isotopes: uranium-234, thorium-230, and radium-226. When activities of these four nuclides are within the same order of magnitude, it is likely that the material has neither gained nor lost any of these isotopes in recent geologic time. Conversely, disequilibrium among these activities is evidence of open-system behavior, where dissolution or precipitation processes have affected the activity of one or more of these isotopes. Because of the dynamic nature of groundwater, with its comparatively short residence time in the aquifer, and because of the differing geochemical properties of the daughters (see Table 1), disequilibrium

TABLE 1. GEOCHEMICAL PROPERTIES OF URANIUM-DECAY SERIES NUCLIDES

Nuclide	Half-Life (yr)	Geochemical Behavior
Uranium-238	4.51×10^9	Insoluble under reducing conditions; readily soluble when oxidized but strongly adsorbed by iron-manganese oxyhydroxides at near-neutral pH
Uranium-234	2.47×10^5	Same as uranium-238
Thorium-230	7.8×10^4	Generally insoluble except at low pH ($< \sim 3$)
Radium-226	5.77	Slightly soluble, especially at low pH; precipitates in the presence of sulfate and carbonate; strongly adsorbed by clays and silica at near-neutral pH

among these isotopes in groundwater was expected. Thus, analysis focused on secular equilibrium in soil samples.

White King Mine

Uranium-decay series activities in White King mine materials for DFRI and pre-RI data are presented in Figures 9 and 10,

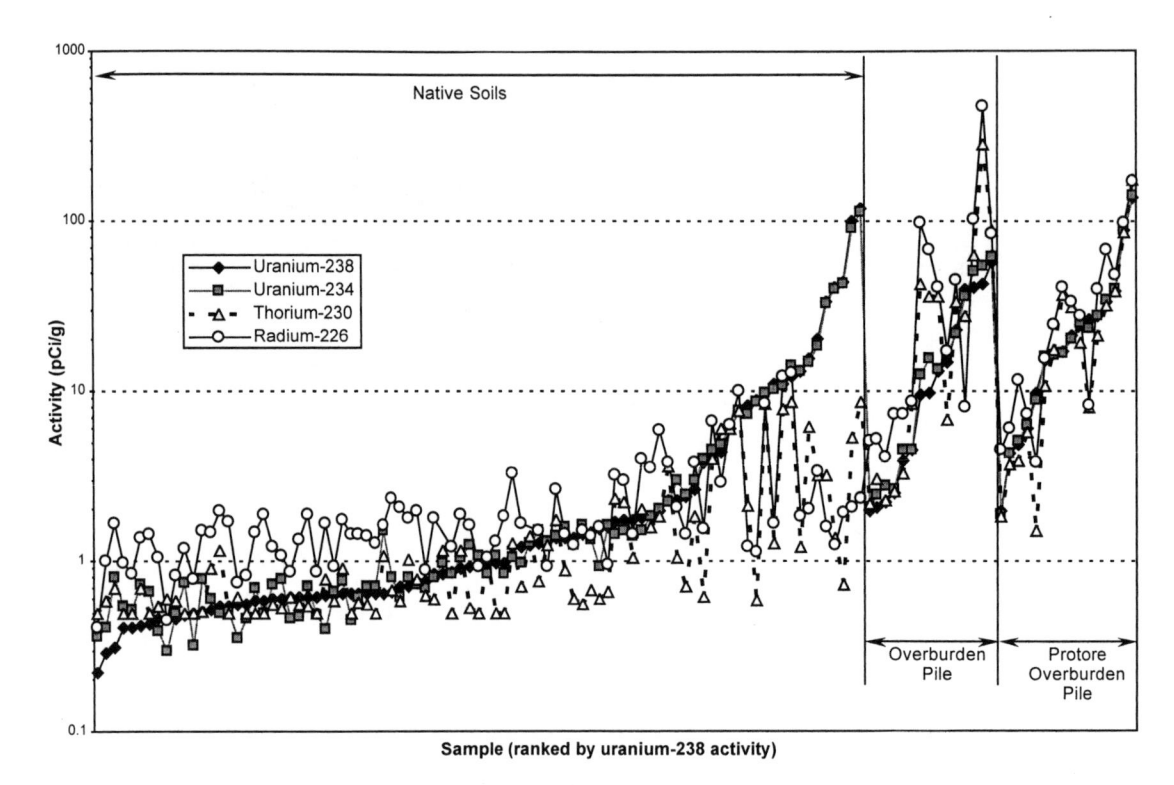

Figure 9. DFRI uranium-decay series activities in White King mine materials.

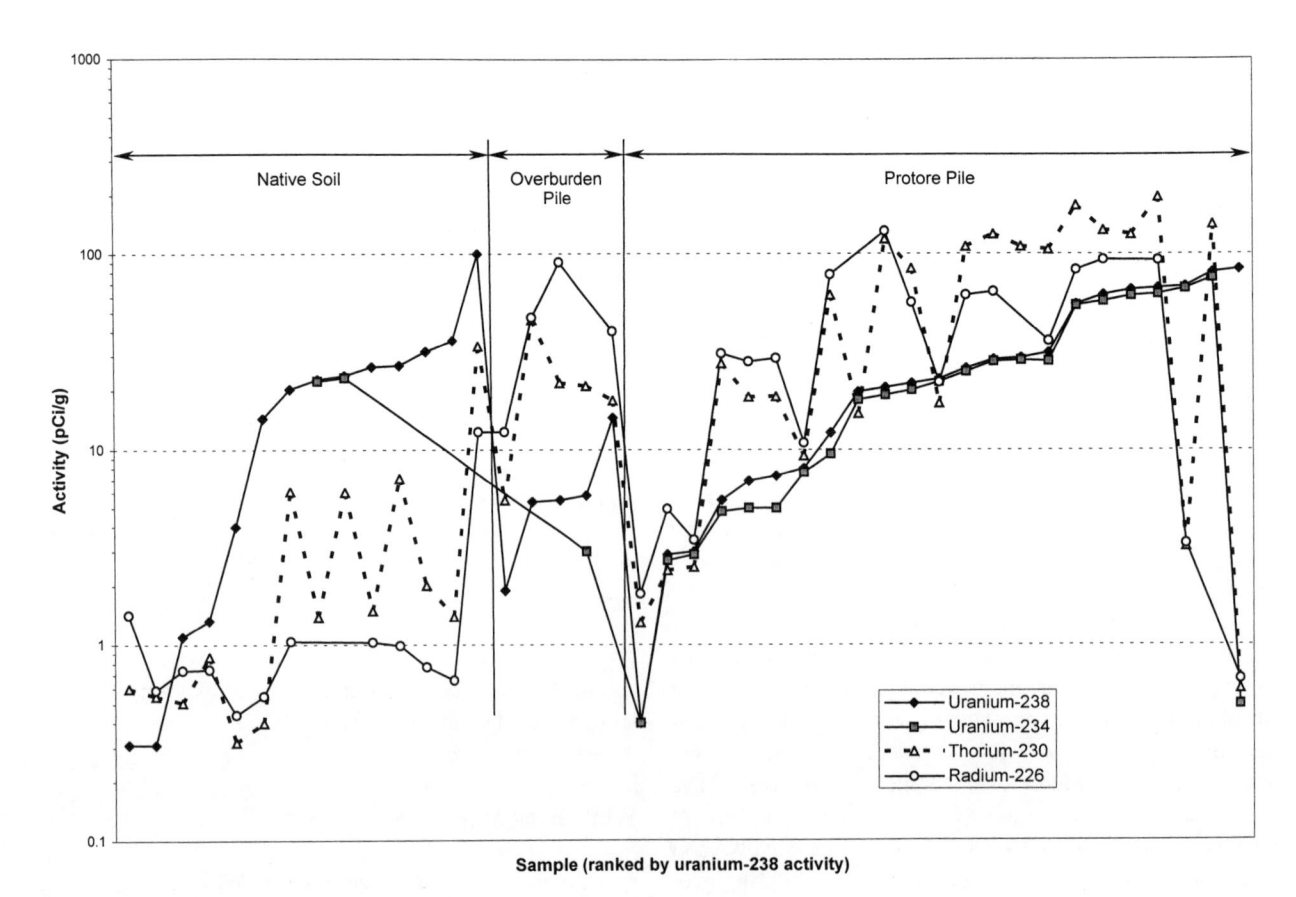

Figure 10. Pre-RI uranium-decay series activities in White King mine materials.

respectively. Samples are grouped according to soil type, including native soils, overburden pile, and protore pile. The DFRI data show that activities are generally low in native soils (<10 pCi/g for each of the radionuclides). The consistently higher activity for radium-226 relative to the other radionuclides in native soil does not follow any discernible pattern and may be a reflection of bias inherent in the analytical methodology. Uranium-234, thorium-230, and radium-226 in native soil exhibit considerable variability relative to uranium-238 at these low activities but in general were considered to be in secular equilibrium. Some native soil samples exhibited elevated levels of uranium-238 and -234 without corresponding levels of thorium-230 or radium-226, consistent with a recent accumulation of uranium. The affected native soil samples were collected adjacent to or beneath the overburden and protore piles, indicating that the piles are the likely source of the uranium. These mining-affected soils do not represent near-surface natural mineralization, which would exhibit secular equilibrium.

In the overburden and protore piles at White King mine, uranium-234 activities closely match parent uranium-238 activities. Thorium-230 and radium-226 are frequently higher than the associated uranium activities, suggesting leaching of the parent uranium from the piles. This is consistent with the observed accumulation of uranium in underlying and adjacent native shallow soils.

The pre-RI uranium-decay series data from the White King mine (Fig. 10) are consistent with the general trends of radiological data presented in the DFRI (Fig. 9). Although some decay-series data are incomplete because uranium-238 activity was not measured, the data are adequate to be included for determining radium-226 background levels at the White King mine.

To establish the appropriate local background level for arsenic at the White King mine, it was necessary to determine whether the high arsenic concentrations in native soils were attributable to leaching from overburden and subsequent precipitation in native soil or were naturally occurring. Unleached soil samples selected on the basis of secular equilibrium in the uranium-decay series should contain naturally occurring levels of arsenic as well because the leaching characteristics of these elements are similar. Arsenic and uranium-238 decay series analyses of the ten highest arsenic soil samples from White King mine are presented in Table 2. All samples are in secular equilibrium except for a slight disequilibrium in samples WK-SS05 and FS-07. Therefore, most of the arsenic in native soils is naturally occurring and is not derived from leaching of overburden.

Lucky Lass Site

Uranium-decay series activities in Lucky Lass mine materials for DFRI and pre-RI data are presented in Figures 11 and 12, respectively. The DFRI data show that activities are generally low in native soils (<10 pCi/g for each of the radionuclides) with a pattern of consistently higher radium-226 similar to that seen in Figure 9 for White King mine materials. Three native soil samples with higher activity exhibit enrichment in uranium-238, uranium-234, and thorium-230 relative to radium-226. Because thorium-230 is relatively insoluble, radium-226 may have been leached from the sample materials. With the exception of the three soil samples with the highest uranium activities, radium-226 in native soils at the Lucky Lass mine is in secular equilibrium with its precursors. The high-activity native soils exhibiting disequilibrium are located in the Lucky Lass meadow in the upper 0.6 m (2 ft), below the overburden pile and Lucky Lass pond. It is likely that a small amount of high-grade ore was deposited on top of the meadow soils by previous mining activities. For this reason, meadow soils were excluded from calculation of local background levels at the Lucky Lass mine.

In the overburden pile at Lucky Lass, all radionuclides were in approximate secular equilibrium, consistent with the geochemistry of the site. In contrast to the White King ore, Lucky Lass ore and overburden contain no sulfide minerals to weather and produce acidity. Because the pH in the overburden pile has remained near neutral, there has been minimal leaching of these radionuclides.

The pre-RI data from the Lucky Lass mine (Fig. 12) are consistent with the general trends of the DFRI data (Fig. 11). Nearly all samples have activity values for all four radionuclides. For the purpose of site characterization, the pre-RI radiological data were combined with DFRI data to provide a more complete view of site conditions and a larger data set for calculating local background levels at the Lucky Lass mine.

Trilinear Analysis of Groundwater Quality

Trilinear analysis of groundwater quality was used to determine the effect of mining and oxidation of sulfide-bearing minerals in overburden on groundwater quality. The major-ion chemistry of groundwater samples from water-bearing units at the White King and Lucky Lass mines is plotted on trilinear diagrams (Figs. 13 and 14), along with surface water from both mines (Fig. 15). Water samples representative of background conditions are tightly clustered on the anionic and combined portions of the trilinear diagrams, indicating the predominance of bicarbonate among the anions. The cationic compositions of these samples are clustered near the center of the cation portion of the trilinear diagram, indicating relatively equal percentages of calcium, magnesium, and sodium plus potassium. The trilinear analysis shows that alluvial, shallow bedrock, and deep bedrock groundwater background samples closely resemble each other, indicating a common origin with no significant geochemical modification. From a geochemical standpoint, regional background groundwater at the White King and Lucky Lass mines is from a single water-bearing unit.

Other monitoring wells at the White King mine define a trend toward sulfate-dominated anionic compositions with elevated total dissolved solids (TDS) and sulfate. Cation compositions in these wells may exhibit increased percentages of calcium and magnesium. Typically, these groundwater samples are from wells completed from within and beneath the protore and overburden piles. These samples have a pH of 6 or less, indicating that the perched water in the piles or the groundwater has been affected by

TABLE 2. ARSENIC AND URANIUM-238 DECAY SERIES ANALYSES FOR THE TEN HIGHEST ARSENIC SOIL SAMPLES AT THE WHITE KING MINE

Sample Date	Units of Measure	Location	Sample Number	Analyte	Qual.	Result	Depth (Meters)
7/11/90	MG/KG	FS-07	SL-LKV-27-00	Arsenic		822	0
7/11/90	PCi/g	FS-07	LKV-27-00	Radium-226		12.3	0
7/11/90	PCi/g	FS-07	LKV-27-00	Thorium-230		33.9	0
7/11/90	PCi/g	FS-07	LKV-27-00	Uranium-234		98	0
7/11/90	PCi/g	FS-07	LKV-27-00	Uranium-238		100	0
7/25/95	MG/KG	WK-SB-01-B	WK-SB-01-B-01	Arsenic		143	1.2
7/25/95	PCi/g	WK-SB-01-B	WK-SB-01-B-01	Radium-226		3.82 ± 0.27	1.2
7/25/95	PCi/g	WK-SB-01-B	WK-SB-01-B-01	Thorium-230		3.57 ± 0.63	1.2
7/25/95	PCi/g	WK-SB-01-B	WK-SB-01-B-01	Uranium-234		2.23 ± 0.41	1.2
7/25/95	PCi/g	WK-SB-01-B	WK-SB-01-B-01	Uranium-238		2.26 ± 0.41	1.2
6/26/95	MG/KG	WK-SB-10-A	WK-SB-10-AD-04	Arsenic		149	4.6
6/26/95	PCi/g	WK-SB-10-A	WK-SB-10-AD-04	Radium-226		5.92 ± 0.51	4.6
6/26/95	PCi/g	WK-SB-10-A	WK-SB-10-AD-04	Thorium-230		1.85 ± 0.65	4.6
6/26/95	PCi/g	WK-SB-10-A	WK-SB-10-AD-04	Uranium-234		2.00 ± 0.37	4.6
6/26/95	PCi/g	WK-SB-10-A	WK-SB-10-AD-04	Uranium-238	J	1.95 ± 0.36	4.6
6/26/95	MG/KG	WK-SB-10-A	WK-SB-10-AD-05	Arsenic		1570	6.1
6/26/95	PCi/g	WK-SB-10-A	WK-SB-10-AD-05	Radium-226		12.6 ± 0.63	6.1
6/26/95	PCi/g	WK-SB-10-A	WK-SB-10-AD-05	Thorium-230		8.59 ± 0.84	6.1
6/26/95	PCi/g	WK-SB-10-A	WK-SB-10-AD-05	Uranium-234		14.1 ± 1.18	6.1
6/26/95	PCi/g	WK-SB-10-A	WK-SB-10-AD-05	Uranium-238	J	12.0 ± 1.08	6.1
6/15/95	MG/KG	WK-SB-12-A	WK-SB-12-AD-01	Arsenic		316	1.5
6/15/95	PCi/g	WK-SB-12-A	WK-SB-12-AD-01	Radium-226		12.1 ± 0.39	1.5
6/15/95	PCi/g	WK-SB-12-A	WK-SB-12-AD-01	Thorium-230		7.89 ± 0.93	1.5
6/15/95	PCi/g	WK-SB-12-A	WK-SB-12-AD-01	Uranium-234	J	10.8 ± 0.87	1.5
6/15/95	PCi/g	WK-SB-12-A	WK-SB-12-AD-01	Uranium-238	J	11.0 ± 0.87	1.5
6/23/95	MG/KG	WK-SB-A	WK-SB-A-04	Arsenic		233	6.1
6/23/95	PCi/g	WK-SB-A	WK-SB-A-04	Radium-226		9.96 ± 0.49	6.1
6/23/95	PCi/g	WK-SB-A	WK-SB-A-04	Thorium-230		7.62 ± 0.80	6.1
6/23/95	PCi/g	WK-SB-A	WK-SB-A-04	Uranium-234		7.61 ± 0.86	6.1
6/23/95	PCi/g	WK-SB-A	WK-SB-A-04	Uranium-238		7.75 ± 0.86	6.1
6/28/95	MG/KG	WK-SB-E	WK-SB-E-03	Arsenic		602	7.6
6/28/95	PCi/g	WK-SB-E	WK-SB-E-03	Radium-226		6.58 ± 0.39	7.6
6/28/95	PCi/g	WK-SB-E	WK-SB-E-03	Thorium-230		3.97 ± 0.55	7.6
6/28/95	PCi/g	WK-SB-E	WK-SB-E-03	Uranium-234		4.53 ± 0.61	7.6
6/28/95	PCi/g	WK-SB-E	WK-SB-E-03	Uranium-238		4.22 ± 0.58	7.6
7/14/95	MG/KG	WK-SS-05	WK-SS-05	Arsenic		199	0
7/14/95	PCi/g	WK-SS-05	WK-SS-05	Radium-226		1.92 ± 0.19	0
7/14/95	PCi/g	WK-SS-05	WK-SS-05	Thorium-230		0.730 ± 0.29	0
7/14/95	PCi/g	WK-SS-05	WK-SS-05	Uranium-234		42.6 ± 2.08	0
7/14/95	PCi/g	WK-SS-05	WK-SS-05	Uranium-238		43.4 ± 2.10	0
7/13/95	MG/KG	WK-SS-09	WK-SS-09	Arsenic		733	0
7/13/95	PCi/g	WK-SS-09	WK-SS-09	Radium-226		8.52 ± 0.44	0
7/13/95	PCi/g	WK-SS-09	WK-SS-09	Thorium-230		8.40 ± 0.81	0
7/13/95	PCi/g	WK-SS-09	WK-SS-09	Uranium-234		9.69 ± 0.98	0
7/13/95	PCi/g	WK-SS-09	WK-SS-09	Uranium-238		9.81 ± 0.98	0
7/12/95	MG/KG	WK-SS-19	WK-SS-19	Arsenic		895	0
7/12/95	PCi/g	WK-SS-19	WK-SS-19	Radium-226		24.7 ± 0.67	0
7/12/95	PCi/g	WK-SS-19	WK-SS-19	Thorium-230		19.3 ± 1.19	0
7/12/95	PCi/g	WK-SS-19	WK-SS-19	Uranium-234		15.2 ± 1.04	0
7/12/95	PCi/g	WK-SS-19	WK-SS-19	Uranium-238		15.2 ± 1.04	0

Note: Data from Weston, 1997.

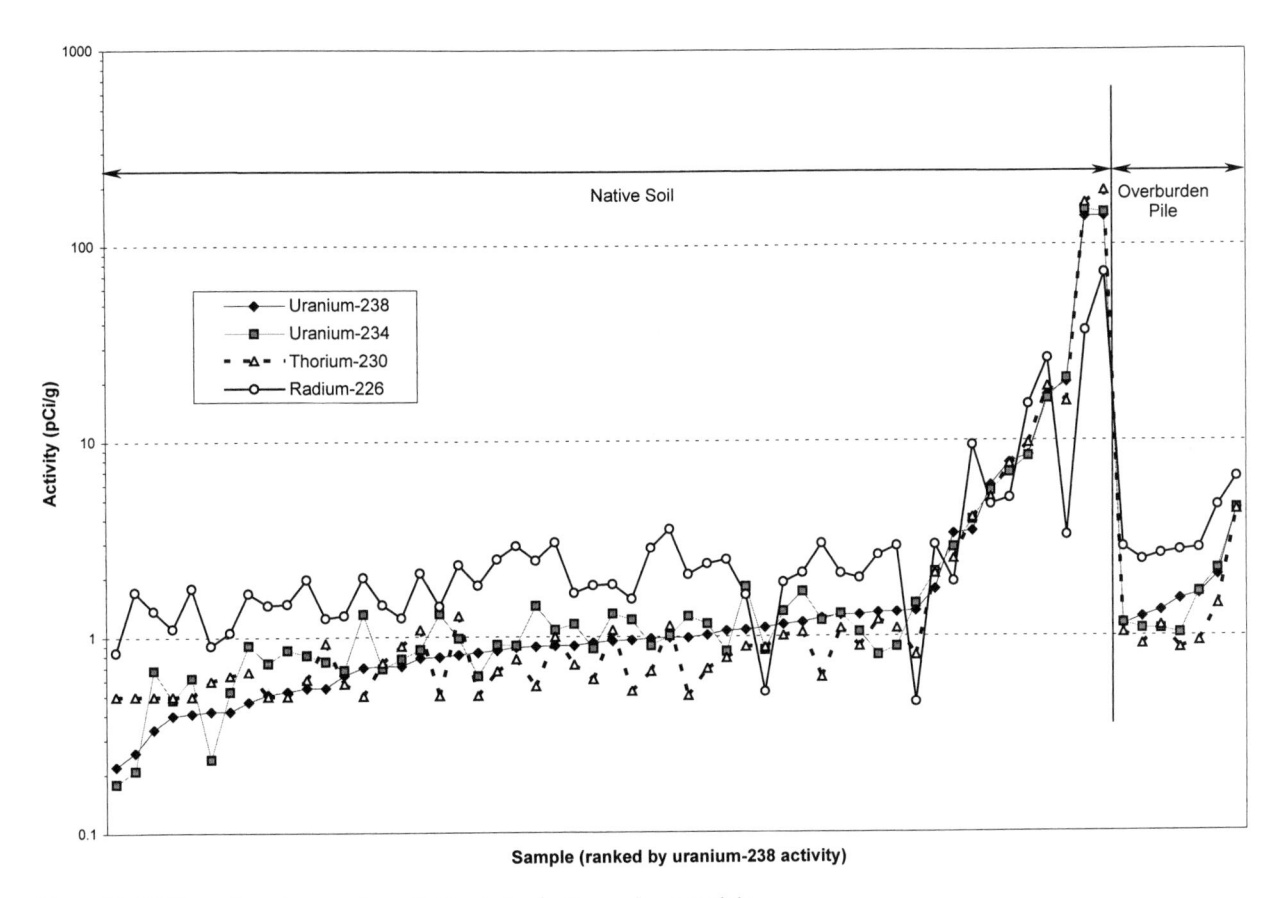

Figure 11. DFRI uranium-decay series activities in Lucky Lass mine materials.

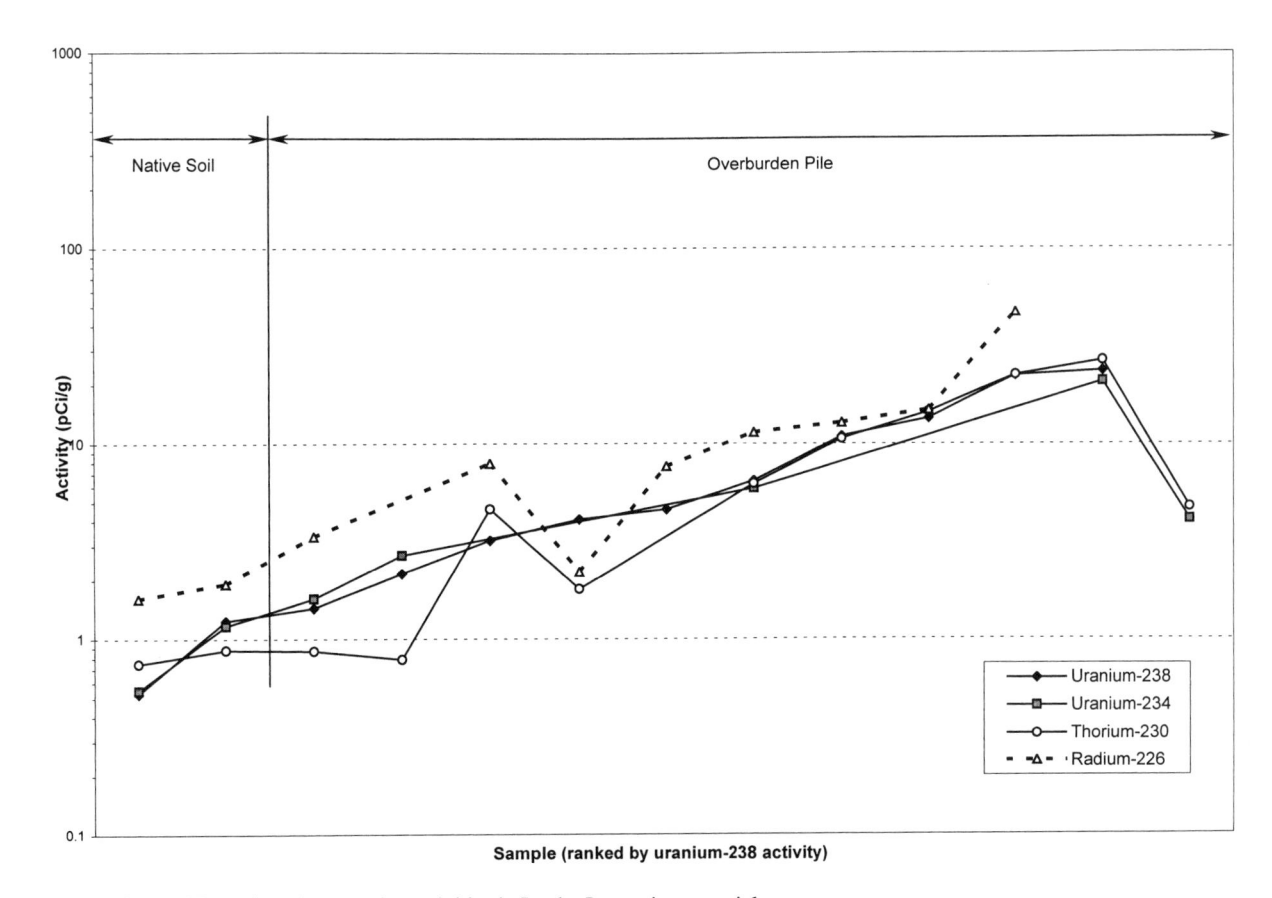

Figure 12. Pre-RI uranium-decay series activities in Lucky Lass mine materials.

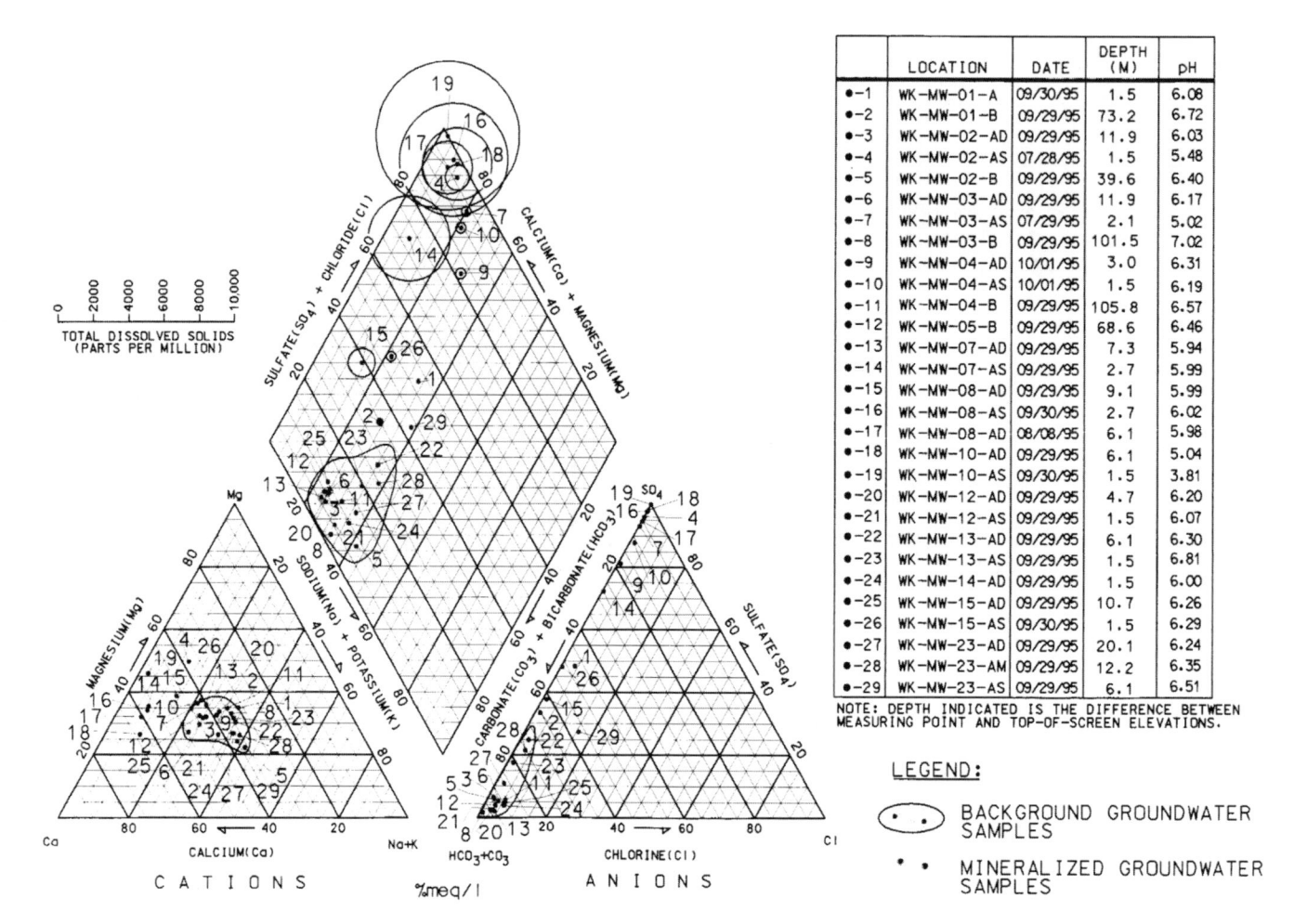

Figure 13. Major-ion chemistry of groundwater at the White King mine.

oxidation of sulfide minerals. These wells plot outside the area defined as background. It is not possible to distinguish mineralized groundwater leached from the piles from ambient mineralization in native groundwater at the White King mine because oxidation of sulfide minerals is the source of sulfate and TDS in both cases. A 240-ft-deep (73 m) monitoring well (WK-MW-01-B) that is completed adjacent to or within the ore body at the White King mine is slightly more mineralized than background groundwater, with elevated levels of sulfate and TDS. Groundwater at this depth is probably unaffected by seepage from the piles and most likely reflects the geochemical influence of the ore body. Monitoring wells exhibiting high TDS in groundwater samples have circles that are proportional to the TDS scale shown on Fig. 13.

Groundwater quality from wells affected by oxidation of sulfide minerals at the White King mine closely resembles that of the White King pond (Fig. 15, location WK-P-SW-01). The White King pond had a pH of 4 before neutralization as compared to a pH of 7 at the Lucky Lass pond (LL-P-SW-01). Other surface water samples from Augur Creek are within the range of background water quality for groundwater indicating the contribution of regional groundwater to stream flow.

POTENTIAL TRANSPORT OF ARSENIC AND URANIUM IN GROUNDWATER AT THE WHITE KING MINE SITE

The potential for transport of arsenic and uranium from the overburden piles to the surrounding soil and groundwater was examined by comparing concentrations of sulfate, sulfite, arsenic, and uranium-decay series radionuclides in groundwater to their concentrations in water from shallow wells completed within or immediately beneath the piles. Measurements of these parameters and well depths at the White King mine are summarized in Table 3.

At the White King mine, sulfate, arsenic, and uranium are soluble in infiltrating water under the oxidizing conditions present in the overburden piles and in the alluvial aquifer. Thorium and radium exhibit only limited solubility under the geochemical conditions in the piles (moderately acidic pH and high sulfate) and have not migrated to groundwater from the piles. Sulfates in groundwater are either derived from weathering and oxidation of sulfides in the piles and subsequent leaching to the groundwater system or from in situ oxidation of ore in the groundwater system.

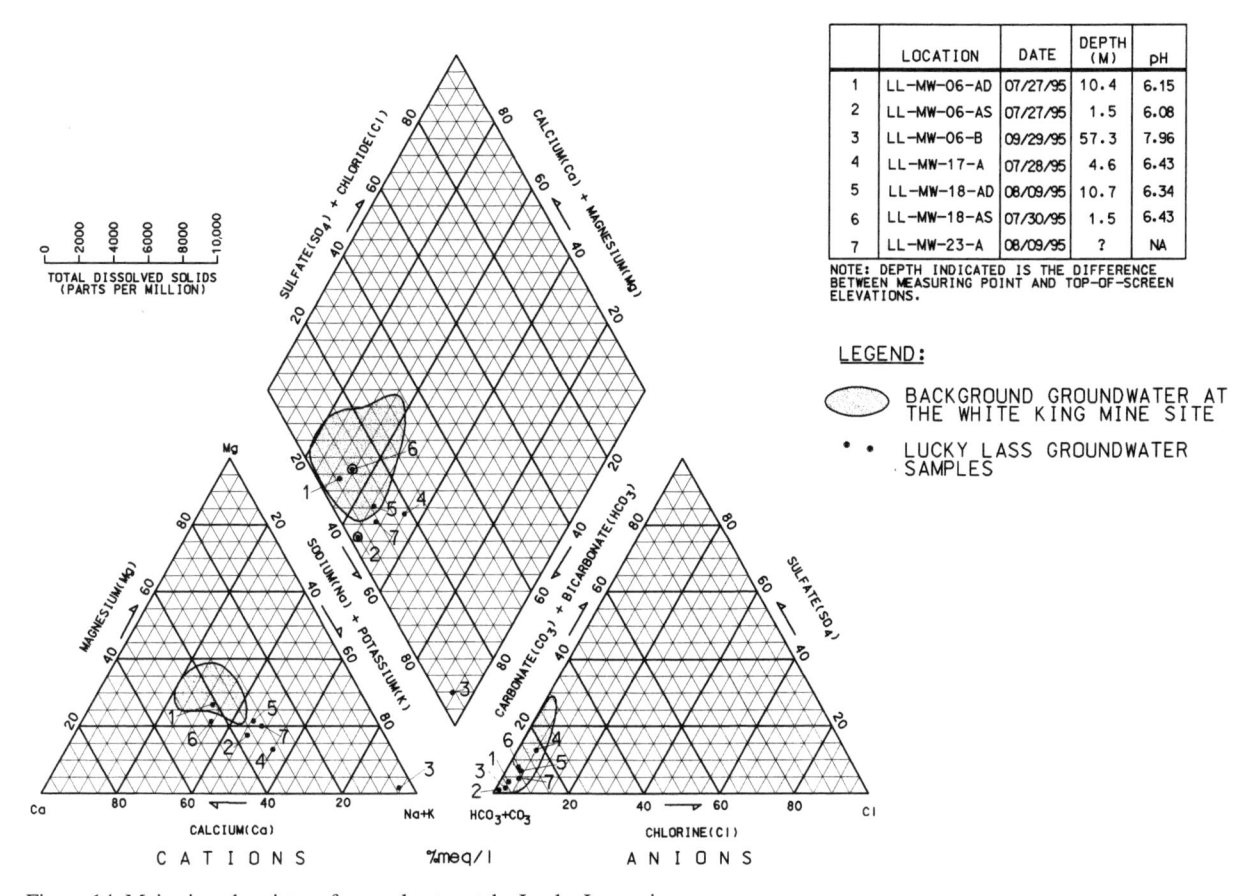

Figure 14. Major-ion chemistry of groundwater at the Lucky Lass mine.

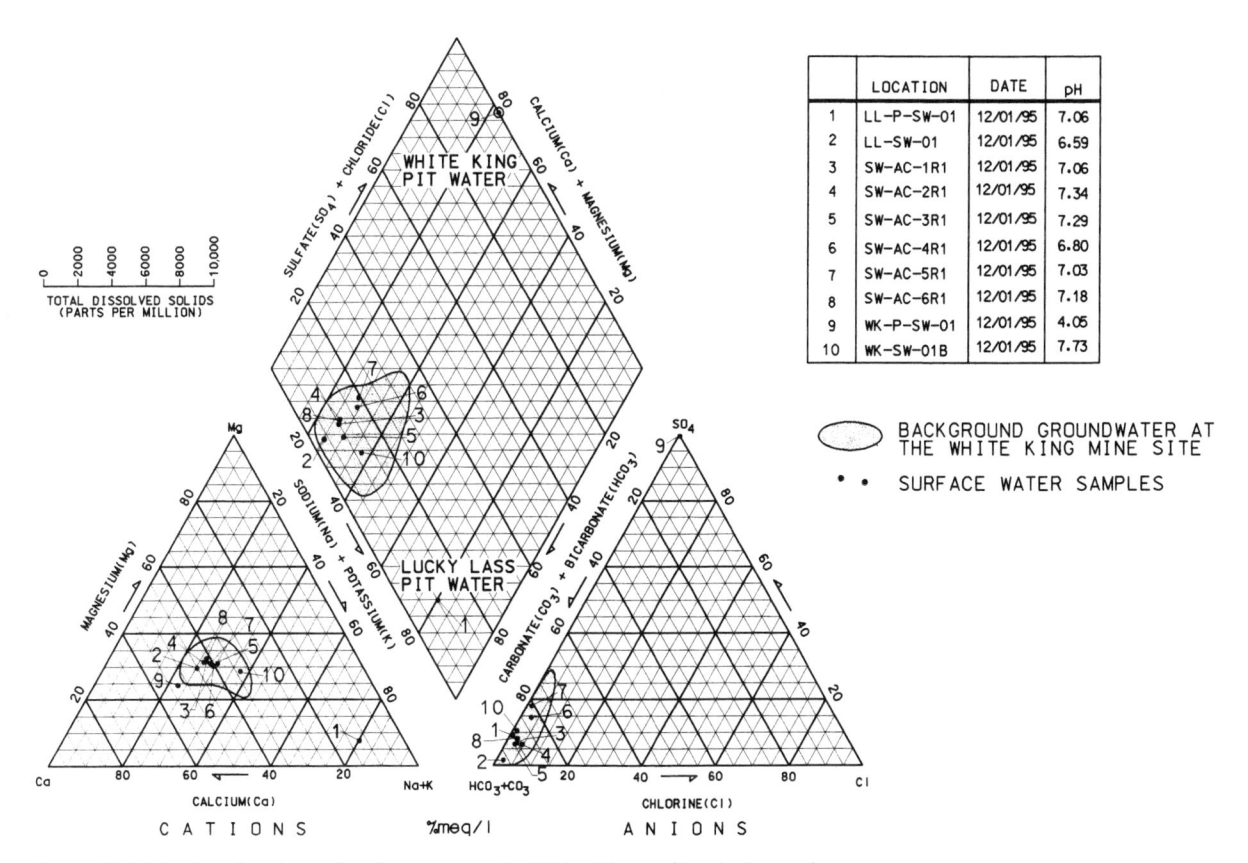

Figure 15. Major-ion chemistry of surface water at the White King and Lucky Lass mines.

TABLE 3. MEASUREMENTS OF SULFATE, SULFITE, ARSENIC, AND RADIONUCLIDES IN GROUNDWATER AT THE WHITE KING MINE

Location	Depth (meters)	pH	Sulfate (mg/L)	Arsenic (μg/L)	Uranium-238 (pCi/L)	Uranium-234 (pCi/L)	Thorium-230 (pCi/L)
WK-MW-01-A	1.5	6.08	140	17.7	ND	ND	ND
WK-MW-02-AS	1.5	5.48	520	4.6	ND	ND	ND
WK-MW-04-AS	1.5	6.19	180	66.5	ND	ND	ND
WK-MW-12-AS	1.5	6.07	5.7		ND	ND	ND
WK-MW-13-AS	1.5	6.81	14	6.7	ND	ND	ND
WK-MW-14-AS	1.5	6.00		1.4	ND	ND	ND
WK-MW-15-AS	1.5	6.29	86	1.3	0.96	0.98	ND
WK-MW-03-AS	1.5	5.02	190	1.3	ND	ND	0.71
WK-MW-04-AD	3	6.31	200	48.4	ND	ND	ND
WK-MW-12-AD	4.6	6.2		1.8	ND	ND	ND
WK-MW-13-AD	6.1	6.3	21	6.7	ND	ND	ND
WK-MW-23-AS	6.1	6.51	14	11	1.59	2.13	0.63
WK-MW-15-AD	10.6	6.26	20	1.7	9.76	9.61	ND
WK-MW-02-AD	11.9	6.03	7.3	2	0.9	1	ND
WK-MW-03-AD	11.9	6.17	16	2.5	2.2	3	ND
WK-MW-23-AM	12.2	6.35	35	13.4	ND	ND	ND
WK-MW-23-AD	20.1	6.24	6	130	0.51	0.68	ND
WK-MW-02-B	39.6	6.40	4.4	38	ND	ND	ND
WK-MW-05-B	68.6	6.46	2.8	5.2	ND	ND	ND
WK-MW-01-B	73.1	6.72	36	10.2	0.73	1.79	ND
WK-MW-03-B	101.5	7.02		2.9	ND	ND	ND
WK-MW-04-B	105.7	6.57	3.4	1.5	ND	ND	ND

Shallow Monitoring Wells Completed Within or Beneath the Overburden Piles

Location	Depth	pH	Sulfate	Arsenic	U-238	U-234	Th-230
WK-MW-10-AS	1.5	3.81	3200	45.5	22,300	21,300	1.86
WK-MW-07-AS	2.7	5.99	1200	2060	8.02	9.77	ND
WK-MW-08-AS	2.7	6.02	2300	35,600	2.8	3.48	2.04
WK-MW-09-AD	6.1	5.98	980	4.4	ND	ND	ND
WK-MW-10-AD	6.1	5.04	1600	23,100	35.6	39.6	1.68
WK-MW-07-AD	7.3	5.94	3.4	8.5	ND	ND	ND
WK-MW-08-AD	9.1	5.99	190	434	1	1.37	ND

Notes: Data from Weston, 1997; ND = no data

Upgradient monitoring wells outside the ore body, not influenced by seepage from the piles, have the lowest sulfate concentrations, between 3.4 and 7.3 mg/L. In the perched water in the piles and in groundwater in the ore body, sulfate concentrations in solution probably reflect a dynamic equilibrium between oxidation of sulfide-bearing minerals and precipitation of sulfate-bearing minerals such as gypsum. Uranium and arsenic may be leached from minerals in the piles but are attenuated by adsorption and precipitation in the native soils beneath the piles and the aquifer matrix. The monitoring-well data show that sulfate, arsenic, and uranium are highest in shallow wells completed within or immediately beneath the piles (depth < 10 ft bgs). Even though groundwater at the White King mine may have been influenced by leaching from the overburden piles, naturally occurring mineralization within the ore body also contributes to poor groundwater water quality. Because there are no sulfide minerals to oxidize in the Lucky Lass mine overburden, groundwater is unaffected by infiltration through these materials. Potential risks to future residents from naturally occurring radon in groundwater prohibit beneficial use of the water for domestic consumption. Groundwater clean-up was not required at either mine due to naturally occurring mineralization and the presence of radon in groundwater.

REMEDIAL ACTIONS

The remedy for the White King mine was to consolidate the overburden pile with off-pile and haul road material at the location of the protore pile. Some high-activity materials from the

Lucky Lass site were relocated to the White King consolidated pile (Golder Associates, Inc., 2005b). The protore pile was recontoured to remove it from the Augur Creek flood plain and achieve compliance with the flood plain and erosion standards of the State of Oregon (ORS 469.375). The final remedial design for the White King mine is presented in Figure 16. The consolidated pile at the White King mine was covered with 2.2 m (7.5 ft) of compacted clay overburden followed by 0.6 m (2 ft) of clean rock and soil. The compacted clay, constructed from overburden with only low levels of arsenic and radium-226, provides an infiltration barrier and freeze-thaw protection for underlying overburden. The rock and soil cover will help promote vegetation and limit the intrusion of small burrowing animals commonly found in the area. Because overburden materials with the most elevated concentrations of arsenic and radium-226 are isolated below a total of 2.8 m (9.5 ft) of cap, the potential for direct exposure and inadvertent human intrusion is essentially eliminated. Consolidation of overburden into one pile at the White King site reduced the total cover area. In addition, consolidation permitted reclamation of a portion of the Augur Creek meadow wetland habitat to pre-mining conditions.

The remedy for the White King pond was in situ neutralization to maintain a consistent neutral pH. Based upon the success of pilot studies begun in 1998, neutralization is an effective and relatively low-cost option for addressing the acidic conditions in the pond. The pond was neutralized by spraying a solution of hydrated lime on its surface. Gabions of crushed limestone deposited in the deepest parts of the pond address a seasonal pocket of low-pH water that develops where acid mine drainage exits a former mine shaft. The pond has been maintained at pH 6 and has become biologically active with diverse species. EPA is monitoring the pond on an annual basis to evaluate the success of

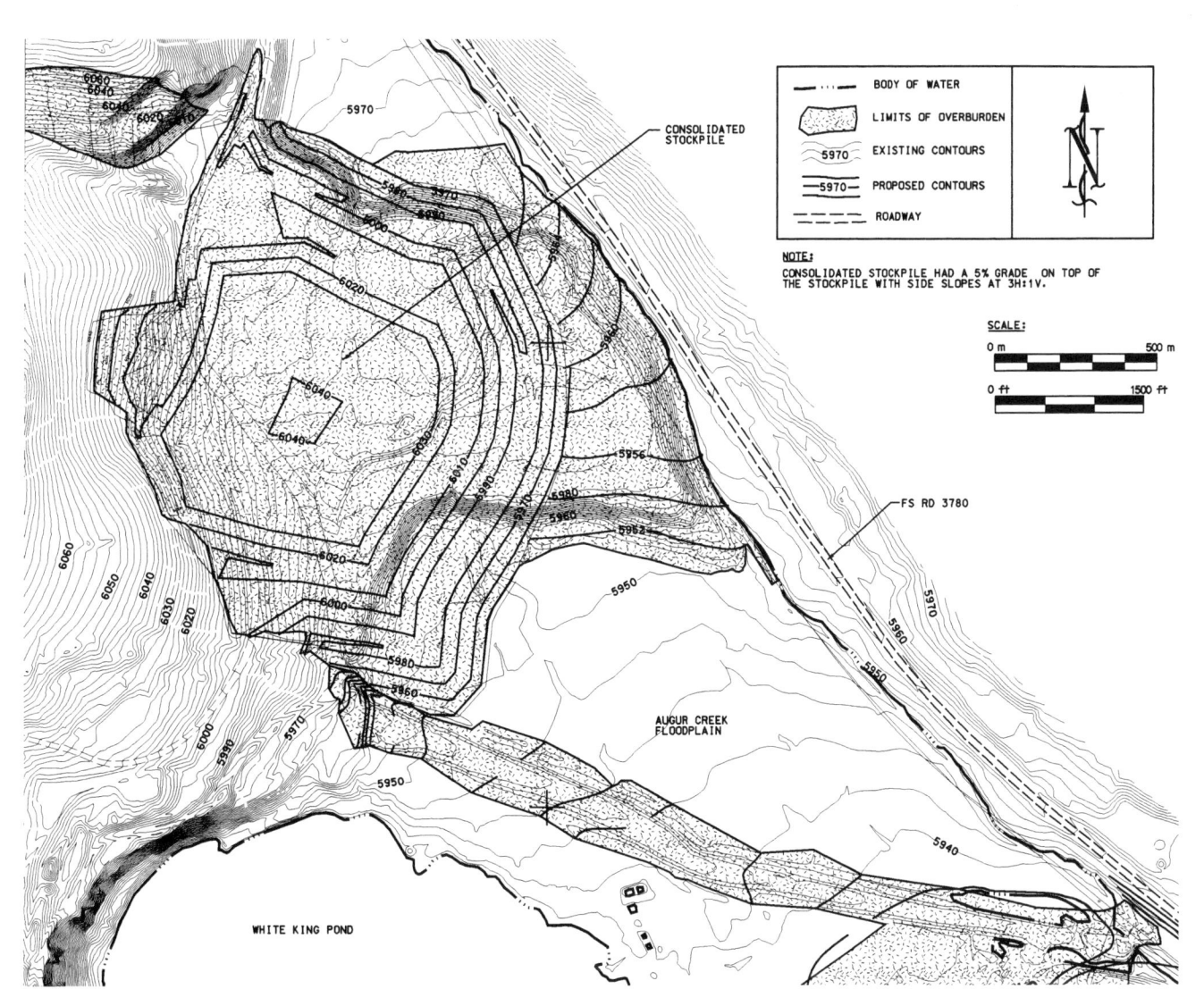

Figure 16. Final remedial design for the White King mine.

neutralization and ensure that the remedy is protective. An ecological study conducted at White King pond determined that a benthic community has been established, providing food for wildlife, and that there is no threat to wildlife from direct sediment exposure or from ingestion of aquatic life from the pond (Golder Associates, Inc., 2005a).

The selected remedy for the Lucky Lass mine was to remove some material with elevated radium-226 activity on the Lucky Lass overburden pile and meadow to the White King mine consolidated pile. The remaining Lucky Lass overburden pile was graded, and soil cover was placed on the pile and disturbed areas.

The remediated consolidated piles will be managed by physical controls such as fencing and institutional controls such as an amendment of the forest management plan to prevent future residential use. (The remediated piles still pose some risk from direct exposure to soils under a residential exposure scenario.) This provides overall protection of human health and the environment as well as long-term effectiveness. Institutional controls will be used at both mines to prevent residential use of the properties and local development of groundwater. Site characterization and the proposed remedy resulted in selection of this Superfund Site for EPA's ROD of the Year in 2001.

REFERENCES CITED

Golder Associates, Inc., 2003, Remedial design work plan for the White King/Lucky Lass Mines Superfund site: Golder Associates, Inc., Redmond, Washington.

Golder Associates, Inc., 2005a, White King Pond and Augur Creek study report on Phase 1 at the White King/Lucky Lass mines Superfund site: Golder Associates, Inc., Redmond, Washington.

Golder Associates, Inc., 2005b, Remedial design report for the White King/Lucky Lass mines Superfund site: Golder Associates, Inc., Redmond, Washington.

Jacobs Engineering, 1998, Independent evaluation report, White King and Lucky Lass mine sites, Lakeview, Oregon: Report prepared by Jacobs Engineering for the United States Forest Service.

U.S. Environmental Protection Agency (EPA), 2001, White King/Lucky Lass Superfund site record of decision: Prepared by Office of Environmental Clean-up EPA Region X.

Weston, R.F. Inc., 1996, Draft remedial investigation report: RI/FS consultants' report prepared by Roy F. Weston, Inc. for Kerr-McGee, White King/Lucky Lass mine sites, Lakeview Oregon.

Weston, R.F. Inc., 1997, Draft final remedial investigation report: RI/FS consultants' report prepared by Roy F. Weston, Inc. for Kerr-McGee, White King/Lucky Lass mine sites, Lakeview Oregon.

MANUSCRIPT ACCEPTED BY THE SOCIETY 28 NOVEMBER 2006

Passive treatment of acid rock drainage from a subsurface mine

Martin Foote
Helen Joyce
Suzzann Nordwick

MSE Technology Applications, Inc., Mike Mansfield Advanced Technology Center, P.O. Box 4078, Butte, Montana 59702, USA

Diana Bless

U.S. Environmental Protection Agency, 26 W. Martin Luther King Dr., Cincinnati, Ohio 45268, USA

ABSTRACT

Acidic metal-contaminated drainages are a critical problem facing many areas of the world. Acid rock drainage results when metal sulfide minerals, particularly pyrite, are oxidized by exposure to oxygen and water. The deleterious effects of these drainages on receiving streams are well known. To address this problem, efforts are being made to use biological processes as an innovative, cost-effective means for treating acidic metal-contaminated drainage. Biological sulfate reduction (BSR) technology can be adapted to diverse site conditions and water chemistry. The Lilly mine near the community of Elliston, Montana, illustrates some of the specific conditions that can challenge effective application of BSR technology.

Keywords: acid drainage, sulfate reduction, metal removal, biological metal contamination

INTRODUCTION

Acidic, metal-laden, aqueous drainages are often associated with mining properties that contain sulfide-based minerals. These acidic drainages are produced by the reaction of sulfide-based minerals, primarily pyrite (FeS_2), with an oxidizing agent and water. The acidic nature of the drainage is produced from the oxidation of sulfide mineral(s) and hydrolysis of the reaction products as shown in the following reactions (Stumm and Morgan, 1981):

$$2FeS_2 + 7O_2 + 2H_2O = 2Fe_2 + 4SO_4^{2-} + 4H^+ \quad (1)$$

$$4Fe^{2+} + O_2 + 4H^+ = 4Fe^{3+} + 2H_2O \quad (2)$$

$$Fe^{3+} + 3H_2O = Fe(OH)_3 + 3H^+ \quad (3)$$

An extensive literature base exists on the subject of acid generation including a treatise on the subject published by the American Chemical Society in 1992.

The metal content in the drainages is produced by numerous processes that include the above reactions and by leaching and dissolution of metals from other minerals that come in contact with the drainage. These drainages emanating from mining properties are a widespread environmental problem in the United States because both the lowered pH and the metal and sulfate content of these drainages cause significant environmental damage. Hunter (1989) reports that ~10,000 miles streams and 29,000 acres of surface water impoundments in this country have been seriously impacted by the affects of such drainages. Conventional treatment of these drainages is often not feasible due to the remoteness of the site, the lack of power, and limited site accessibility. For such sites, there is a need for a low-cost passive remedial technology to immobilize metals and increase the pH of the drainage prior to the drainage's entering a surface water receiving stream.

This report describes the results of the implementation of such a treatment method at the Lilly mine near Elliston, Montana, by the Mine Waste Technology Program (MWTP). The MWTP is a federal program funded by the U.S. Environmental Protection Agency (EPA) and jointly administered by the EPA and the U.S. Department of Energy (DOE) at the Western Environmental Technology Office (WETO) through an interagency agreement. MSE Technology Applications, Inc. is the implementing entity for the program.

SITE DESCRIPTION

The Lilly mine is a small subsurface mine located ~11 miles south of the community of Elliston, Montana, within the Elliston Mining District of Powell County.

The mine was first opened in the late 1890s (Aiken, 1950) and was worked intermittently until the 1960s. The subsurface workings consist of a 250-foot shaft, six horizontal drifts, two known stopes located to the northeast of the shaft, and a small winze. All of the openings to the horizontal workings of the mine are now collapsed. The portal of the 74-foot level (74 adit) is the primary drainage path for acidic water from the mine. A cross section of the subsurface workings of the mine is shown in Figure 1.

Because the subsurface workings of the mine are not accessible, characterization of the site prior to the implementation of the technology was limited primarily to monthly sampling of the water (chemical constituents and flow rates) discharging from the portal of the 74-foot adit. These data did not detail any seasonal variations in the concentrations of species in the mine water. No data relating to the flow regime of the water within the mine could be acquired. As shown by the data in Table 1, the 74-foot adit portal discharged seasonally variable flows of water. The seasonal variation of flow from the portal has continued throughout the operation of the project.

The concentration of a number of constituents and the pH of the waters associated with the mine prior to the treatment implementation are shown in Table 2.

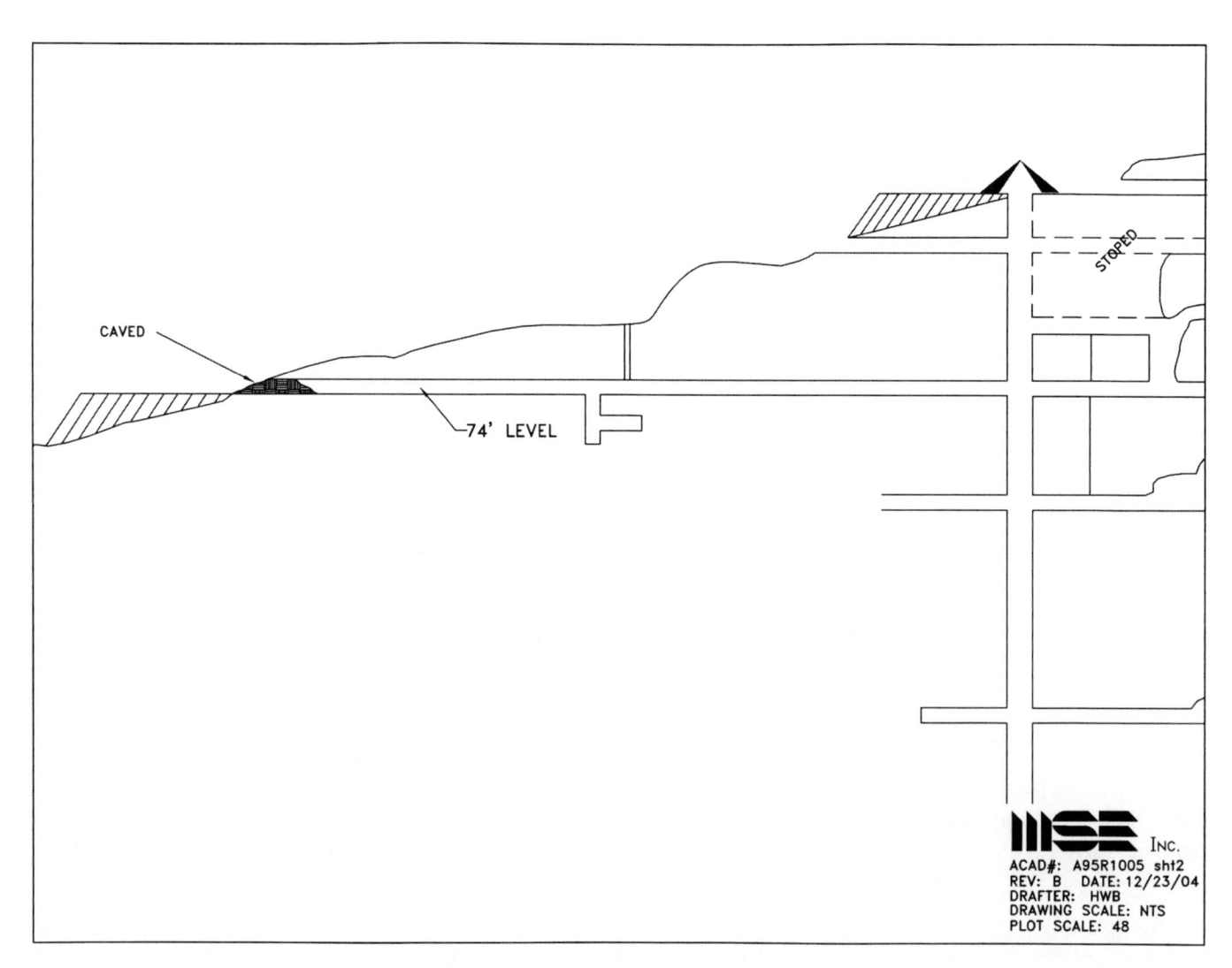

Figure 1. Mine works at Lilly mine near Elliston, Montana.

TABLE 1. REPRESENTATIVE RATES OF FLOW FROM THE LILLY MINE AS GALLONS PER MINUTE

Jan.	Feb.	March	April	May	June	July	Aug.	Sept.	Oct.	Nov.	Dec.
0.89	0.84	0.88	1.74	5.47	4.62	2.18	1.79	1.30	1.10	0.86	0.88

TABLE 2. PRETREATMENT ANALYSES OF DISSOLVED CONSTITUENTS IN THE LILLY MINE WATERS

Date	Al mg/L	As mg/L	Cd mg/L	Cu mg/L	Fe mg/L	Mn mg/L	Zn mg/L	SO_4 mg/L	pH
Adit Portal									
5/20/1993	4.02	0.21	0.14	0.30	10.90	2.62	11.00	n.d.	2.9
9/7/1993	8.09	0.02	0.24	0.24	4.87	6.28	21.60	n.d.	n.d.
10/1/1993	5.97	0.08	0.20	0.16	16.90	4.77	18.00	214	3.0
5/18/1994	7.88	0.12	0.27	0.51	15.00	4.87	19.10	n.d.	n.d.
6/14/1994	7.95	0.07	0.30	0.50	11.80	5.85	20.10	236	n.d.
7/21/1994	8.38	0.04	0.27	0.24	13.60	5.80	20.20	n.d.	n.d.
8/8/1994	9.06	0.04	0.25	0.20	18.40	6.62	22.10	n.d.	3.4
Upper Shaft									
10/1/1993	9.06	0.29	0.31	0.26	28.30	6.75	26.4	262	n.d.
8/8/1994	9.71	2.60	0.37	0.41	26.60	5.01	24.6	n.d.	3.1
Shaft Well									
9/1/2003	0.030	1.93	0.033	0.002	9.71	4.39	2.65	94	5.4
5/1/2004	0.091	1.03	0.011	0.002	6.00	3.05	1.26	57	2.72
Drift Well									
9/1/2003	26.7	75.00	1.23	0.23	218.00	5.17	84.40	667	2.85
5/1/2004	30.7	76.90	1.29	0.03	223.00	4.13	99.20	706	1.56

In addition to these early samples, two monitoring wells were drilled in 2003 to sample waters from the deeper portions of the shaft (deep shaft well), and the waters in the rock adjacent to the upper portion of the shaft (shallow shaft well). The data yielded by these samples are also shown in Table 2.

Data from Table 1 show that the flow rate from the 74 adit peeks in the May–June time frame of each year to a flow of ~5 gallons per minute. Subsequently, the flow diminishes to a low flow rate of ~0.8 gallons per minute in February.

As can be seen from the Table 2 data, the water in the upper portions of the shaft and the water from the deeper portions of the shaft (deep-shaft well) are distinctly different in dissolved metallic constituent concentrations. The deeper shaft water is generally characterized by lower dissolved concentrations of metallic constituents when compared to the water in the upper portions of the shaft. This phenomenon is believed to be caused by a mixing of the upwelling, deeper shaft water and the water flowing into the shaft from the surface areas surrounding the shaft (near-surface shaft well). The shallow-shaft well waters are higher in concentration of dissolved metallic constituents than any of the other waters in the shaft. The orientation of the subsurface workings of the mine (Fig. 1) allows the near-surface waters in the vicinity of the shaft to enter the upper portion of the shaft and combine with the waters found there. In the fall and winter of each year, the flow into the upper reaches of the shaft is dominated by the flow from the deeper shaft as the quantity of surface water diminishes due to winter conditions. During the spring of the year the flow into the upper portion of the shaft from the surface increases and changes the character of the water in the upper portion of the shaft.

Data from Table 2 show that in the fall of the year the concentrations of the metallic constituents in the water draining from the 74 adit are generally lower than the concentrations observed in the upper shaft water. This lowering of the metallic concentrations is believed to be due to the oxidation of the water and the subsequent precipitation of ferrihydrate from the 74 adit water. Such a precipitation would directly lower the iron concentration, and adsorption onto the precipitated particles would lower the concentrations of several other elements.

Lastly, since the characteristics of the 74 adit portal water and the upper shaft water are similar, it is probable that no large source of metal-laden water exists between the shaft and the adit portal. Therefore, the upper shaft water is the best representation that can be made to approximate the composition of the mine waters prior to treatment.

Table 3 depicts the load of metals, in terms of kilograms per year, emanating from the Lilly mine prior to the implementation of the treatment technology. These data have been calculated from the flow rates and concentration of the water emanating from the adit portal prior to the implementation of the treatment technology.

TECHNOLOGY DESCRIPTION

The water-treatment technology implemented at the Lilly mine is known as biological sulfate reduction (BSR). This technology exploits the actions of sulfate reducing bacteria (SRB) to produce soluble sulfide and alkalinity to treat the acidic metal-laden water. The primary function of this technology can be described in terms of the following reaction:

$$SO_4^{2-} + 2CH_2O = H_2S + 2HCO_3^- \tag{4}$$

The process requires an organic carbon source to serve as an electron donor, dissolved sulfate to serve as an electron acceptor, and specific environmental conditions including a pH of 5–8 and a redox potential below 100 mV to ensure the optimal growth of the bacteria (Cohen and Staub, 1992). The primary treatment of acidic metal-laden water involves the two products of reaction 4 and the reduction in the concentration of sulfate due to the formation of soluble sulfide. Soluble sulfide (H_2S) reacts with several metals to produce insoluble metal sulfides (reaction 5), and bicarbonate ions (HCO_3^-) react with hydronium ions to raise the pH (reaction 6).

$$H_2S + M^{+2} = MS + 2H^+, \text{ where } M = \text{a metal} \tag{5}$$

$$HCO_3^- + H^+ = H_2CO_3 \tag{6}$$

A number of secondary treatment processes are known to occur in conjunction with the aforementioned primary processes. Among these secondary treatment processes are ion exchange and adsorption functions involving the removal of metals from the drainage by the organic-rich substrate, precipitation of metal hydroxides and carbonates, and the adsorption of metals by the solid metal hydroxides.

TECHNOLOGY IMPLEMENTATION

A BSR water treatment process was implemented at the Lilly mine in September 1994 by placing two masses of organic substrate into the subsurface workings of the mine.

The first of these masses was placed on top of two permeable platforms suspended within the shaft of the mine, one in each compartment of the shaft, at a position 30 feet below the static water level of the mine shaft. Organic substrate formed of a mixture of compressed cow manure, straw, and sawdust was piled on top of each platform to a level approximately 3 feet above the static water level of the shaft. The position of this mass of substrate forces the upwelling water from the deeper portion of the shaft and a portion of the water entering the shaft from the surface to pass through the organic substrate and come in contact with the growing bacteria and the nourishing substrate.

The second mass of organic substrate (in the form of two unsupported piles) was placed into the horizontal drift on the 74-foot level of the Lilly mine at a position ~75 feet prior to the

TABLE 3. METAL LOADING FROM THE LILLY MINE AS KILOGRAMS PER YEAR

Al	As	Cd	Cu	Fe	Mn	Zn
25.4	0.3	0.8	1.1	46.3	17.3	63.3

shaft. That portion of the drift is S curved. Each mass was placed into a leg of the S curve by injection into the working via two injection wells drilled for that purpose. Again, the organic substrate is comprised of a mixture of compressed cow manure, straw, and sawdust. The placement of the masses allows the partially treated water flowing down the drift from the shaft to again pass through the organic substrate and come in contact with the growing bacteria and the nourishing substrate. The treated water subsequently flows out of the mine through the 74-foot drift portal. A third well used to monitor the process was drilled into the 74-foot level drift ~125 feet down drift from the S-curve section of the workings. A cross-sectional view of the underground workings of the mine showing the placement of the masses is shown in Figure 2.

The placement of the substrate masses within specific segments of the subsurface workings of the mine has produced a number of positive effects: It has insulated and thus protected the chemical and biological processes from rapid fluctuations in temperature and freezing; it has protected the reactors from the effects of human interference; and it has protected the reactors from large influxes in flow by allowing such large increases to partially bypass the reactors.

Previous experience has shown that large influxes in water flow and temperature can be detrimental to the operation of a BSR reactor (MSE, 1993; Zaluski et al., 2000).

RESULTS OF TREATMENT

Sampling of water subsequent to the implementation of the treatment process began in September 1994 and has continued to the present. Samples have been taken on a monthly and/or quarterly basis at the 74 adit portal, the previously described monitoring well, and one of the two injection wells used to place the aforementioned substrate piles in the 74-level drift. Analytical data have included flow rate, dissolved and total metal concentrations (aluminum, arsenic, cadmium, copper, iron, manganese and zinc), alkalinity, temperature, dissolved oxygen, pH, E_H, sulfate, sulfide, biochemical oxygen demand, chemical oxygen demand, and volatile fatty acids.

Due to the inaccessibility of the subsurface workings, direct comparisons of pre- and posttreatment water quality are not possible. Therefore, inferences related to knowledge of the subsurface flow regime and the geochemical character of the water will be used to derive these comparisons.

The mean concentrations of the posttreatment time frame for the injection well sampling site, the monitoring well sampling site, and the portal sampling site are contained in Table 4.

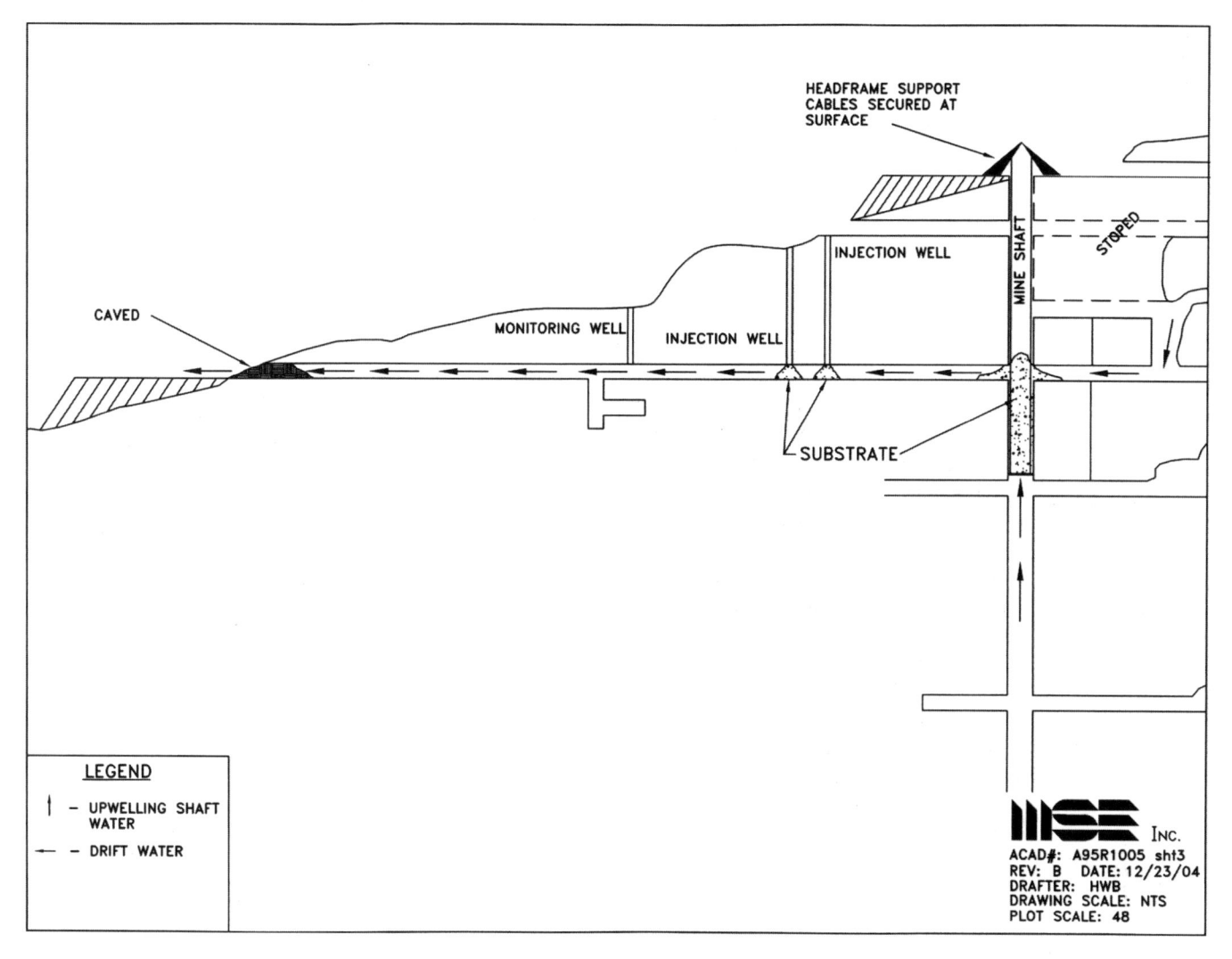

Figure 2. Location of substrates and other components used in the BSR treatment at Lilly mine.

Injection Well

Data gathered at the injection well are used to monitor the interstitial water from a portion of one of the drift substrate piles. These samples represent an area of the process that is relatively isolated and not representative of the entire flow at any point in the process. These data serve to determine the viability of the BSR operation internal to the substrate mass. As can be seen from the data presented in Table 4, the dissolved metallic concentrations for the water from this well are much lower than either of the feed water (the upper shaft water or the shallow shaft well water) concentrations shown in Table 2. In addition, these data are distinctly stable through the reporting period (September 1994–September 2003) with variations of less than 0.05 mg/L for all monitored species. The dissolved concentrations of Cd, Cu, and Zn are reduced by at least two orders of magnitude below the concentrations found in the feed waters. The dissolved concentrations of Al and As are also reduced below the concentrations found in the feed waters. The dissolved Fe concentration in the samples taken from the injection well were not reduced below the concentra-

TABLE 4. POSTTREATMENT AVERAGES OF DISSOLVED CONSTITUENTS IN THE LILLY MINE WATERS

Sampling Site	Al mg/L	As mg/L	Cd mg/L	Cu mg/L	Fe mg/L	Mn Mg/L	Zn mg/L	pH
Injection Well	0.08	0.18	0.004	0.005	24.59	3.15	0.061	7.1
Monitoring Well	2.20	1.55	0.065	0.151	18.5	5.90	10.0	4.55
74 Portal	3.40	2.27	0.143	0.201	18.0	6.01	15.5	4.75

tions found in the feed water because the treatment of iron by the BSR process is less effective due to kinetic complications. These results demonstrate that the BSR treatment processes have continued to occur throughout the time frame of the project within the organic substrate.

Monitoring Well

The sample data gathered from the monitoring are used to ascertain the effectiveness of the treatment process on the entire flow within the 74-foot level drift. It should be stated that the monitoring well was installed approximately one year after the treatment process was initiated.

As shown in Table 4, the dissolved concentrations of the monitored species are higher in the monitoring well when compared to the injection well. However, the dissolved metal concentrations found in the monitoring well are generally lower than the concentrations found in either of the feed sources. These phenomena are due to a portion of the water from the feed sources bypassing the treatment process and not being treated. The treated and untreated waters later combine prior to being sampled at the monitoring well.

The data from the monitoring well also differ from injection well data in that a number of the monitored species display a distinct seasonal variation in concentration. These seasonal variations for a number of dissolved metals as well as pH and E_H are shown in Table 5. Additionally, the average dissolved concentrations of iron and zinc for the period January 1996 through January 2000 at both the monitoring well and the adit portal are shown in Figures 3 and 4.

As can be seen from the data presented in Table 5 and the two figures, the dissolved concentration of Al, Cd, Cu, and Zn increase in the spring of the year and subsequently decrease throughout the remaining portion of the year reaching a minimum in the late winter. The dissolved concentrations of Fe and As show a substantial decrease in the concentration of each metal in the spring that continues for several months. The concentrations of these two elements reach a minimum in the fall of any year and

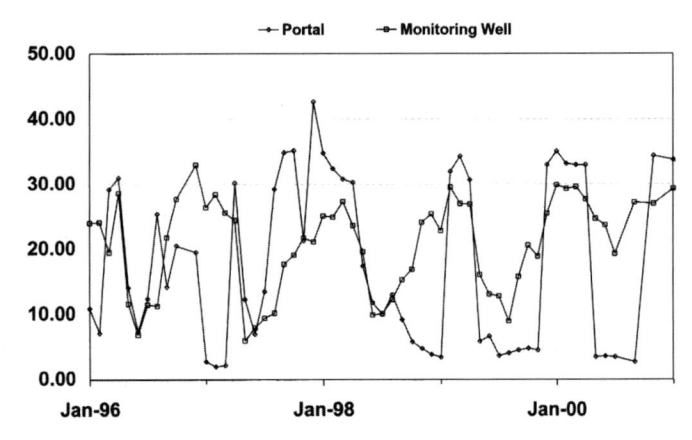

Figure 3. Graph comparing dissolved iron concentrations from 1996 to 2000 measured in water at the mine portal and monitoring well.

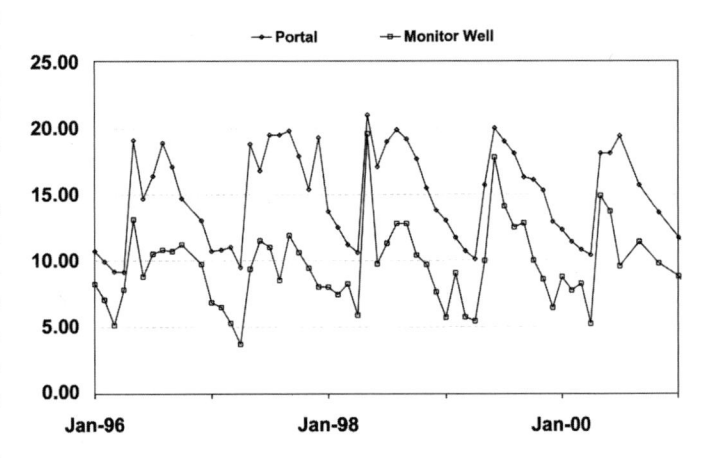

Figure 4. Graph comparing zinc concentrations from 1996 to 2000 measured in water at the mine portal and monitoring well.

subsequently increase throughout the remaining portion of the year reaching a maximum in the late winter (Fig. 3). The pH and E_H of the monitoring well water show the effects of treatment and display a seasonal variation (Table 5). The pH is generally higher

TABLE 5. SEASONAL VARIATIONS OF DISSOLVED CONSTITUENTS IN THE LILLY MINE WATERS

Sampling Site	Al mg/L	As mg/L	Cd mg/L	Cu mg/L	Fe mg/L	Zn mg/L	pH	E_H
Injection Well								
Spring–Summer	0.074	0.174	0.004	0.005	24.9	0.63	6.38	−123
Winter	0.083	0.185	0.004	0.005	24.2	0.59	7.82	−146
Monitoring Well								
Spring–Summer	4.00	0.1	0.125	0.300	9.00	15.0	3.3	517
Winter	0.40	3.0	0.004	0.002	27.0	5.0	5.8	−22
74 Adit Portal								
Spring–Summer	6.5	0.03	0.250	0.400	32.0	4.0	3.5	480
Winter	0.30	4.25	0.035	0.002	13.0	32.0	6.0	50

than that of either feed source, whereas the E_H of the monitoring well waters is generally lower than the untreated waters.

Adit Portal

The sample data collected at the portal of the 74-foot level adit are relatively similar to the data from the monitoring well and can also be used to determine the effectiveness of the treatment process on the entire flow within that adit. In addition, these data can be used to establish the posttreatment discharge quality of the water emanating from the mine.

As was the case with the samples taken at the monitoring well, the dissolved concentrations of a number of the sampled parameters from the adit portal display similar seasonal trends (Table 5 and Fig. 4). It should be noted that the dissolved concentrations of the metallic constituents of the adit portal samples are higher than those of the monitoring well for the majority of any year. Much like the samples taken from the monitoring well, the pH and E_H of the adit portal water show the effects of treatment with the aforementioned seasonal variation. The pH is generally higher than the values obtained prior to the implementation of the treatment process, whereas the E_H is generally lower than the untreated waters.

The total concentration of the metallic species analyzed within the water from the drift portal displays similar seasonal trends to the dissolved concentrations of the same metals. The total concentrations of all the elements are only slightly higher or equal to the associated dissolved concentrations.

Posttreatment Summary

Soon after the implementation of the treatment technology at the Lilly mine, the water emanating from the drift portal changed significantly when compared to the water emanating from that mine opening prior to the treatment implementation. The dissolved concentrations of Al, Cd, Cu, and Zn dropped significantly and remained lower than the historic concentration throughout the year. The dissolved concentrations of Fe and As in the adit portal water increased soon after the treatment process began to levels higher than the historic concentrations. In the spring of the first year of treatment, the dissolved concentrations of Fe and As dropped to levels lower than had been known prior to treatment. Subsequently, this cycle has been repeated each year.

Two calculations have been performed to compare the concentrations of metals in the posttreatment water at the drift portal to the pretreatment concentrations at the same sampling site. Those calculations are a dissolved concentration removal percentage and a dissolved discharge load calculation.

The removal percentage is calculated as the difference in concentration of a metal in the posttreatment drainage compared to the concentration in the pretreatment drainage. The average monthly posttreatment concentrations were used for this calculation whereas the pretreatment concentrations were from the monthly samples taken at the portal prior to the treatment implementation. The removal percentage results (both average and monthly range) are shown in Table 6.

The treatment process has been able to lower the concentration of the dissolved metals Al, Cd, Cu, and Zn discharging from the mine portal since the inception of the project. However, the treatment process has been able to lower the concentration of Mn, Fe, and As discharging from the mine portal only at specific times during the project. The times when the concentrations of these metals have been lowered are during the summer months. However, the discharge concentration of these three metals has been increased above that which was historically discharged during the fall and winter months. Overall, the average monthly removal percentages for these metals are negative.

In terms of the load of metals exiting the 74-level adit portal and entering the receiving stream subsequent to the implementation of the treatment process, the amount of aluminum has been reduced by ~12 kg per year, while the loads of zinc, copper and cadmium have been reduced by 2.3, 0.4, and 0.3 kg per year, respectively. The loads of iron, manganese, and arsenic exiting the portal of the mine have increased since the inception of the project. However, these metals are captured as oxidized precipitates soon after leaving the portal and prior to entering the creek. The chemical parameters of the partially treated water increases the rate of oxidation of the iron in the portal drainage and allows for the precipitation of the oxidized precipitates. The solid precipitates adsorb the dissolved arsenic and other species in the water thus preventing their entering the creek.

CONCLUSIONS AND LESSONS LEARNED

A number of conclusions can be reached that relate to the BSR treatment project at the Lilly mine. The most important conclusion is that the biological reactors have continued to operate throughout the duration of the project in the manner in which they were designed. In mine sites such as the Lilly, where seasonal

TABLE 6. METAL REMOVAL PERCENTAGES FOR THE 74 ADIT PORTAL SAMPLING SITE

	Al mg/L	As mg/L	Cd mg/L	Cu mg/L	Fe mg/L	Mn mg/L	Zn mg/L
Yearly Average Removal Percentage	67.7%	–8900%	53.3%	67%	–199.7%	–52.2%	5.8%
Monthly Range of Removal Percentage	95%-Apr. 13%-May	–20,000%-Apr. –50%-July	82%-April 22%-May	98%-April 1%-July	34%-April –282%-April	18.8%-Sept. –112%-Dec.	26%-Sept. 25%-May

flow variations show large fluctuations, a means of capturing the increased flow and treating it over a longer time period would normally be used. However, this option was not possible at the Lilly because the substrate was placed subsurface without reopening the mine workings. As such, the substrate placement was made into portions of the mine where the increase in flow would be able to bypass and not damage the function of the process. The pile of substrate in the 74-level drift has continued to operate without interruption as shown by the samples of water taken from the injection well. These samples show effective water treatment that has varied very little since the inception of the project. Therefore, the large and rapid increases of water through the mine have not had a deleterious effect on the processes occurring in the substrate piles inside the Lilly mine.

However, the placement of the reactors means that only a portion of the water flowing through the mine would be treated and that the amount of water treated would vary seasonally. In addition, the treated water exits the substrate masses within the subsurface workings of the mine. These workings contain precipitated solids, some of which are readily dissolved in the treated water. As such, the water being sampled at the monitoring well and the drift portal is not representative of the treated water but is a combination of treated water, untreated water, and treated water that has become contaminated by the dissolution of solids within the subsurface mine workings.

The dissolved concentration of a metal species is higher at the portal than the monitoring well, whereas the suspended fraction is higher at the monitoring well. This is due to the flow of water-suspending particles, which then may dissolve during the passage through the subsurface workings, contributing to the concentration of metals within the water flow. The source of the solid particles is not known but may be material historically deposited in the mine or particles produced by the action of the treatment process. The treatment process reduces the dissolved concentration of metals in the treated flow by producing solid particles that precipitate. The Cd, Cu, and Zn are removed as sulfides due to the formation of soluble sulfide by the BRS reactor; Al is removed as a hydroxide precipitate.

The concentrations of iron and arsenic exiting the drift portal have been increased above the historic norms by the implementation of the treatment process. This occurrence can be explained by examining the chemical processes that occur within the mine adit. Prior to the implementation of the treatment process, ferrous iron in the mine water was oxidized by atmospheric oxygen as the water flowed through the mine drift. The oxidized ferric iron would then hydrolyze to form insoluble ferric hydroxide (ferrihydrate), which would precipitate and adsorb substantial quantities of arsenic (Robins and Huang, 1988). Subsequent to the implementation of the treatment technology, the soluble iron in the adit (per the monitoring well data) is reduced ferrous iron. The implementation of the treatment process reduced the oxidation potential of the treated water flow to the point where the iron in that flow is not appreciably oxidized. As such, less of the iron oxidizes during the flow out of the mine workings, and therefore, more dissolved iron and arsenic reach the drift portal. In addition, it is probable that some quantity of the previously precipitated ferrihydrate is dissolved by the more reducing conditions, thus contributing to both the iron and arsenic concentrations in the water as shown by the increase in dissolved iron concentration between the monitoring well and the adit portal. During the spring time frame, the flow rate increases and a lower percentage of the water is treated. Therefore, the reduction in oxidizing potential is not significant, and thus, the concentration of both iron and arsenic in the water flow is reduced during those months.

The BSR treatment process has been operating at the Lilly mine in a passive manner for more than 10 years. During that period, the BSR treatment process has increased the pH and alkalinity of the water exiting the adit portal and significantly reduced the concentration and load of several contaminant metals from that water. In addition, this extended operation has been very valuable in increasing the knowledge base related to the use and operation of these processes for the treatment of acidic metal-laden drainages. Among these lessons learned is that the placement of the organic substrate within the subsurface workings does provide a number of advantages but also leads to other difficulties. Among those difficulties are that the organic substrate must be protected from large variations in flow rate. Therefore, systems must be developed that provide that protection while still treating the majority of the water. Providing these systems in a subsurface scenario can be problematic and require that the mine workings be open and enterable. Lastly, the pH and alkalinity of the treated water have been raised above that of the water that historically drained from the adit portal. Because the rate of oxidation of dissolved iron and other species is directly proportional to pH (Hustwit et al., 1992), the dissolved iron and other species in the Lilly mine drainage oxidize faster after emerging from the mine than was seen historically. This oxidation allows the iron to hydrolyze and precipitate while adsorbing substantial quantities of other dissolved species, thus removing these species from the water. Therefore, it would be advantageous to place an oxidation system and catch basin external to the adit portal to facilitate these secondary treatment processes. Because the iron concentration is high in the Lilly drainage and other similar drainages, this method of enhancing iron oxidation, precipitation, and adsorption could be a significant addition to a multistep BSR treatment process.

ACKNOWLEDGMENTS

The research associated with this document was funded under Interagency Agreement No. DW899388-70-01-1 between the U.S. Environmental Protection Agency and the U.S. Department of Energy and was conducted by MSE Technology Applications, Inc., through the National Environmental Technology Laboratory (DOE Contract No. DE-AC22-96EW96405) at the Western Environmental Technology Office located in Butte, Montana.

REFERENCES CITED

Aiken, W., 1950, The Lilly mine of Powell County, Montana: B.S. thesis, Montana School of Mines, Butte, Montana.

American Chemical Society, 1992, Environmental chemistry of sulfide oxidation: ACS Symposium Series No. 550, Alpers, C.N., and Blowes, D.W., eds., Washington, D.C.: ACS.

Cohen, R.R.H., and Staub, M.W., 1992, Technical manual for the design and operation of a passive mine drainage treatment system: prepared for the U.S. Bureau of Reclamation by the Colorado School of Mines, Golden, Colorado.

Hunter, R.M., 1989, Biocatalyzed partial demineralization of acidic metal sulfate solutions: Ph.D. thesis, Montana State University, Bozeman, MT.

Hustwit, C.C., Ackman, T.E., and Erickson, P.M., 1992, Role of oxygen transfer in acid mine drainage treatment: U.S. Bureau of Mines, Report of Investigations, RI 9405.

MSE, Inc., 1993, Activity 1, Volume 1, Appendix C, Issues identification and technology prioritization for sulfate reducing bacteria: MWTP 15, Butte, Montana, May.

Robins, R.G., and Huang, J.C.Y., 1988, The adsorption of arsenate ion by ferric hydroxide: Proceedings of the Arsenic Metallurgy Symposium, TMS/AIME Annual Conference, Phoenix, Arizona, January.

Stumm, W., and Morgan, J.J., 1981, Aquatic chemistry: An introduction emphasizing chemical equilibria in natural waters (2nd edition): John Wiley and Sons, New York.

Zaluski, M., Trudnowski, J., Canty, M., and Harrington Baker, M., 2000, Performance of field-bioreactors with sulfate reducing bacteria to control acid mine drainage: Proceedings of the Society for Mining Engineers, the Fifth International Conference, ICARD 2000, Denver, Colorado, USA.

MANUSCRIPT ACCEPTED BY THE SOCIETY 28 NOVEMBER 2006

ND# Management of mine process effluents in arid environments

Christopher Ross*
Bureau of Land Management Nevada State Office, P.O. Box 12000, Reno, Nevada 89507, USA

ABSTRACT

Modern large-scale gold mining by cyanide leaching of low-grade ore generates a large volume of process fluids. Reduction and disposal of these fluids presents unique challenges. Leaching solutions, tailings dewatering, and even postmining pit lakes must be managed both in the immediate short term and over decades or longer. Methods for reducing influx to these sources with covers and capillary breaks as well as attenuating, reducing, and disposing of them via above and subsurface land application, evaporation, and vegetation, both xeric and in engineered wetlands, among other techniques, are an evolving art still requiring an adequate base of data and observable experience. Predictive modeling of fluid volume and behavior has proved very inaccurate over both shorter and longer time intervals. Climatic extremes and intensity of precipitation events compound the problem in arid areas. Ecological risk assessment is used to estimate exposure to contaminants of concern. Experience has demonstrated the inadequacy of predictions about process fluid management postclosure, and the need for comprehensive fluids bonding both for short-term contingencies such as bankruptcy and for long-term effluent disposal maintenance and monitoring.

Keywords: metal mining, process fluid, pit lake, tailings, evaporative cells, engineered wetlands, land application, ecological risk assessment, cyanide, fluid management, evapotranspiration, capillary break, effluent, bonding, revegetation, heap leach

INTRODUCTION: PROCESS OVERVIEW, ISSUES, AND DEFINITIONS

Modern precious metal mining often involves leaching of low-concentration ores with solutions of cyanide in water adjusted to a relatively high pH. As fluids are circulated and recirculated, they accumulate not only dissolved target metals but other metals and salts. These are concentrated by evaporative loss. Such process fluids pose problems for disposal that persist long after active mining ceases. Process fluids include leach solutions, both "pregnant" (with target metals in solution) and "barren" (stripped of target metals but still containing other metals, salts, cyanide, etc.), tailings, and other waste water. Pit lakes that form after dewatering of mines ceases are discussed separately because such waters are generally different from native ground water as a result of their exposure to the mining process. Waste rock drainage and the management of acidity and other contamination issues are not discussed here, although these issues certainly affect surface and groundwater at many sites.

Disposal of process fluids other than tailings and pit waters usually involves attempts to attenuate contaminants, reduce volume, and then dispose of the fluids on site. Decontamination may be accomplished by recirculation and chemical addition to break down cyanide and evaporate fluids, followed by discharge either to evaporation cells, evaporation basins, engineered wetlands, reinjection underground, or land application. Evaporation is facilitated not only by reapplication to heaps but also by spraying, sometimes with modified snow-making machines, absorptive

---

*cross@nv.blm.gov

Ross, C., 2007, Management of mine process effluents in arid environments, *in* DeGraff, J.V., ed., Understanding and Responding to Hazardous Substances at Mine Sites in the Western United States: Geological Society of America Reviews in Engineering Geology, v. XVII, p. 163–169, doi: 10.1130/2007.4017(10). For permission to copy, contact editing@geosociety.org. ©2007 The Geological Society of America. All rights reserved.

evaporative fabrics, and wetland vegetation. Long-term disposal by underground leach fields or injection and surface land application have environmental impacts and are now generally discouraged in favor of "zero discharge" systems. Pit waters are not amenable to disposal, and long-term treatment of water-quality issues is hampered by the difficulty of accurate predictive modeling of water quality and quantity and by the financial impact of long-term monitoring and treatment of large volumes. Temporary closure plans address events that might cause temporary closure, such as natural disasters, and how fluids will be managed to prevent discharge to the environment. However, experience with bankruptcies has demonstrated the need for rapid response to interim fluid management in order to prevent overflow of process ponds and other containment when power to recirculation pumps ceases.

EVALUATION OF PROCESS WATER QUALITY

During production, process water quality is not an issue for the Bureau of Land Management (BLM) and the Nevada Division of Environmental Protection (NDEP) as long as the operator is processing within the guidelines established in the approved federal Plan of Operations and state permits. However, evaluation of process solutions becomes an integral component of operating plans/permits when an operator proposes to enter a temporary or permanent closure phase.

The state of temporary closure as defined by Nevada NAC 445A.382 means "the cessation of the operation of a process component for more than 30 days as a result of a planned or unplanned activity." In general, under a temporary closure, operators must maintain process solutions per their operating permit or must begin neutralization or stabilization of the fluids. Stabilization must then meet the standards outlined in the final permanent closure plan.

Final permanent closure plans must ensure that no mined areas may release contaminants that have the potential to degrade the waters of the state. Spent ore from heap leach pads must be rinsed until weak acid dissociable (WAD) cyanide levels in the effluent are less than 0.2 mg/L, the pH is between 6.0 and 9.0, and any other contaminants in the effluent will not degrade waters of the state. The state does allow variances from these standards if it can be demonstrated that spent ore can be stabilized to inhibit contaminant transport, meteoric waters can be inhibited from migrating through the material, and mobilizing contaminants and stabilization of spent ore will not have the potential to degrade waters of the state. Tailings impoundments must be stabilized "so as to inhibit the migration of any contaminant that has the potential to degrade the waters of the state." The state requires postclosure monitoring for up to 30 years or until it can be demonstrated that the process component has been stabilized.

Nevada BLM policy for water-quality standards is summarized in Instructional Memorandum No. NV-2004-064 (June 2, 2004). This policy basically states that, at a minimum, state standards must be met. The goals of the policy are to ensure the health of the land and water resources, to use good science in making decisions, and to collaborate with appropriate federal, state, local, and tribal agencies and other affected interests. Essentially, BLM utilizes the NEPA process to set water quality standards for effluents from process components. These standards may be variable depending on the type of closure technique proposed as discussed later.

ECOLOGICAL RISK ASSESSMENT

Water quality, especially of projected pit lakes, tailings, and sludges, may be evaluated by the ecological risk assessment (ERA) process, which analyzes the likelihood that adverse ecological effects may be caused by exposure to contaminants. The Bureau of Land Management has established criteria for metal contaminant levels affecting humans and wildlife (Ford, 2004). The U.S. Fish and Wildlife Service also establishes toxicity benchmark values. Pathways of exposure, exposure scenarios, and ranges of expected levels of contaminants of concern (COCs) are modeled and evaluated against the risk management criteria (RMC) for each metal. Exposure scenarios within a given ERA model might include residents drinking the water, ATV drivers being exposed to dust, surveyors, fishermen eating their catch, etc. Where the combination of level of COC and exposure potential leads to an intake that exceeds the RMC, adverse effects on humans or wildlife can be expected, and water can be treated or otherwise managed to prevent problems. This process can be applied to aquatic systems, using the following procedure. Resource values and potential risks are identified. The source(s) of risks are identified and quantified; then the fate, transport, and accumulation of chemicals are defined to determine risk to specific receptors. Assessment endpoints are characterized in terms of ecological effects to receptors. From this risk, endpoints can be characterized and risk assessed. (Cardwell et al., 1993). In Nevada, a screening-level ERA, using conservative assumptions and simple assessments, is typically conducted, and the results of this determine whether a full-scale ERA is warranted.

TYPES OF PROCESS FLUIDS AND CHEMICAL/MONITORING ISSUES

Traditional attention to process fluid has centered on heap effluents, both the pregnant and barren solutions that are used in metal extraction during mining, and on postmining effluent, or "draindown." Cyanide solutions involved in actual metal extraction circuits are well contained, typically in double-lined ponds and channels with leak detection drains and monitoring ports between the layers and storm event overflow capacity. Thus, spills and leaks are rare at active mines. However, both pregnant and barren solutions first received intense environmental scrutiny in the 1980s when massive kills of migratory waterfowl and mammals, especially bats, occurred on widely scattered operations (Ross, 1992a). Imaginative methods of hazing, or scaring, of birds including radio-controlled model boats and planes, laser shows, rafts blasting high-decibel hard-rock music, and propane cannons all became quickly passé to the birds. In the face of these violations of the Migratory Bird Treaty Act, the Nevada Division of Wildlife issued guidance that pond permits would henceforth require that wildlife either be excluded, typically by nets or floating balls, or exposure to cyanide be at low levels. Recent studies suggest that even low levels of cyanide in solution have detrimental effects on many vertebrates.

Interim Fluid Management

Following the 1999 bankruptcy of Alta Gold it became clear that economic issues could affect fluid management as much or more than accidents or natural disasters. Alta Gold declared bankruptcy and essentially walked away from Olinghouse, leaving the state to maintain process fluids on containment while BLM began legal proceedings to collect the bond from the surety company. The state spent approximately $177,000 maintaining fluids and had to write off nearly $77,000 as uncollectible (Nevada Appeal, 2004). As a result, the state environmental commission adopted regulations providing for surety bonding specific to maintenance of short-term fluid management in a trust fund. Such funding typically allows operation of pumps and other circulation equipment for six months to allow the time it takes to collect long-term bond/surety funds (Nevada Administrative Code Chapter 519A.360 Adopted Regulation R120-00). However, reclamation bonds must still have an interim fluid management component equal to six months of fluid management costs, which are used to repay the state fund if needed.

Heap Effluent

Meteoric water infiltration and discharge are poorly understood under many arid land conditions, even for parameters as important as water supply to large urban areas. For example, results presented at the 2004 Geological Society of America Annual Meeting in Denver demonstrated that long-held assumptions about snow melt contribution to surface water supply in Colorado were radically incorrect (Associated Press, 2004). In the 1980s and early 1990s, fluid modeling and environmental analysis of heap draindown rates and times routinely assumed that large heaps in arid areas would cease emitting fluid in perhaps six months or so after cessation of fluid application. These estimates were made without benefit of experience because large-scale heap leach mining was a new technology that had not yet provided any examples of heaps in the closure stage of the mining lifecycle. The first heaps closed demonstrated that such assumptions were unrealistic and led to major revisions in hydrologic modeling and methods to reduce influx of meteoric water to heaps. The HELP model, originally used for many heap predictions, is now regarded by many as useful only for evaluating variation among different cover designs, not for actual prediction of outflow (Van Zyl, 2001). Because heaps have complex, often heterogeneous physical geometry and operate in an unsaturated state, seepage models are typically inadequate (Fredlund and Thode, 2003). It is now assumed that heaps will drain, at least at low flow rates, for years or longer, even in arid sites experiencing drought conditions. For example, at Nevada's Bald Mountain mine, draindown rates and times were seriously underestimated initially, mirroring experiences at other sites (Parshley et al., 2003). Furthermore, despite the fact that undisturbed sites in arid rangelands show little deep infiltration, heaps show sudden increases years after closure in response to precipitation events (Tyler, 2001). This has forced the development of techniques for long-term, low-maintenance disposal of effluent. The BLM in Nevada, in cooperation with contractors, the state, and the Desert Research Institute, is installing lysimeters at some heaps that are being reclaimed after corporate bankruptcy to monitor heap draindown output in relation to draindown and meteoric input on inactive heaps. These are simple liner/collection systems buried in a representative portion of the heap with cover material identical to the rest of the heap.

ENGINEERED AND EVAPOTRANSPIRATIVE COVERS AND CAPILLARY BREAKS

Obviously it is preferable to intercept rain and snow before they infiltrate heaps and contribute to effluent outflow. Such evapotranspirative (EVT) covers are currently accepted practice for new heap closure plans. However, they remain largely untested, and several important constraints are routinely ignored in the planning process. One is that most infiltration models were not developed with the arid West in mind. In Nevada, almost all precipitation occurs in winter, when plants are dormant and in many cases leafless. The cover must, therefore, store the water for perhaps six months until the weather is warm and plants are actively transpiring. Summer precipitation is likely to be intense and episodic, overwhelming storage capacity of the cover. Clay barriers and capillary layers are added to assist this, but they will develop preferential infiltration pathways over time as plant roots and rodents invade them and as freeze-thaw and wet-dry cycles occur. Plant root depths are usually drastically underestimated in cover designs, many of which rely on only one foot or less of covering growth media. Many desert plants, even annuals, send down roots five meters or more, which is far below the capillary barrier and well into potentially saline/toxic heap material. Even if this does not cause plant mortality, it may result in metals and other heap contaminants being introduced into the food chain through plant uptake. At Florida Canyon in northern Nevada, initial modeling by UNSAT2D, TAR2DU, Hydrus2d, and SoilCover models gave the "typical answer of 'zero' net infiltration." Subsequent experience, here as elsewhere, did not bear these predictions out. In addition to missed processes, such as root distributions, soil changes, fines in the capillary break, and construction issues, Niccoli et al. (2003) blame the lack of calibration data, pointing out that they attempted to predict effects before construction and real data, and that they were thus calculating as opposed to modeling. Another complication is that many taxa actively bioaccumulate uranium, selenium, arsenic, lead, and other toxic metals from water and soil.

Extreme precipitation events, especially in arid areas and early in the revegetation stage, often cause serious erosion, mass wasting, and alteration of drainage topography on heap covers. In a number of cases in Nevada, heap and landfill covers have been seriously damaged shortly after installation. Thus bonds should be released cautiously and the possibility of significant additional work considered even after initial dirtwork has been completed. The AA leach pad at Barrick's Goldstrike operation on the Carlin trend is probably the most sophisticated heap design in Nevada, with extensive run-on and run-off design and state of the art revegetation. Very shortly after the cover was completed, and before plants were well established, a storm dumped 1.6 inches of rain in 20 minutes. This cover survived with minimal damage due to its advanced design (Myers et al., 2003), but in several other cases at active and inactive mines as well as landfills in Nevada, the results have been much more destructive.

Models generally assume that no vegetation will be harvested. This is in conflict with most stated postmining land uses, such as grazing and wildlife habitat. Furthermore, to ensure this condition, it is necessary to fence the heap cover. Fencing requires frequent monitoring and maintenance/repair, mandating bonding in perpetuity to ensure fence integrity. The potential of vegetation loss from wildfire carries similar financial constraints. Plant succession will alter the intentionally installed communities over time, sometimes quite rapidly and to types less suitable for evapotranspiration, such as annual species or taxa with high water-use efficiency. Climatic change and pathogens may also impact the successional trajectory. For example, wide-scale shrub die-offs have occurred repeatedly in western states for reasons still largely unknown (Nelson et al., 1990). Fire is increasingly common on arid western ranges and almost invariably causes a shift to annual plant communities with very different EVT characteristics. Any sort of cover that relies on vegetation will require indefinite monitoring and financing to ensure that if undesirable changes occur they can be reversed through site alteration and reseeding/replanting if needed.

EFFLUENT MANAGEMENT

While effluent management systems can be classified as either short-term active or long-term passive, it is increasingly apparent that even so-called passive systems such as leach fields and engineered wetlands have a long-term maintenance component. Process water leaving the leach circuit may contain arsenic, antimony, selenium, cyanide, sulfate, and salts, among numerous other problematic chemistry. Nevada regulations mandate that such effluents must not degrade the waters of the state. Thus, disposal options, especially in areas of shallow groundwater, may be constrained (Bowell, 2003). BLM, on the other hand, must be concerned with the vadose zone and potential receptors through metals uptake.

Land Application

(Active, end-of-mine life.) Land application has been used at a number of sites in Nevada. It consists of flooding or spraying heap effluent onto the surface of the ground. Because of the high concentrations of salts, metals, and other components in such fluid, it often results in death of vegetation at the application site, as well as creating a salty, contaminated surface attractive to wildlife. It is rarely permitted now in Nevada, which favors zero-discharge of fluids from containment.

Leach Fields

(Passive, maintenance and monitoring required.) Where depths to groundwater are great, and especially where groundwater quality is already low, shallow leach fields similar to septic tank systems have been used to dispose of heap draindown. Often these are equipped with a dosing tank system that retains fluid, then releases a pulse into the field so that the entire field is used rather than just the first portion of each line that would absorb a gradual trickle of fluid. Obviously, such systems require monitoring and maintenance and are prone to scale buildup and freezing. Like land applications, leach fields are discouraged.

Evaporation

(Active and passive.) Reduction of fluid volume by evaporation may take place either on or off the heap itself. In its simplest form, the heap solution is just pumped repeatedly back onto the heap, where it evaporates and concentrates salts. This process is sometimes sped up by either spraying the solution with standard irrigation equipment or by the use of snow-making equipment, which uses compressed air or fans to blast a fine spray into the atmosphere. These techniques may also be used at the ponds or elsewhere off-heap but on containment (Fig. 1).

Other off-heap fluid reduction systems involve evaporation basins and/or wetlands where plants are to transpire the fluid. These often use the existing pregnant, barren, and overflow (or storm event) lined ponds, which are filled with gravel or other coarse material, then filled to a level that facilitates evaporation from the substrate surface but does not allow animals to contact the solution directly. Pilot projects are underway to use turbine ventilators, often used on outbuildings, to draw air through perforated pipe buried in the gravel to increase air flow and evaporation. All of these require monitoring, maintenance, and overflow capacity in redundant backup systems.

Engineered Wetlands

(Passive, monitoring and maintenance required.) Recently, engineered wetlands have been proposed as better systems for evaporating heap effluent and other process solutions. A number of serious obstacles exist, though. Most of these are the same as or similar to those that affect EVT covers on heaps, with similar con-

Figure 1. Heap solution evaporation by snow-making equipment, Candelaria silver mine, Nevada. This often results in salt concentrations on the heap surface, which interfere with revegetation.

straints on long-term performance and bonding requirements. An important addition to this list is the fact that as fluids continue to flow into such wetland cells and are evaporated or transpired from them, the salinity and TDS content of the wetlands will increase dramatically. Because plants tolerate only certain ranges of these parameters, the species that initially thrive will eventually (in some cases soon) find their tolerances exceeded. Many mine sites are in very remote arid areas where it is unlikely that more tolerant taxa will invade naturally. Nor is it likely that years down the road there will be attention to replanting the cells with vegetation that can tolerate the evolved conditions. In addition, shed leaves and other dead vegetation on the surface of the wetland growth media radically reduce airflow across evaporative surfaces and insulate from solar input, further reducing evaporation potential. They will probably turn into outflow systems unless they are periodically rejuvenated through a long-term maintenance program.

BONDING ISSUES

There are several components to most reclamation bonds reflecting not only the complexity of modern mines and their components but also the need to be able to mobilize manpower and equipment quickly to deal with fluid management issues in unexpected mine shutdowns. Unanticipated premature mine closure fluid issues are now covered by the interim fluid management (IFM) component in the Nevada bond cost estimate. The state IFM fund covers short-term response while that bond is being collected. The reclamation bond can cover both short- and long-term draindown. A long-term trust may cover long-term passive and active treatment and disposal of draindown. What were once perceived as passive (short- and long-term) closure issues are now recognized to be potentially a very long-term closure responsibility that requires a long-term trust fund for heap leach and tailings fluid management.

TAILINGS

Older tailings impoundments were often unlined or not equipped with leak detection and, thus, leak to the vadose zone and, perhaps, groundwater. Newer impoundments are lined with underdrains to collect fluids for treatment by any of the means discussed above. Like heaps, tailings impoundment hydraulics are difficult to predict. Because of their high water retention, large impoundments may take anywhere from decades to thousands of years to completely drain (Fig. 2; Milszarek et al., 2004).

Because of their size, liners, and the mechanical and chemical characteristics (acidity) of tailings, EVT covers dependent on vegetation are often not a viable option unless very thick covers are applied. Thickeners and wick drains may speed drying, as can a compressive overburden. Alternate placement of tailings into pits or underground operations may be feasible in a few cases where chemical and water issues do not preclude them. Closure goals at Barrick's Goldstrike large AA block tailings facility are typical objectives (Henderson et al., 2004):

- Design for chemical and physical stability.
- Maintain hydraulic separation between tailings and surface/groundwater.
- Return the site to a productive postmining land use.
- Minimize closure time.
- Limit environmental liability to all parties.

PIT LAKES

Hydrogeochemical modeling is widely used to predict water quality of postmining pit lakes. However, variables such as evapoconcentration, oxidation/reduction of compounds in newly exposed rock, biological activity, adsorption, and ligand exchange/complexation add greatly to the complexity and unreliability of predictive modeling of pit water quality (Ross, 1992b). Predictive

Figure 2. Tailings impoundment, Paradise Peak mine, Nevada. After drying, tailings are often high in metals and salts and are prone to moving off containment by wind erosion.

accuracy decreases as pit lake age increases. Great uncertainties in modeling of pit lake levels, chemistry, and turnover contribute to debate about future quality and exposure to hazards. Although addition of lime to control pH and precipitate COCs may be useful in the short term, perpetual treatment is generally not considered acceptable and would in any case be very expensive and difficult to bond (Fig. 3).

In addition, unrealistic expectations that such lakes in very arid areas will not be stocked with fish or used recreationally in coming years, while common in environmental impact analyses, are of concern to those responsible for public health and physical safety. Although the state appears to require that pits be reclaimed, regulations offer exemptions if there would be an economic impact or if evidence of mineralization would be obscured. In effect, this eliminates all pits from required reclamation beyond reseeding of the pit floor and berming/fencing of the highwall. Pits are rarely backfilled except where haul logistics offer an economic advantage or where poor water quality is expected. The ledges and steep sides of pits make them especially dangerous even when there is no lake below. Where water does exist or is projected, rock falls and ledges above and below the water line are certain to result in accidents after mines are no longer active. Experience with quarries in eastern states suggests that falls and drownings may be fairly common.

SUMMARY

The rapid development of large-scale cyanide heap leach mining on an unprecedented scale has created many situations in which the use of untested techniques for dealing with process fluids in closure is widespread. In many such situations, assumptions about closure methodology and duration have been proven inadequate. In at least one western state, industry has recently proposed the elimination of heap rinsing and rinse standards for cyanide and pH, preferring to go directly to capping and covering heaps and dealing with the effluent as it emerges. With increasing numbers of mines approaching closure, certain precautions will help in dealing with these uncertainties.

Heaps, waste rock dumps, and tailings features should be fully instrumented during the closure process and, perhaps, during initial construction as well. Simple lysimeters to monitor fluid influx and outflow, coupled with sampling ports and automated weather stations, will help determine effectiveness of covers and provide data with which to refine and validate predictive models. Monitoring periods should be extended over a much longer time, with financial assurance that monitoring will be conducted.

Bonding for extended maintenance and/or repair of covers and for revegetation in the event of loss due to fire, grazing, or climatic extremes should be in place. In most cases, this should include support for long-term fence maintenance. Bonds should be regularly updated to reflect changes in labor, equipment, and fuel costs.

Planners should accept that characterizations of storm events, such as "100-year storm," are based on very limited data sets. It is common, especially in the arid West, to experience multiple "100-year events" in a human lifetime. Thus, planning should be for longer temporal horizons and events of greater intensity.

Realistic assessment of the ability of engineered wetlands to adapt to climatic changes and especially evapoconcentration of salts and other process fluid elements should be incorporated into the planning and closure cycle.

Figure 3. Pit lake, Paradise Peak mine, Nevada. This mine is in bankruptcy, with inadequate bonding to reclaim the site. The pit lake is acidic and high in metals, with obvious failures of the high walls.

Figure 4. Betze-Post pit at Barrick/Goldstrike mine, Carlin trend, Nevada. After dewatering ceases, this will fill with groundwater to become the largest lake wholly within the boundaries of Nevada.

Planners must acknowledge that pit lakes and other bodies of water associated with mining will be stocked with fish and will be used recreationally in the future, especially in areas where other water sources are scarce. Where it is not possible or desirable to backfill pits to eliminate lakes, pits should be reconfigured to create more stable walls and to eliminate hazardous ledges both above and below the range of potential water levels (Fig. 4).

REFERENCES CITED

Associated Press, 2004, Study: Snowmelt only responsible for a portion of runoff: Reno Gazette Journal Online, November 7.

Bowell, R., 2003, Defining the purpose of water quality assessment as part of a cyanide heap leach closure project: Proceedings, Heap Leach Closure Workshop, Mining Life Cycle Center, University of Nevada Reno, March 25–26, 2003.

Cardwell, R.D., Parkhurst, B.R., Warren-Hicks, W., and Volosin, J.S., 1993, Aquatic ecological risk: Water Environment and Technology, v. 5, p. 47–51.

Ford, K.L., 2004, Risk management criteria for metals at BLM mining sites, Technical Note 390, Revised: BLM/RS/ST-97/001+1703), Denver, Colo. 80225.

Fredlund, M.D., and Thode, R., 2003, Seepage modeling a heap leach pad: Proceedings, Heap Leach Closure Workshop, Mining Life Cycle Center, University of Nevada Reno, March 25–26, 2003.

Henderson, M., Kowalewski, P., Espell, R., and Giraudo, J., 2004, Closure options for the Goldstrike tailings facility: Proceedings, Tailings Impoundment Closure Workshop, Mining Life-Cycle Center, University of Nevada, Reno, June 8–9, 2004.

Milszarek, M., Yao, M., and Thompson, T., 2004, The effect of tailings characteristics on mine reclamation and closure, 2004: Proceedings, Tailings Impoundment Closure Workshop, Mining Life-Cycle Center, University of Nevada, Reno, June 8–9, 2004.

Myers, K., Espell, R., and Zhan, J., 2003, Hydrologic, erosional, and revegetation performance of the cover system on the AA heap leach pad at Barrick's Goldstrike mine: Proceedings, Heap Leach Closure Workshop, Mining Life Cycle Center, University of Nevada Reno, March 25–26, 2003.

Nelson, D.L., Harper, K.T., Boyer, K.C., Wever, D.J., Austin Haws, B., and Marble, J.R., 1990, Wildland shrub dieoffs in Utah: An approach to understanding the cause, in Durant, E.D., et al., eds., Proceedings, symposium on cheatgrass invasion, shrub die-off, and other aspects of shrub biology and management: GTR. INT-276, Ogden Utah. USDA USFS Intermountain Research Station, 351 p.

Nevada Appeal, 2004, State writes off costs from mine restoration: Appeal Capitol Bureau, Dec. 8, www.Nevadaappeal.com/article/20041208/NEVADA/112080030.

Niccoli, W.L., Marinelli, F., and Carlson, B., 2003, Predictions of long-term flow from vegetative covers: Why are we SO far off?: Proceedings, Heap Leach Closure Workshop, Mining Life Cycle Center, University of Nevada Reno, March 25–26, 2003.

Parshley, J., Buffington, R., and Rykaart, M., 2003, Lessons learned from the closure of the Yankee heap leach pad: Proceedings, Heap Leach Closure Workshop, Mining Life Cycle Center, University of Nevada Reno, March 25–26, 2003.

Ross, C., 1992a, Leach heap operations and groundwater: Groundwater Monitoring and Remediation 11:2, p. 88–90.

Ross, C., 1992b, Gold mining affects ground water: Ground Water Monitoring and Remediation 12:4, p. 100–102.

Tyler, S., 2001, Hydrologic behavior of heaps: Estimation of leachate quality, caps and covers: BLM NTC Course Number 1703-17, Dec. 12, Reno, Nevada.

Van Zyl, D., 2001. Caps and covers: BLM NTC Course Number 1703-17, Dec. 12, 2001, Reno, Nevada.

Manuscript Accepted by the Society 28 November 2006

Sampling and monitoring for closure

Virginia T. McLemore*
New Mexico Bureau of Geology and Mineral Resources, New Mexico Institute of Mining and Technology, Socorro, New Mexico 87801, USA

Kathleen S. Smith†
U.S. Geological Survey, M.S. 964, Denver Federal Center, Denver, Colorado 80225-0046, USA

Carol C. Russell‡
U.S. Environmental Protection Agency, 999 18th Street, Denver, Colorado 80202, USA

Sampling and Monitoring Committee of the Acid Drainage Technology Initiative—Metal Mining Sector

ABSTRACT

An important aspect of planning a new mine or mine expansion within the modern regulatory framework is to design for ultimate closure. Sampling and monitoring for closure is a form of environmental risk management. By implementing a sampling and monitoring program early in the life of the mining operation, major costs can be avoided or minimized. The costs for treating mine drainage in perpetuity are staggering, especially if they are unanticipated. The Metal Mining Sector of the Acid Drainage Technology Initiative (ADTI-MMS), a cooperative government-industry-academia organization, was established to address drainage-quality technologies of metal mining and metallurgical operations. ADTI-MMS recommends that sampling and monitoring programs consider the entire mine-life cycle and that data needed for closure of an operation be collected from exploration through postclosure.

Keywords: mine-life cycle, mine closure, mine drainage

INTRODUCTION

Acidic drainage (AD) results from sulfide oxidation in rocks exposed to air, water, and microorganisms and can occur naturally or as a result of mining-related activities. Acidic drainage resulting from mining commonly is referred to as acid mine drainage (AMD). Acid rock drainage (ARD) is a more general term that encompasses both naturally occurring and mining-related AD. In discussing mine drainage, we prefer the term *mining-influenced waters* (MIW; Schmiermund and Drozd, 1997) because acidic, neutral, and alkaline waters all can transport metals and other contaminants.

The Acid Drainage Technology Initiative (ADTI), a coalition of government agencies, industry, academia, and consulting firms, was created in 1995 by the National Land Reclamation Center, National Mining Association, Office of Surface Mining, U.S. Environmental Protection Agency, Bureau of Land Management, U.S. Geological Survey, and Interstate Mining Compact Commission. ADTI has a coal-mining sector (ADTI-CMS) and a metal-mining sector (ADTI-MMS). Its overall mission is to identify, evaluate, and develop cost-effective and practical AMD technologies and to address other drainage-quality issues related to mining (Hornberger et al., 2000; Williams, 2003; van

*ginger@gis.nmt.edu
†ksmith@usgs.gov
‡russell.carol@epa.gov

Zyl et al., 2006). In addition to these responsibilities, the ADTI-MMS disseminates information about cost-effective, environmentally sound methods and technologies to manage metal-mine wastes and related metallurgical materials from active, abandoned, inactive, and future mines (ADTI-MMS, 2006). It also addresses MIW drainage-quality issues related to metal mining and metallurgical operations, including pit lakes, and has assembled a multidisciplinary team of experts to identify current prevention and remediation technologies and to propose promising technologies for research. For more information on ADTI-MMS, see http://www.unr.edu/ mines/adti/.

Geological and environmental variations between mines have fostered a broad range of techniques to sample, monitor, predict, model, mitigate, and control acid and MIW related to mining, metallurgy, and materials processing. At present, there is no single authoritative guide available for evaluating the various techniques for assessing MIW. ADTI-CMS members wrote two handbooks on AMD related to coal mines (Skousen et al., 1998; Kleinmann, 2000), focusing on avoidance, prediction, remediation, and treatment technologies. Similarly, ADTI-MMS members are developing a handbook dealing with drainage from metal mines and related metallurgical facilities. The purpose of this paper is to outline the site-specific sampling and monitoring activities and strategies needed throughout a mine's life and provide examples of their importance. This paper and an earlier paper (McLemore et al., 2004) are a condensed version of the sampling and monitoring chapter of the ADTI-MMS Handbook that is under development.

PHILOSOPHY OF SAMPLING AND MONITORING

More than 30 years have produced only limited descriptions of the philosophy of sampling and monitoring. The holistic philosophy taken in this paper is that sampling and monitoring programs should be incorporated into all aspects of the mine-life cycle on a site-specific basis. Designing for ultimate closure of new mines or mine expansions is a relatively new concept because miners and regulators of the past rarely planned for closure. Closure planning ideally begins during the exploration and feasibility stages of mining and is augmented throughout development, mining, and milling. Additional sampling and monitoring continues through closure and postclosure stages to ensure that previous plans are working.

WHAT IS A SAMPLING AND MONITORING PROGRAM?

A sampling and monitoring program uses sample analyses to guide decision making, especially with respect to environmental and financial risks. Sampling and monitoring plans document procedures to ensure that the resulting samples and analytical data meet the goals and objectives of the program. This approach contributes to informed decisions. These plans, especially those undertaken early in the mining process, are not conducted merely to fulfill the regulatory process but to ensure a profitable and environmentally responsible undertaking.

Sampling strategies should be tailored to site characteristics, such as environmental conditions, geology, available sample media, and access, and should consider the costs for undertaking sampling versus the costs of not having the data when needed. Sampling and monitoring programs can be designed to address specific issues. When several sampling and monitoring plans are used during the life of a mine, it is important to record the objectives and methods of each plan so the resulting data can later be evaluated relative to why the data were needed and what decisions were being made based upon them.

Monitoring and data assessment can address many questions, such as those suggested by Harrison (1998):

- What are the potential issues?
- What are the baseline conditions?
- What are the potential sources of problems?
- What are the pathways affected, and are people or the environment impacted?
- What are the desired results of remediation, reclamation, or restoration?
- Who will determine if the program is successful?
- Was the sampling and monitoring plan successful?
- Is the site or will the site be in compliance with environmental laws?

Data quality objectives (DQOs) for monitoring should be tailored to each specific site in order to answer these questions. A 12-step DQO process is summarized in Figure 1.

THE MINE-LIFE CYCLE

The mining sequence, or the mine-life cycle, can be divided into several stages including:

- Exploration
- Mine development
- Operations
- Closure/postclosure

Figure 2 illustrates the progression of these stages. The following sections briefly explain each stage of the mine-life cycle and its relationship to sampling and monitoring for closure.

Exploration

The *Mining Engineering Handbook* describes the first stage of mining, the prospecting stage, as the search for a mineral deposit (Hartman, 1992). A *mineral deposit* is any occurrence of a valuable commodity or mineral that is of sufficient size and grade (concentration) for potential economic development under past, present, or future favorable conditions. An *ore deposit* is a well-defined mineral deposit that has been tested and found to be of sufficient size,

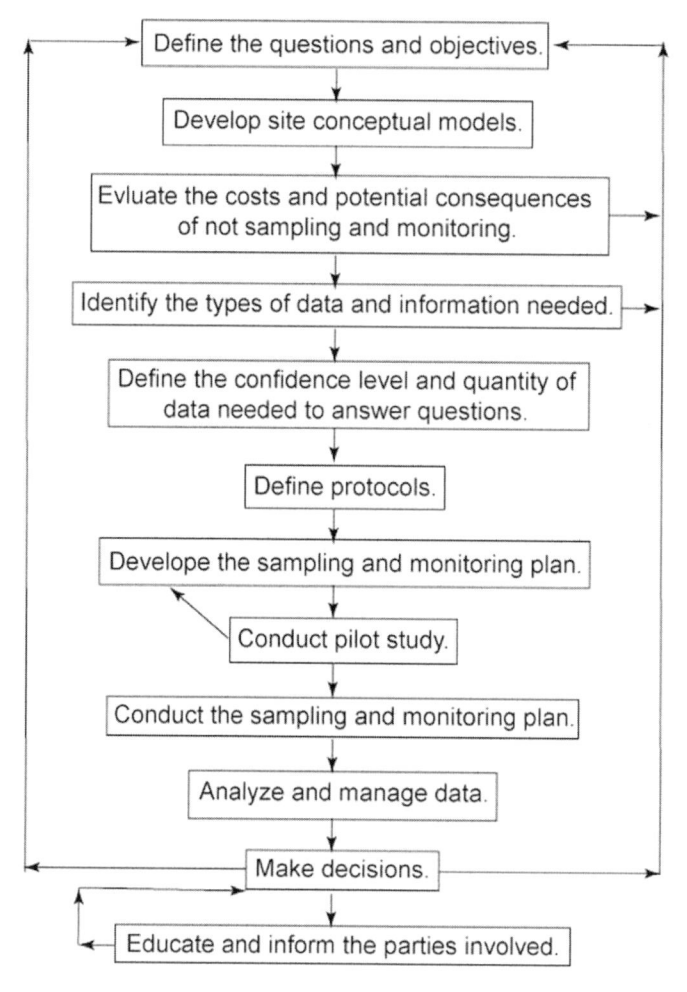

Figure 1. Data quality objective steps for implementing a sampling and monitoring program.

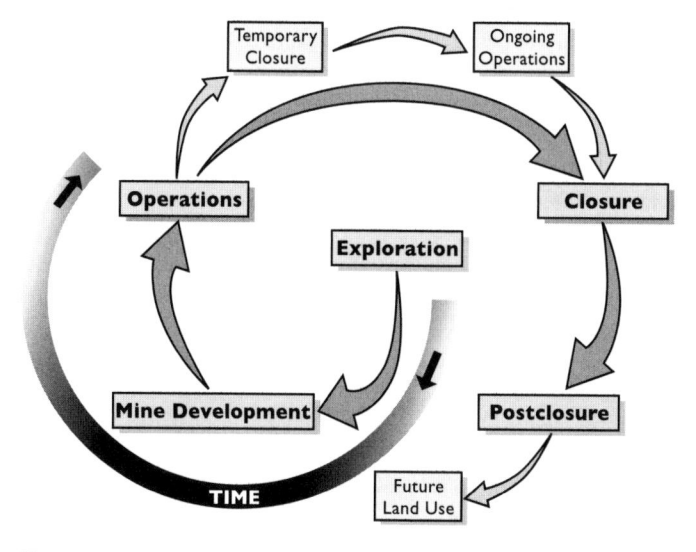

Figure 2. Illustration showing the stages of the mine-life cycle. Future land use can include further exploration and ongoing mining; other economic uses, such as renewable energy or grazing; and wildlife habitat (from van Zyl, 2005).

grade, and accessibility to be extracted (i.e., mined) and processed at a profit at a specific time. Most exploration programs never evolve into an operating mine because deposit size is too small, grade is too low, or environmental risks are too high to be profitable. The economics include the quality of the mineral deposit and the potential for environmental costs, both of which dictate the need for sampling and monitoring at this stage.

Mineral exploration typically requires 10 years or more of study before a decision is made to develop or to abandon a potential mineral deposit. Exploration can include literature searches; geological mapping; geochemical sampling of rock, soil, water, or biological media; geophysical and remote sensing surveys; aerial photography; and drilling. A tremendous amount of data collected during exploration can be useful later in identifying, predicting, and mitigating AD or MIW.

Exploration can disturb the baseline conditions, and typically these disturbed areas need to be reclaimed. The effects of exploration can range from habitat disruption to pollution from abandoned fuel caches. Exploration commonly requires trenching or drilling to retrieve rock samples for assay or other chemical analyses. Exploration roads, drill pads, and trenches can damage water and drainage quality from the erosion of sediment alone. Trenches and drill sites and their access roads must be reclaimed after exploration unless further development of the deposit is planned. Geochemical sampling, road construction, and drilling, which typically occur during the exploration stage, can continue into the mine-development stage.

Acidic drainage can result where outcrops of sulfide-bearing rock are disturbed. One common misconception is that ores near the surface contain no sulfide minerals because they have been oxidized. Whether sulfide minerals exist at the surface depends on local climate and its effects on weathering rates, mineral type and composition, length of exposure to oxygen (time since uplift or exposure), location of rocks above the local water table, and amount and rate of water flow through the rocks. Generally, more accessibility of sulfide minerals to moisture and oxygen promotes oxidation, which is why sulfide-bearing rocks tend to weather more slowly in dry climates than in wet climates. Sulfides are not present in all cases, but the possibility of AD should be considered in both shallow and subsurface exploration.

In the past, prospectors panned in streams and sampled at the surface to identify potential areas for continued exploration. Once a likely area was identified, the prospector would typically dig pits, shafts, or adits to test the mineral occurrence and determine the grade and extent of the deposit. Commonly, the prospector proceeded directly into mining as long as minerals were visible. Many such abandoned prospect sites have AD because pyrite, the nearly ubiquitous indicator of most types of ore, was disturbed at the surface and broken into chunks that exposed more pyrite to oxygen and water. This exposed pyrite then weathers to form AD.

Both traditional and modern exploration technologies, such as satellite and other remote-sensing imagery, geologic maps, geochemical and geophysical studies, and geologic reports, can be used in environmental studies. Traditional exploration samples from stripping, drilling, trenching, bulk sampling, and even panning can provide useful environmental insights into potential water- and air-quality problems. Generally, with just a few extra analyses, sediment samples and biological materials can be used for baseline environmental monitoring data. Geophysical techniques that explorationists use routinely to predict the depth and breadth of mineral deposits can be used to characterize the probable extent of natural AD sources prior to mining. Geologic maps, the foundation of mineral exploration, can provide fundamental information useful in interpreting baseline environmental implications of geochemical analyses.

Environmental evaluations in mining are useful not just for site characterization but for determining if a deposit is economically feasible. The economic feasibility determination must consider baseline or ambient conditions and the costs of closure. Where mineral deposits associated with sulfide minerals occur near the surface, it is likely that current soil productivity, natural drainage quality, and biological conditions are affected or impaired by minerals associated with that deposit. Furthermore, these unique environmental conditions may support rare or endangered species that have colonized these unusual habitats. Defining these conditions before mining begins can be critical in evaluating the economic and environmental liabilities of mine development.

Another economic assessment consideration includes the proximity to wilderness areas, national parks, or other valuable or fragile environments. The presence of threatened or endangered species or their habitats can make or break a project. Knowing these features in advance of or during exploration commonly determines the "go or no-go" decision. A significant gold deposit in Montana near the border of Yellowstone Park drew unprecedented opposition from citizens concerned about how mining might affect the park. Their opposition was perhaps of key importance in the company's decision to abandon its plans for this deposit (Russell, 2006).

Regulations governing exploration programs vary from jurisdiction to jurisdiction. Some exploration projects are so extensive that operational and reclamation plans must be developed prior to drilling. Permitting is required for exploration in most areas in the United States and elsewhere, and reclamation of disturbances caused by exploration is required.

Sampling and Monitoring During the Exploration Stage

Information and data collected during exploration can be applied to later stages of the mine-life cycle. By slightly expanding data collection during this stage, exploration sampling and monitoring can become a useful foundation for later mining stages, especially mine planning, development, and closure. As the first to set foot on potential mine sites, exploration geologists have a chance to develop insights and fill many future data needs provided they sample, analyze, and preserve data and information with an eye toward the later mining stages and ultimate closure.

By analyzing more ground than is ever developed, explorationists can produce a regional geochemical and environmental picture that can be applied to environmental planning and monitoring. For example, by noting areas prone to flooding or landslides, explorationists can lead mine planners away from such dangerous or unstable features. Recent studies near Lake City, Colorado, show that head scarps of major landslides lie in hydrothermally altered rock that is unstable due to rapid weathering caused by pyrite oxidation (Diehl and Schuster, 1996). This alteration is related to nearby mineral deposits of gold and sulfide veins. Exploration alone might have located these mineral deposits, but exploration work integrated with soil chemistry and physiographic analyses should be used in a case like this to identify unstable areas to avoid waste rock piles and other mine structures.

By recording observations of media other than rocks, such as plant life, soil characteristics, and water chemistry, the explorationists lend insights into native environmental conditions and set the stage for monitoring programs and features that can minimize impacts to the environment. Hydrologic features, such as wetlands, seeps, and iron bogs, can indicate where MIW will surface during mining or postclosure. Subsidence and historical mining structures tend to affect how subsequent mines are planned and developed, and because explorationists tend to discover such features through ordinary ground studies, their observations can be useful for mine planners.

Explorationists tend to concentrate on analyzing potentially valuable elements, such as gold, copper, and zinc, and on studies of "pathfinder elements" to assess large areas and focus on smaller drill targets. If, in addition to these limited analyses, the explorationist analyzes a broader suite of chemical parameters, data thus produced can be useful for studies as diverse as geochemical modeling, weathering, and slope stability. Complete chemical analyses of water, soil, sediment, rock, and vegetation commonly are necessary to determine the background or baseline geochemistry and biochemistry of a site. If explorationists integrate exploration and environmental sample results, there is a better chance of identifying problems that mining will encounter. For example, at the Kendall mine in Montana, thallium was found to be a problem long after mining had begun (W. McCollough, pers. comm., June 2003). The difference in cost between complete chemical analyses and limited analyses of only pathfinder elements can be insignificant compared with the potential consequences of inadequate documentation of premining conditions. Also, it is important to keep in mind that analytical limits of detection adequate for determining premining environmental conditions can be lower than those needed for exploration. Whenever possible, it is good practice to retain all data and samples, even if they give negative results for exploration purposes. Properly archived data and samples could be essential during reclamation and closure because they may provide the only premining information.

Efforts should begin early in the mine-life cycle to record climatic conditions in areas of exploration interest. Parameters of interest include annual and seasonal precipitation rates, daily temperature range, relative humidity, predominant wind direction, wind speed, and evaporation rates. Seasonal data are necessary and supplementary climatological data can be required to adequately characterize the site. Weather stations can seem like a waste of exploration dollars, but the data thus developed serve as a basis for water-balance calculations that will be needed throughout the mine-life cycle. Moreover, on-site climatic data can be far more accurate and, thus, more applicable for mine design, than data collected from the nearest weather monitoring stations, which may be distant in location or climatic conditions. For example, at Summitville, Colorado, where climatological data came from a site 75 km away, the site-wide hydrologic balance underestimated the precipitation, which led to critical shortcomings in the water budget and the site's water-management needs (Pendleton et al., 1995). Many climatic stations can be required at some sites to adequately characterize different microclimates due to elevation and topographic changes. Such is the case at the Questa molybdenum mine, New Mexico, where elevations range from 7800 to 10,000 feet and slope aspects vary from north- to south-facing, which places some slopes directly in the sun where evaporation/sublimation is the greatest while other slopes are in the shade where snow pack is present much later in the season.

Surface and subsurface rock samples can provide essential mining and environmental information, and closure/postclosure needs ought to be considered in the choice of collection methods. Surface-rock samples are collected with either manual tools or a backhoe. Rocks from below the surface can be collected with a variety of drilling methods, including churn, hammer, reverse circulation, and diamond drilling. The former three drilling methods produce rock chips or powders, whereas the diamond-drill method produces a continuous core of the rocks it traverses. Diamond-drill cores provide a record of the original stratigraphy of overburden, waste rock, and mineralized rock and are considered the most useful for general purposes. Some exploration drill holes can be converted to monitoring wells to provide samples for baseline conditions or other monitoring needs.

Mine Development

Once an economic mineral deposit is discovered and its limits defined, mining can proceed to the second stage of the mine-life cycle, mine development. If baseline environmental conditions were not described during the exploration stage, they should be described during the earliest part of development, prior to ground disturbances. This is because disturbances during development and mining typically are environmentally indistinguishable from naturally degraded conditions in near-surface mineral deposits. It is important to incorporate appropriate natural conditions into the site plan to set realistic reclamation goals. Permitting, operational, and mine-closeout plans generally are developed at this stage. Mine development of sites typically requires reserve drilling to determine three-dimensional features of the deposit to sufficiently develop plans for excavating the deposit. This phase, which is commonly conducted by a contractor or a junior mining company, also includes construction of mine access and mine-haul roads; rail lines and terminals (if they are required); installation of electrical, gas, and other utilities; and construction of mine buildings, support structures, and sometimes, housing facilities. In remote or undeveloped areas, facilities construction can include housing and social structures, such as schools, churches, and shopping centers. Excavation requires the use of heavy machinery and the use of petroleum and other industrial products for operation and maintenance. If the site is not pristine prior to exploration, organic chemical scans can be useful to establish baseline conditions.

Before mine-waste repositories, roads, buildings, or other structures are placed on a site, condemnation drilling should occur; that is, closely spaced wells should be drilled to ascertain that viable minerals and historical mine workings do not lie beneath the proposed structure. Information collected through such drilling can be applied to geochemical and hydrological balance studies.

The mine-development stage also includes construction of metallurgical-processing facilities, such as mills, pipelines, conveyance facilities, and leach pads. Metallurgical testing and development are conducted on-site at metallurgical pilot plants. For operations requiring cyanide or acid leaching, leach pads, processing ponds, and their related infrastructure are constructed during the development stage and usually continue into the operational stage, especially at large, multiyear operations.

Metallurgical testing is one of the most overlooked areas for developing environmental data in the mining industry. While metallurgical testing produces assays and geochemical analyses similar to many that are run during exploration, this stage of mining describes the rocks and minerals at a more detailed mineralogical and physical scale. During this phase of testing, trace-element composition of minerals is examined, and from the environmental perspective, this testing can be important. For instance, cadmium is commonly present as a trace element in sphalerite (ZnS), but sphalerite in some deposits is not of sufficient quantity to warrant recovery. In such cases, the sphalerite could be specially handled by placing it in a specific, more isolated area of the waste rock facility than other gangue minerals (noneconomic minerals, such as quartz or calcite, that are part of an ore deposit). Also, additional environmental protections can be put in place to minimize the distribution of cadmium and zinc into the environment. Such detailed mineralogical observations typically are made during metallurgical testing.

Material strength properties, which are typically determined during mine development, can provide useful environmental and closure information. Whereas mine engineers examine rocks to determine the maximum height or depth of mine benches, they also can elect to determine the conditions under which mine faces will fail after closure. For example, when a mine pit fills with water after mine closure, pit walls can become unstable. Slope failures after mining can be as much a problem for mine engineers as are slope failures during mining.

Sampling and Monitoring During the Mine-Development Stage

Mine development is the transition stage between exploration and mine operation. Hence, many issues considered during exploration also need to be addressed during this stage of mining. If collection of on-site climatological data did not begin during exploration, it is essential that this information be collected early in the mine-development stage. In order to establish successful revegetation at the time of mine closure, observations of native vegetation, soil depth, slope and aspect, and other pre-existing conditions must be documented. This information also is useful during the operation stage for planning placement of waste materials; it is more economical to move waste only once. For example, during mine development/expansion, Kennecott at the Bingham Canyon facility redesigned the waste repository system to more efficiently and effectively deal with waste rock. During the early stages of development it is important to consider reclamation. An example of this practice is when Nevada Placer Dome hired the Duckwater/Shoshone tribe to collect seeds and grow plants needed for revegetation (B. Upton, pers. comm., September 2003).

During the mine-development stage, metallurgists evaluate detailed mineralogy. Concurrently, others are developing a water balance for the site and defining baseline environmental conditions. It can be extremely useful at this stage to set aside an area for and initiate predictive tests, such as acid-base accounting, meteoric water mobility, static and kinetic tests, or other long-term weathering tests, conducted under natural local conditions. Air monitoring also should begin in this stage or earlier to develop baseline conditions prior to ground disturbance. These data are useful in the prediction of acid generation and water management requirements at the site. For example, the acid-generation characteristics of the waste rock at the Zortman-Landusky mine changed over time (Miller and Hertel, 1997) due to the buildup of ferric iron, a strong oxidant, in the sulfidic waste rock piles and the heap. This unanticipated change in mineralogy led to the eventual bankruptcy of the parent company (Russell, 2006).

Permitting, which generally takes place during the mine-development stage, can require significant environmental data to accommodate the permitting process. The mine-development stage is a good time to design and implement compliance-monitoring programs. Lysimeters can be installed at this time to measure depth to water and to aid in evaluating infiltration rates. Pilot studies, such as test heap-leach pads and site waste rock weathering columns or test pads, can be constructed at this time, and monitoring can begin. It is good practice to continue these site weathering and closure pilot studies throughout the remainder of the mine-life cycle to provide information for closure. For example, when research on pilot heap closures was undertaken at the Golden Sunlight mine in Montana with the U.S. Environmental Protection Agency's Office of Research and Development and MSE Technology Applications, Inc., the operator (Placer Dome) found pilot-testing results to be very useful and is now undertaking similar tests at its operations in Nevada (B. Upton, pers. comm., September 2003).

Transport and pathways of waters and potential contaminants need to be identified during this stage. Detailed geological maps, including structural information, should be refined. Potential biological receptors should be evaluated and indicator organisms established.

Disturbance from road building and other structures should be sampled and characterized for use during closure. It is important to keep in mind that closure can occur any time during mine development if feasibility studies determine that continued development or mining cannot be economical. If closure occurs, all disturbed ground will need to be reclaimed.

Operations

The operations stage of mining includes the process whereby ores and waste rock are removed from the ground, milled and/or leached, and concentrated into a saleable product, such as a metal sulfide concentrate or high-grade gold bullion. This generally is followed by smelting and refining. Sampling of the waste materials is necessary in order to know what potential contaminants exist in the rock mass, and monitoring of ground or surface water can be necessary for environmental compliance. Air-pollution monitoring also can be needed.

Excavation

The initial and usually ongoing phase of the operations stage is *excavation,* which includes the extraction of ores by drilling, blasting, and mucking (removing the broken material from the mine). During excavation, as rock is removed from the ground, excess water must be pumped or drained from the area of underground and surface workings in order to accommodate miners and equipment. At open-pit mines, water is typically pumped from wells adjacent to the pit, a process that lowers the water table below the bottom of the pit. In underground mines, water is usually pumped from sumps that are constructed in the lower reaches of the excavation and piped upward through vertical mine shafts to the surface. At many older underground mines located in steep terrain, such as in Colorado, underground mining was made possible through construction of an adit (horizontal shaft) or tunnel that drained groundwater from one or more mines to a nearby stream. Such drainage tunnels, which were constructed miles underground, could dewater many square miles and reduced the need for expensive pumping of water from mine workings. However, drainage-tunnel water commonly is of poor quality and requires treatment.

Whenever the water table is lowered to allow mining below the water table, or fluctuates (such as occurs seasonally or during storm events), there is a chance that water quality will degrade due to oxidation of sulfide minerals. When this happens, acid water or its derivatives will form wherever there are sulfide minerals that had resided below the water table. Monitoring these waters during mining, both inside and outside the mine area, will be important in planning for mine closure. Dewatering that occurs during the operation stage may have to continue after excavation ceases in order to control environmental impacts. Once a mine floods, resulting MIW can require treatment or other special management.

Milling, Concentrating, and Leaching

After ores are excavated from the ground, they generally are reduced to a nominal size to allow ore minerals or leachable metals to be segregated from waste rock or gangue minerals. Crushing and milling generate a rock mass with a surface area that is many times greater than the original rock mass. Once exposed to air, crushed and milled rock will generate acid if it contains acid-generating sulfide minerals, such as pyrite. Sulfide ores typically are milled to a fine grain size (generally less than 100 µm, although this varies). This allows particles of ore minerals to be segregated from gangue minerals or from other ore minerals via flotation, gravity, electromagnetic, or other means. Other types of ores, including many gold and copper oxide ores, need only be reduced to chunks (typically 1–5 inches) so they can be piled on a liner and treated with a leaching solution to remove the economic metals. However, some gold ores are milled to a fine size to expose gold-bearing minerals to leaching solutions or are roasted/autoclaved to oxidize the sulfidic minerals. Tailings are finely ground waste from metal extraction and are managed by dewatering.

Mineral-separation techniques have changed over time, and the sampler must be aware of these historical and modern technologies because different technologies can lead to potentially different environmental consequences (Jones, 2007). For example, stamp mills commonly were used at gold mines during the late 1800s and early 1900s. These mills ran gold ore through a number of different levels with cam-driven hammers pounding the ore into finer and finer particles. Gold was separated from the finest particles using mercury amalgam. Waste materials from stamp mills commonly contain fine-grained sulfide minerals that can result in acidic metal-rich drainage, and mercury contamination can be found in the vicinity of some historical stamp mills.

Leaching is a common technique in which valuable, leachable metals are removed from the gangue, typically along with other less desirable metals. The metals are then extracted from the high-metal "pregnant" solution via electrowinning. Leaching chemicals, such as cyanide and sulfuric acid, are potential environmental contaminants.

Smelting and Refining

Although not always the case, metal smelters and refineries today typically are large, specialized facilities that cannot be profitable to operate at all mining sites. Consequently, many are located in populated areas away from mining districts. Smelting, which involves heating in the presence of chemicals, is employed to melt the mineral concentrates and segregate metals from their combined elements (e.g., oxygen or sulfur). Refining is a more advanced process to segregate one metal from another and produce a higher-grade refined metal or metallic compound that can be sold for use in metal applications. Refining of the recovered metal can involve pyrometallurgical, chemical, and electrochemical processes that must be properly managed to prevent negative environmental impacts from wastes or emissions. Smelting, especially, and refining produce a waste slag (glass) that generally requires special handling. These processes also produce flue gases or smokestack emissions that need to be sampled and monitored. Sulfur dioxide and smelter dusts can be produced, which must be properly managed to prevent air and drainage-quality impacts. Sulfuric acid is commonly produced and sold as a byproduct of sulfide smelting and roaster operations.

Mine Expansion

Many mines undergo one or more expansions during their operating life as economic conditions change as a result of changes in metal prices and the development of new technologies and new reserves. The challenge is to keep the existing mine operating efficiently while the permitting process to expand mining moves forward. Ultimately, the ease in permitting expansions depends upon the ownership of land, past compliance record, the attitudes of the general public, and the policies of regulating agencies.

Standby/Inactive Mines

Many mines will be forced to temporarily cease production for a variety of reasons, mostly related to economic conditions. Monitoring is highly recommended during temporary closure. At this time, few people are on-site to visually monitor potential problems, and the local community generally is not as focused toward the mining operation. The possibility of oversight by regulatory agencies increases because this is the mine's time of highest vulnerability. Mine sites can be inactive for many years depending upon government regulations and economic conditions. Funding for continuation of sampling and monitoring should be maintained during these standby periods.

Sampling and Monitoring During the Operations Stage

The operational units of mining and metallurgical processing can include the mine, overburden storage facilities, development waste materials, ore stockpiles, tailings, crushing facilities, concentration facilities, leaching facilities, and other operational facilities. Each of these units can require monitoring, depending on its potential contaminants and its handling and closure characteristics.

Potential environmental effects of mineral production include impacts to wildlife and fisheries; habitat loss; changes in local water balance; sedimentation; introduction of potential contaminants in tailings ponds or leaching-solution impoundments; formation of unstable dams and highwalls; construction of tailings ponds or leaching pads and ponds; potential acid generation from ore, waste rock, pit walls, and tailings; metal leaching by acid, neutral, and alkaline drainage; cyanide-solution contamination at gold and silver metallurgical recovery operations; windborne dust; and gas and dust emissions from metallurgical operations. Most of these impacts are routinely addressed or resolved as part of the development and production stages.

An important aspect of the operations stage is compliance monitoring. One key to compliance monitoring is to have complete chemical analyses with analytical detection limits and other quality-assurance/quality-control documentation. Complete analytical data can be used to verify predictions made during exploration and mine development and can provide a basis for modifying plans during the operations stage (Kuipers et al., 2005). For example, selective handling of waste material can become necessary during the operations stage. Modeling of groundwater flow can require chemical data that were obtained during the exploration or mine-development stages. Similarly, chemical analyses of bench samples of blast holes during open-pit production can be useful in characterizing the mine-waste piles for reclamation. It is good practice to set site-specific threshold levels that trigger additional sampling or monitoring activities to ensure that compliance levels are not exceeded. Although many mines are not required to monitor nitrogen compounds or organics in water, it is good practice to include nitrate analysis to monitor the effects of blasting, and to include trichloroethylene (TCE) analyses to monitor residues from degreasing of equipment.

It can be important to characterize all materials, including backfill materials and dewatering fluids, using chemical, physical, and mineralogical techniques and to maintain detailed records of sampling and analyses. At the Pinto Valley mine in Arizona, a tailings impoundment failed when waste rock was placed on tailings (Kiefer, 1998). Supplementary characterization and monitoring of the physical properties of the tailings and impoundment may have prevented this failure.

The water balance determined during the mine-development stage needs to be continuously updated during the operations stage as new data become available (O'Kane and Waters, 2003; Marcoline et al., 2003). The U.S. Environmental Protection Agency found that a common thread among recently active Superfund mining sites is that they did not have an accurate water balance (Russell, 2006).

Expansions and temporary closures can occur during the operation of the mine. Frequent reexamination of the sampling and monitoring programs and revision of sampling plans are ways to lessen the impacts of these circumstances. It is good practice to maintain monitoring programs during temporary mine closures and to maintain adequate staff and funding to be able to address unforeseen problems.

Closure/Postclosure

Closure is the period following excavation during which activities must be conducted to complete mineral extraction and return lands to their intended postmining use. Wherever possible, closure activities should be planned concurrently with operations. This overcomes the cost of moving material more than once and is, thereby, cost-effective because operations handle some of the closure costs. Closure plans usually require mine structures be removed, land be graded and revegetated, openings be sealed or safeguarded, and lands be generally returned to a use other than mining. Postclosure is that period during which local environmental equilibration takes place (e.g., groundwater levels return to normal and vegetation successions reestablish). During both the closure and postclosure stages, monitoring can be required, and continuing monitoring after closure is proactive.

Today, numerous mines and metallurgical facilities in many countries have mine-closure plans in place prior to operation. Some closure activities can take place concurrent with mine production in areas where ore has already been depleted or when leach pads, spoils piles, mine-waste rock piles, and tailings facilities have reached their ultimate capacity. The cost of closure should be accounted for in the overall feasibility of mining the deposit, or else the economics of the deposit will be inaccurate. Because mining and water-quality regulations tend to change over time, closure plans should be periodically reexamined to account for changing metal prices and evolving postmining liabilities and regulatory requirements. Depending on corporate policies, local regulations, planned land use, and landowner requirements, closure can involve recontouring of pit walls and mine waste-rock piles, covering and revegetating reactive tailings and waste-rock piles, decommissioning roads, dismantling buildings, reseeding and planting of disturbed areas, ongoing monitoring, and possibly water treatment and management for an extended period after closure.

Problems that most commonly appear after mining include erosion, high-wall failure, surface subsidence, seepage of contaminated solutions into ground and surface water from AD and soluble salts, slope failure, and changes due to weathering. Acidic metal-rich drainage can increase after production, and impacts can affect wildlife; fisheries; habitat; vegetation; air quality; and agricultural, recreational, or drinking-water supplies, and in some cases can generate physical hazards to humans and wildlife. Plugging of shafts and adits needs to be carefully evaluated because, in some cases, plugging can cause changes in the groundwater flow and contribute to groundwater contamination.

Sampling and Monitoring During the Closure/Postclosure Stage

Closure plans typically establish a postmining use of the mine site commensurate with the desires of the landowner and adjacent local land uses. Some sites can require long-term monitoring. Important factors in closure/postclosure monitoring are to maintain monitoring and not to stop monitoring too soon. For example, Placer Dome plans to monitor Las Cristinas mine in Venezuela for potential AMD for at least 20 years after mine closure because there are extensive underground and surface pathways for AD, and any problem would have to be mitigated quickly (Dahlberg, 1999). Problematic conditions can arise during or years after closure. At the San Luis mine in southern Colorado, a metal-rich seep developed adjacent to a backfilled pit during closure. Even though operations had ceased, there were personnel and equipment on-site to handle the problem, and mitigation was able to begin quickly. Continued monitoring can lend insight into source-control possibilities in lieu of perpetual treatment of MIW.

Continued characterization of solid materials moved during contouring and reclamation can avert unexpected results during closure. Also, predictive tests need to be reevaluated to assess their continued reliability. Some postmining uses, such as property sales or sale of nonacidic waste rock for construction purposes, could generate income for continued monitoring of the site.

SUMMARY

Today's mines are being designed to take into account conditions from exploration through postclosure. Sampling and monitoring programs should be started as early in the mine-life cycle as practical. Data required for the closure stage are obtained throughout the mine-life cycle, and baseline and background conditions are determined prior to mining. Data management and documentation of sampling and monitoring programs are important to successful closure. Because mines can close or go on standby status any time during the mine-life cycle, it is important to maintain monitoring programs and personnel during these periods in order to prevent or minimize detrimental impacts to the environment.

ACKNOWLEDGMENTS

We would like to acknowledge the members of the Sampling and Monitoring Committee of ADTI-MMS for their contributions to this report. We thank Harry Posey, Doug Peters, and Bob Kleinmann for their insightful reviews, and Charles Bucknam, Terry Chatwin, and Jim Crock for reviewing an earlier version of this manuscript. Kathy Smith would like to acknowledge the U.S. Geological Survey Mineral Resources Program for funding her contributions to this report. Virginia McLemore would like to acknowledge the New Mexico Bureau of Geology for funding this work.

REFERENCES CITED

ADTI-MMS, 2006, Acid Drainage Technology Initiative Metal Mining Sector Web site, http://www.unr.edu/mines/adti/ (accessed on April 16, 2006).
Dahlberg, K., 1999, Fourteen steps to sustainability: Mining report card, Placer Dome, Inc.: Mineral Policy Center, Washington, D.C., 39 p. (Available at http://www.mineralpolicy.org/publications.cfm?pubID=17.)
Diehl, S.F., and Schuster, R.L., 1996, Preliminary geologic map and alteration mineralogy of the main scarp of the Slumgullion landslide, in Varnes, D.J., and Savage, W.Z., eds., The Slumgullion earth flow: A large-scale natural laboratory: U.S. Geological Survey Bulletin 2130, chapter 3. (Available at http://pubs.usgs.gov/bul/b2130/.)
Harrison, J.E., 1998, Key water quality monitoring questions: Designing monitoring and assessment systems to meet multiple objectives, in Proceedings, National Water Quality Monitoring Council National Monitoring Conference, Reno, Nevada, July 7–9, 1998. (Available at http://www.nwqmc.org/98proceedings/Papers/17-HARR.html.)

Hartman, H.L., 1992, Introduction to mining, in Hartman, H.L., ed., Mining engineering handbook: Society for Mining, Metallurgy, and Exploration, Inc., Littleton, Colorado, p. 3–42.
Hornberger, R.J., Lapakko, K.A., Krueger, G.E., Bucknam, C.H., Ziemkiewicz, P.F., van Zyl, D.J.A., and Posey, H.H., 2000, The Acid Drainage Technology Initiative (ADTI), in Proceedings, Fifth International Conference on Acid Rock Drainage, Denver, Colorado, May 21–24, 2000: Society for Mining, Metallurgy, and Exploration, Inc., Littleton, Colorado, p. 41–50.
Jones, W.R., 2007, History of mining and milling practices and production in San Juan County, Colorado, 1871–1991, in Church, S.E., von Guerard, P., and Finger, S.E., eds., Integrated investigations of environmental effects of historical mining in the Animas River watershed, San Juan County, Colorado: U.S. Geological Survey Professional Paper 1651 (in press).
Kiefer, M., 1998, Cleaning the creek: A mining company mucks out its own mess at Pinto Creek: Phoenix New Times, May 7, 1998. (Available at http://www.phoenixnewtimes.com/issues/1998-05-07/feature.html.)
Kleinmann, R.L.P., ed., 2000, Prediction of water quality at surface coal mines: West Virginia University, Morgantown, West Virginia, The National Mine Land Reclamation Center, 239 p. (Available at http://wvwri.nrcce.wvu.edu/programs/adti/pdf/adti_predictions.pdf.)
Kuipers, J., Maest, A., MacHardy, K., and Lawson, G., 2005, Evaluation of the National Environmental Policy Act (NEPA) process for estimating water quality impacts at hardrock mine sites, in Proceedings, Society for Mining, Metallurgy, and Exploration Annual Meeting and Exhibit, Salt Lake City, Utah, February 28–March 2, 2005: Society for Mining, Metallurgy, and Exploration, Inc., Littleton, Colorado.
Maest, A.S., Kuipers, J.R., Travers, C.L., and Atkins, D.A., 2005, Predicting water quality at hardrock mines: methods and models, uncertainties, and state-of-the-art: Buka Environmental and Kuipers & Associates, 77 p.
Marcoline, J.R., Beckie, R.D., Smith, L., and Nicol, C.F., 2003, Mine waste rock hydrogeology: The effect of surface configuration on internal water flow, in Proceedings, Sixth International Conference on Acid Rock Drainage, Cairns, Queensland, Australia, July 14–17, 2003: Australasian Institute of Mining and Metallurgy, Victoria, Australia, p. 911–918.
McLemore, V.T., Russell, C.C., Smith, K.S., and the Sampling and Monitoring Committee of the Acid Drainage Technology Initiative—Metals Mining Sector (ADTI—MMS), 2004, Sampling and monitoring for closure, in Preprints of the Society for Mining, Metallurgy, and Exploration Annual Meeting and Exhibit, Denver, Colorado, February 23–25, 2004: Society for Mining, Metallurgy, and Exploration, Inc., Preprint No. 04-62, Littleton, Colorado, 7 p.
Miller, R.A., and Hertel, T.M., 1997, Mine rock characterization: Zortman and Landusky Mines, Little Rocky Mountains, Phillips County, north-central Montana, in Proceedings, Fourth International Conference on Acid Rock Drainage, Vancouver, B.C., Canada, May 31–June 6, 1997, p. 515–532.
O'Kane, M., and Waters, P., 2003, Dry cover trails at Mt. Whaleback: A summary of overburden storage area cover system performance, in Proceedings, Sixth International Conference on Acid Rock Drainage, Cairns, Queensland, Australia, July 14–17, 2003: Australasian Institute of Mining and Metallurgy, Victoria, Australia, p. 147–153.
Pendleton, J.A., Posey, H.H., and Long, M.B., 1995, Characterizing Summitville and its impacts: Setting the scene; in Posey, H.H., Pendleton, J.A., and van Zyl, D., eds., Proceedings, Summitville Forum: Colorado Geological Survey, Special Publication 38, p. 1–12.
Price, W.A., 2005, List of potential information requirements in metal leaching and acid rock drainage assessment and mitigation work, MEND Report 5.10E: Natural Resources Canada, 23 p.
Russell, C., 2006, Mine drainage challenges from A to Z in the United States, in Proceedings, Seventh International Conference on Acid Rock Drainage, St. Louis, Missouri, March 26–30, 2006: Society for Mining, Metallurgy, and Exploration, Inc., Littleton, Colorado, p. 1785–1802.
Schmiermund, R.L., and Drozd, M.A., eds., 1997, Acid mine drainage and other mining influenced waters (MIW), in Marcus, J.J., ed., Mining environmental handbook: London, Imperial College Press, Chapter 13, p. 599–617.
Skousen, J., Rose, A., Geidel, G., Foreman, J., Evans, R., Hellier, W., and members of the Avoidance and Remediation Working Group of the Acid

Drainage Technology Initiative (ADTI), 1998, Handbook of technologies for avoidance and remediation of acid mine drainage: West Virginia University, Morgantown, West Virginia, National Mine Land Reclamation Center, 131 p. (Available at http://wvwri.nrcce.wvu.edu/programs/adti/publications/adti_handbook.html.)

van Zyl, D., 2005, Sustainable development and mining communities, *in* Price, L.G., Bland, D., McLemore, V.T., and Barker, J.M., eds., Mining in New Mexico: The environment, water, economics, and sustainable development: New Mexico Bureau of Geology and Mineral Resources, Decision-Makers Field Guide, p. 133–136.

van Zyl, D.J.A., Parsons, S., McLemore, V., and Hornberger, R.J., 2006, Acid Drainage Technology Initiative: Ten years of mining industry, government agencies, and academia collaboration in the metal and coal mining sectors in the USA, *in* Proceedings, Seventh International Conference on Acid Rock Drainage, St. Louis, Missouri, March 26–30, 2006: Society for Mining, Metallurgy, and Exploration, Inc., Littleton, Colorado, p. 2159–2169.

Williams, R.D., 2003, The Acid Drainage Technology Initiative: An evolving partnership, *in* Proceedings, Sixth International Conference on Acid Rock Drainage, Cairns, Queensland, Australia, July 14–17, 2003: Australasian Institute of Mining and Metallurgy, Victoria, Australia.

Manuscript Accepted by the Society 28 November 2006

Printed in the USA